Chris Bruhn

New York Chicago San Francisco Athens London Madrid
Mexico City Milan New Delhi Singapore Sydney Toronto

3 4 5 6 7 8 9   LCR   24 23 22 21

ISBN   978-1-260-45431-4
MHID   1-260-45431-2

e-ISBN   978-1-260-45432-1
e-MHID   1-260-45432-0

Interior design by Steve Straus of Think Book Works.
Cover and letter art by Kate Rutter.

McGraw-Hill Education books are available at special quantity discounts to use as premiums and sales promotions or for use in corporate training programs. To contact a representative, please visit the Contact Us pages at www.mhprofessional.com.

About the Author

Chris Bruhn began his career as an aerospace engineer before figuring out that teaching physics was his calling. Since becoming an educator, he has taught all varieties of physics, in all types of high schools, and has won several educational awards. Chris is an educational trainer. He likes to create and share curriculum and educational resources as well as lead institutes and study sessions for teachers and students around the country. Chris enjoys sports, travel, painting, movie marathons with his family, and generally having fun. He likes building things—and taking things apart to see how they work!

Contents

Part One: The Mechanics of Physical Objects

*Yes, we're covering Newton's Third Law first!

Part Two: The Physics of Nonsolid Behavior

Part Three: Thermodynamics

12 Circuits 266

13 Magnetism 298

14 Induction 323

Part Five: Waves

Appendix

This book is dedicated to everyone who has ever found physics mystifying and bafflingly beyond their reach. I hope this book helps you see the wonder and beauty of how the universe works.

Acknowledgments

I'd like to thank all the students over the years who have taught me at least as much as I ever taught them. Thanks for learning with an open mind and heart.

Thanks to my daughter, Alina, who is taking AP Physics 1 as I write this book. She has checked my "physics" and given me many great reminders about how beginning physics learners view the material and what would be confusing to them.

Thanks to my son, Cole, in college as a math major, who is checking to make sure I have not made too many mistakes.

Thanks to my wife, Karen, who has made sure that I haven't butchered the English language too much. *And* has put up with me barricaded in the office working on this book for about five months straight.

Introduction

Welcome to your new physics book! Let us try to explain why we believe you've made the right choice. This probably isn't your first rodeo with either a textbook or other kind of guide to a school subject. You've probably had your fill of books asking you to memorize lots of terms (such is school). This book isn't going to do that—although you're welcome to memorize anything you take an interest in. You may also have found that a lot of books jump the gun and make a lot of promises about all the things you'll be able to accomplish by the time you reach the end of a given chapter. In the process, those books can make you feel as though you missed out on the building blocks that you actually need to master those goals.

With *Must Know High School Physics,* we've taken a different approach. When you start a new chapter, right off the bat you will immediately see one or more **must know** ideas. These are the essential concepts behind what you are going to study, and they will form the foundation of what you will learn throughout the chapter. With these **must know** ideas, you will have what you need to hold it together as you study, and they will be your guide as you make your way through each chapter.

To build on this foundation you will find easy-to-follow discussions of the topic at hand, accompanied by comprehensive examples that show you how to apply what you're learning to solving typical physics questions. Each chapter ends with review questions—more than 300 throughout the book—designed to instill confidence as you practice your new skills.

This book has other features that will help you on this physics journey of yours. It has a number of sidebars that will both help provide helpful information or just serve as a quick break from your studies. The BTW sidebars ("by the way") point out important information as well as tell you what to be careful about physics-wise. Every once in a while, an IRL sidebar ("in real life") will tell you what you're studying has to do with the real world; other IRLs may just be interesting factoids.

In addition, this book is accompanied by a flashcard app that will give you the ability to test yourself at any time. The app includes more than 100 "flashcards" with a review question on one "side" and the answer on the other. You can either work through the flashcards by themselves or use them alongside the book. To find out where to get the app and how to use it, go to the next section, The Flashcard App.

Before you get started, though, let me introduce you to your guide throughout this book. Chris Bruhn both teaches AP physics and is an AP physics consultant for the College Board. He has a clear idea about what you should get out of a physics course and has developed strategies to help you get there. Chris also has seen the kinds of trouble that students can run into, and he is an experienced hand at solving those difficulties. In this book, he applies that experience both to showing you the most effective way to learn a given concept as well as how to extricate yourself from traps you may have fallen into. He will be a trustworthy guide as you expand your physics knowledge and develop new skills.

Before we leave you to Chris's sure-footed guidance, let us give you one piece of advice. While we know that saying something "is the *worst*" is a cliché, if anything is the worst in physics, it's force diagrams. Let Chris introduce you to the concept and show you how to apply them confidently to your physics work. Mastering force diagrams will leave you in good stead for the rest of your physics career.

Good luck with your studies!

The Editors at McGraw-Hill

The Flashcard App

his book features a bonus flashcard app. It will help you test yourself on what you've learned as you make your way through the book (or in and out). It includes 100-plus "flashcards," both "front" and "back." It gives you two options as to how to use it. You can jump right into the app and start from any point that you want. Or you can take advantage of the handy QR Codes near the end of each chapter in the book; they will take you directly to the flashcards related to what you're studying at the moment.

To take advantage of this bonus feature, follow these easy steps:

Search for **Must Know High School** App from
either Google Play or the App Store.

↓

Download the app to your smartphone or tablet.

↓

Once you've got the app,
you can use it in either of two ways.

↙   ↘

Just open the app and you're ready to go.	Use your phone's QR code reader to scan any of the book's QR codes.
You can start at the beginning, or select any of the chapters listed.	You'll be taken directly to the flashcards that match your chapter of choice.

↘   ↙

Get ready to test your physics knowledge!

Author's Note

Many, if not most, people find physics mysterious and intimidating. It seems to be secret knowledge that is beyond their grasp. This is unfortunate, because physics is the simplest, most basic foundation of all science. Physics is the basis on which scientific thought and engineering are built. Most importantly, it is a knowledge that is attainable by everyone.

You may not be a scientist, but you already comprehend much about how the world works. Since you were a child, you have been learning physics. In fact, every child is a little physicist who is constantly experimenting in an insatiable quest of discovery. For example, as a child you learned:

- Things fall down. (Gravity)

- Falling on concrete hurts, but jumping on the bed is fun. (Impulse and Momentum)

- Hot chocolate cools off and warms your hands. (Thermodynamics)

- You have to peddle to get a bike moving, but when you stop peddling, the bike keeps going. (Forces and Inertia)

- Touching a 9-volt battery to your tongue makes it tingle. (Electricity)

- Sound travels around a corner, but you can't see things around a corner. (Waves)

- And the list goes on and on …

You have already discovered many of nature's hidden rules and structure. You are already a physicist. All you need is to take what you already know, discover some new ways of looking at the world that you have not yet considered, and organize it all into the framework we call physics. Then, you will grasp the beautifully simple structure of how it all fits together and interconnects.

You can learn physics. You were born to be a physicist!

In this book, we are going to look at physics from a big-picture point of view, so we can see the overall beauty and structure of how the universe operates. We are not going to learn everything. Instead we'll concentrate on the building blocks of physics, that is, the foundations that all physics is built on. We are going to ignore unnecessarily confusing details and extra mathematical complications that can grind us into the dirt. Once you learn the big picture of physics, you can dig as deep and get as dirty as you want.

This book is for:

- Students looking to get a head start in a physics class

- Those already taking a physics course but need help putting it all together

- Anyone who is intimidated by physics

- Everyone who wants to know how physics works without being overburdened by science speak or number crunching

- Anyone who wants to sound smart at a party

Each chapter starts with the **must know** ideas to understand a physical phenomenon, followed by examples in easy-to-understand language. Each topic is extended into a fuller and more mathematical representation. Since physics can be challenging for first-time learners, you will need a lot of practice. Each chapter has lots of examples to help you bridge the gap between "watch me do it" and "do it yourself." Finally, every chapter ends with review questions that will encourage you to put together everything you've learned.

Let's get started!

PART ONE

The Mechanics of Physical Objects

Mechanics shows us the behavior of solid objects, like where they are in space and time, and what they are doing. It explains how objects interact with each other through forces and how forces can change the motion of objects. Amidst all this moving about and changing, mechanics reveals that momentum and energy are conserved.

Motion

MUST ⚡ KNOW

⚡ Position tells us where we are, distance tells us how far we have traveled, and displacement tells us how far and in which direction we are from where we started.

⚡ Speed tells us how fast we are going. Velocity tells us how fast and in which direction we are going.

⚡ Acceleration tells us if we are changing speed and/or direction.

Before we can go very far in physics, we need to know where things are at and how they are moving. The study of motion is called **kinematics**. In this chapter, our goal is to learn the terminology of motion and how to describe motion with several different representations: pictures and words, dot diagrams, graphs, and equations.

Position, Distance, and Displacement

Where are you right now? Your house or school? In physics we need to be a little more exact. Pull out your phone or computer and go to your maps app. Drop a marker at your location, and you will get the latitude and longitude grid coordinates of your exact location on the Earth. In physics, we will set up a coordinate system that is convenient for us to use. The figure shows an **x-axis** that runs horizontally left and right. This is called a **number line** in math, but in physics our axis has units so we know where things are really at. Our x-axis is measured in units of meters (m).

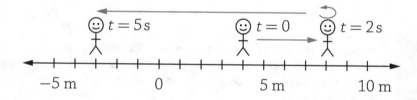

A person is moving along our axis. We want to know the person's position, distance, and displacement.

Position is the exact location of the object at a specific time measured in meters (m). This is a vector. This means it has a magnitude and a direction. For instance, there is a difference between 10 m to the east as opposed to 10 m to the west.

Distance is how far the object has traveled in total measured in meters (m) during a period of time. This is a scalar quantity, meaning it has a magnitude but not a direction.

Displacement Δx is the change in position: $\Delta x = x_f - x_i$ where x_f is the final position and x_i is the initial position. Displacement is only concerned about the starting and ending positions. It tells us how far and in what direction an object is from where it started during the time period of travel. This is a vector quantity because it has a magnitude and direction.

> ▶ Let's practice finding these for our person on our axis.
>
> ▶ At the start, the person is at position 4 m. At a time of 2 s, the person is at position 8 m. At 5 s, the position is −3 m. For the entire 5 s, the person has traveled 4 m to the right and then 11 m to the left for a total distance of 15 m.
>
> ▶ The displacement is $\Delta x = x_f - x_i = ((-3 \text{ m}) - 4 \text{ m}) = -7 \text{ m}$.
>
> ▶ Notice how the displacement is negative. This tells us the person is to the left of where they started.

IRL Cars have an odometer and your phone has GPS. What do these two devices measure: position, distance, or displacement? An odometer indicates how far the car has traveled, so it measures distance. GPS indicates where you are, so it measures position.

Speed and Velocity

Speed tells us how fast the object is traveling over a time period. Speed = distance divided by time interval. **Velocity** tells us how fast, but also gives the direction of travel because it is a vector. Since velocity is a vector, we need to keep in mind that geometry and trigonometry are needed to add velocities that are in perpendicular directions. It is the displacement divided by the time interval: $v = \dfrac{\Delta x}{\Delta t}$. Let's look at our person on the x-axis again.

Let's take a look at our moving stick figure:

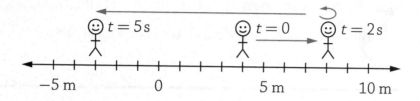

The speed from 2 s to 5's is $\dfrac{11\,\text{m}}{3\,\text{s}} = 3.7\,\text{m/s}$. The velocity over the same time interval is almost the same, but it has a direction associated with it:

$$v = \frac{\Delta x}{\Delta t} = \frac{x_{5s} - x_{2s}}{3\,\text{s}} = \frac{(-3\,\text{m}) - 8\,\text{m}}{3\,\text{s}} = \frac{-11\,\text{m}}{3\,\text{s}}$$

$$= -3.7\,\frac{\text{m}}{\text{s}} = 3.7\,\frac{\text{m}}{\text{s}} \text{ to the left}$$

Notice how velocity contains the direction information in the negative sign, which tells us "in the negative direction" or "to the left."

EXAMPLE

Using the same figure just provided, let's find the speed and velocity of the person over the entire 5 seconds.

$$\text{Speed} = \frac{\text{distance}}{\text{time}} = \frac{15\,\text{m}}{5\,\text{s}} = 3\,\frac{\text{m}}{\text{s}}$$

$$v = \frac{\Delta x}{\Delta t} = \frac{x_{5s} - x_{0s}}{5\,\text{s}} = \frac{(-3\,\text{m}) - (4\,\text{m})}{5\,\text{s}} = \frac{-7\,\text{m}}{5\,\text{s}}$$

$$= -1.4\,\frac{\text{m}}{\text{s}} = 1.4\,\frac{\text{m}}{\text{s}} \text{ to the left}$$

> Notice how speed and velocity are not the same. This is because the person changed direction and didn't travel at the same rate the whole time. This is an average speed over the entire time, while the velocity is telling us the net effect of the journey. If the person had simply walked $1.4\frac{m}{s}$ to the left for 5 seconds, they would have ended up at the same result.

IRL   Cars have a speedometer. Is it really a speed-ometer or is it a velocity-o-meter? Since the meter only tells the speed and not direction, it is named correctly. It is a speed-o-meter. Does the speedometer measure average or instantaneous values? Every time the car is stopped, the speedometer reads zero, so it is reading how fast the car is going at that instant. Some cars also have a travel computer that indicates the average speed of the trip.

Acceleration

When you step on the gas, a car will speed up. In physics, we call this **acceleration**. When you step on the brake and the car slows down, physics still calls this acceleration. Positive acceleration means changing velocity toward the positive direction, while negative acceleration means changing the velocity toward the negative direction. This can be confusing.

Look at the figure on the following page. The top car has a positive velocity and acceleration, which means the car is moving to the right and accelerating to the right. So, the car is speeding up and goes faster and faster. The second car has a positive velocity and a negative acceleration, which means the car is going to the right but accelerating to the left. So, this car is slowing down.

Can you tell what is happening to the bottom two cars? The third car is moving to the left and slowing down because the velocity is to the left and the acceleration is to the right. The fourth car is moving to the left and speeding up because the velocity and acceleration are in the same directions.

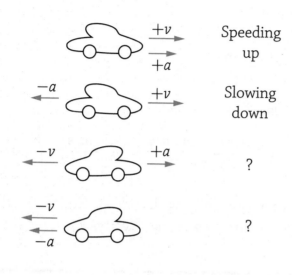

Acceleration is velocity divided by time: $a = \dfrac{\Delta v}{\Delta t}$.

Acceleration is a vector with units of meters per second squared.

EXAMPLE

Suppose we are late for an appointment. You jump in your car and stomp on the gas. You accelerate from zero to 50 m/s in 14 s but then you see a police motorcycle and hit the brakes, slowing to 30 m/s when the clock hits 19 s. See the following figure.

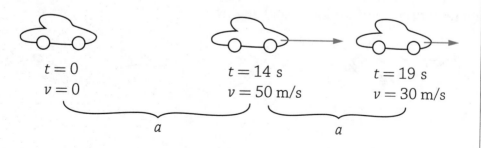

▶ Notice that your acceleration is to the right for the first part of the journey but to the left when you are slowing down. The acceleration for the first part of the trip is:

$$a = \frac{\Delta v}{\Delta t} = \frac{50\,\text{m/s}}{14\,\text{s}} = 3.6\,\text{m/s}^2$$

▶ After you hit the brakes, the acceleration is:

$$a = \frac{\Delta v}{\Delta t} = \frac{v_f - v_i}{t_f - t_i} = \frac{30\,\text{m/s} - 50\,\text{m/s}}{19\,\text{s} - 14\,\text{s}} = \frac{-20\,\text{m/s}}{5\,\text{s}} = -4.0\,\text{m/s}^2$$

▶ You had a greater acceleration when hitting the brakes. Notice that the acceleration is in the opposite direction of the velocity, which means you are slowing down.

IRL Smartphones have measuring devices inside called accelerometers that are used to measure the phone's changes in movement. They let the phone know what orientation it is in and flip the screen for you and know when you are walking so it can count your steps for the day.

Motion Diagrams

In the following figure a dog is running. Drawing all these pictures to represent the dog's motion is challenging.

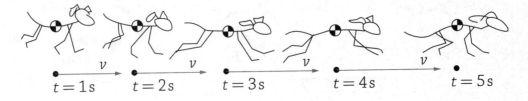

v v v v v

$t = 1\,\text{s}$ $t = 2\,\text{s}$ $t = 3\,\text{s}$ $t = 4\,\text{s}$ $t = 5\,\text{s}$

In physics, we simplify the picture by only drawing one point to represent the entire object. Usually, this point is the **center of mass** represented by the symbol: ◗. Below the dog is a dot to represent the position of the dog as it runs. This is called a **dot diagram** or a **motion diagram**. Times and arrows can be added to show when the dog was at each location and what velocity it was traveling.

EXAMPLE

▶ Look at the three motion diagrams presented next. Each dot is drawn one-tenth of a second apart. A ruler with major marking every centimeter is shown for scale.

▶ Which object is traveling faster: A, B, or C? We can tell that object C is moving the fastest because the dots that represent the object are spaced farther apart. By the same reasoning we can tell that B is the slowest because it's dots are more closely spaced.

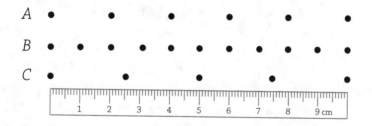

▶ Calculate the speed of the fastest and slowest object:

B travels 1 cm in 0.1 s for a speed of 10 cm/s.

$$\text{Speed} = \frac{\text{distance}}{\text{time}} = \frac{1\,\text{cm}}{0.1\,\text{s}} = 10\frac{\text{cm}}{\text{s}}$$

C travels 10 cm in 0.4 s for a speed of 25 cm/s.

$$\text{Speed} = \frac{\text{distance}}{\text{time}} = \frac{10\,\text{cm}}{0.4\,\text{s}} = 25\frac{\text{cm}}{\text{s}}$$

▶ In the next figure, notice how objects D and E travel the same distance in the 7 s, but the distances between the dots for object D are getting farther and farther apart. Object D started slow and accelerated to the right, while E is traveling at a constant velocity the entire time.

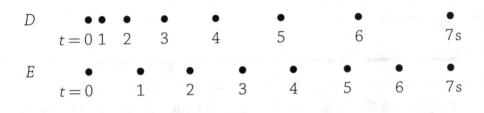

▶ Can you tell when the two objects are traveling the same velocity? Both will have the same velocity when they cover the same distance in the same amount of time. Between 3 s and 4 s both appear to travel the same distance in 1 second and must be traveling the same velocity.

Graphs

Dot diagrams help us visualize motion in a simple way, but sometimes we need more detail. For this we use graphs. There are three graphs you need to know in kinematics: position-time (x-t), velocity-time (v-t), and acceleration-time (a-t) graphs.

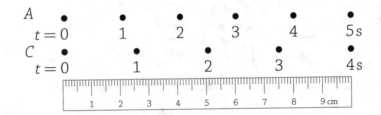

Two objects, *A* and *C*, are moving in the motion diagram shown next. With this information, we can create a data table and a position-time graph.

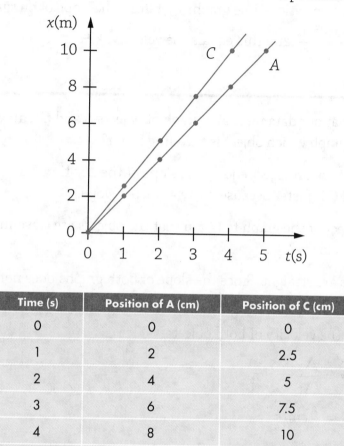

Time (s)	Position of A (cm)	Position of C (cm)
0	0	0
1	2	2.5
2	4	5
3	6	7.5
4	8	10
5	10	

To find the position of the object at any time, we simply need to read the graph. But how can we find how fast the objects are going? For this we need to calculate the **slope of the line**. From math class, remember that slope is the rise over the run, or slope $= \dfrac{\Delta y}{\Delta x}$. But our graph is not an *x-y* graph—it is a position-time graph.

To find the position of the object at any time, we simply need to read the graph—but how can we find how fast the objects are going? For this, we need to calculate the **slope of the line**. From math class we know that slope is the rise over the run, or slope $= \dfrac{\Delta y}{\Delta x}$. But our graph is not an x-y graph; it is a position-time graph. Therefore, the slope of our graph becomes $\dfrac{\Delta \text{position}}{\Delta t}$ and this equals the velocity: $v = \dfrac{\Delta x}{\Delta t}$.

EXAMPLE

▶ Look back at our data table and graph of objects A and C. Can you tell from the graph which object is traveling faster?

▶ The velocity of an object equals the slope of the position-time graph. Therefore, C is faster because it has a steeper slope.

▶ What aspect of the graph tells you that the objects are traveling at a constant velocity?

▶ Both lines are straight. Since the slope of both graphs does not change, both are traveling at a constant velocity.

▶ Calculate the velocity of both objects.

$$v = \frac{\Delta x}{\Delta t} = \text{slope}$$

$$v_A = 2\,\text{cm/s}$$

$$v_C = 2.5\,\text{cm/s}$$

Let's look at another example.

EXAMPLE

▶ Look at the motion diagram for object D presented next. Describe the motion. What is object D doing? (The object is moving to the right and speeding up.) Let's check your description by turning this motion diagram into a graph. Notice that the slope of the graph is getting bigger and bigger, indicating that the object is speeding up. This graph is curved because object D is accelerating. If the object is accelerating at a constant rate, the curve will be a **parabola**.

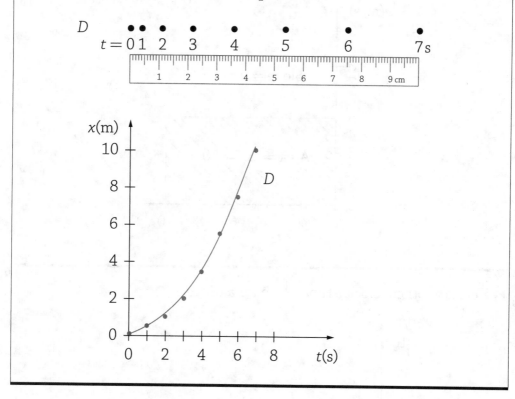

Now let's consider velocity-time (v-t) graphs. Like the name implies, these graphs tell us how fast and in what direction an object is going at any time. But velocity-time graphs also tell us the acceleration and displacement of the object. Look at the graph in the following figure. Describe the motion—but be careful! Don't confuse this graph with a position-time graph. The object is moving in the positive direction at a constant velocity of 5 m/s. The slope of the velocity-time graph is $\dfrac{\Delta v}{\Delta t}$, which equals the acceleration. We do not know where the object started, but we can tell how far it has gone by calculating the **area of the graph**. See the shaded area in the figure. This is true because the area equals length times width, which equals $v\Delta t = \Delta x$ on the v-t graph.

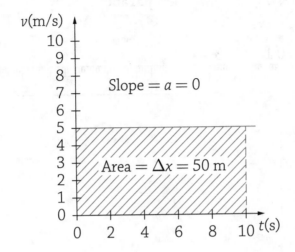

▶ Here is another velocity-time graph for a toy car.

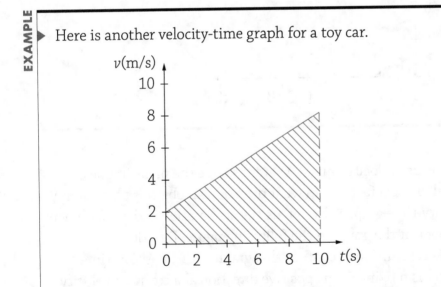

▶ Can you figure out the initial and final velocities of the toy car? Since this is a v-t graph, all we need to do is read the velocities off the graph. At $t = 0$, the car is traveling at 2 m/s and at $t = 10$ s the car has sped up to 8 m/s.

> What is the toy car's acceleration? Acceleration: $a = \text{slope} = \dfrac{\Delta v}{\Delta t} =$
> $\dfrac{(8\,\text{m/s}) - (2\,\text{m/s})}{10\,\text{s}} = 0.6\dfrac{\text{m}}{\text{s}^2}$.

> How far did the car travel over the entire 10 seconds? $\Delta x = $ area of the graph $= 50$ m. If you don't remember the formula for the area of a trapezoid, just break the shape into a rectangle and a triangle.

Our final graph is the acceleration-time graph (*a-t* graph). By reading this graph, we find the acceleration of the object at any time. Calculating the area of the graph will give us the change in velocity (Δv) of the object. Putting these three graphs together gives us a complete picture of three very common types of motion: **constant position**, **constant velocity**, and **constant acceleration**.

Imagine a car sitting in a parking lot. The car has a constant position. It does not move, or in physics terms, it is at **rest**. Let's plot the motion of this stationary car on our three graphs, which are shown next. The *x-t* graph is a constant horizontal line at the original location of the object. Notice how both the *v-t* and *a-t* graphs are zero.

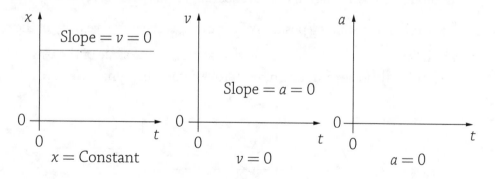

Now plot the motion of a car traveling at a constant velocity on the highway. The *v-t* graph has a constant velocity of v_0. The *x-t* graph will be a straight line because the constant slope equals the constant velocity of the car. Since the velocity is not changing, the acceleration will be zero.

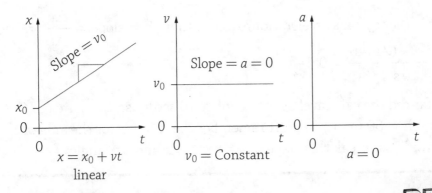

$$x = x_0 + vt$$
linear

$$v_0 = \text{Constant}$$

$$a = 0$$

Now imagine the car rolling slowly backwards and the driver applying the gas so that the car accelerates forward at a constant rate. The *a-t* graph has a constant positive value. The *v-t* graph begins with an initial negative value, but has a constant slope that equals the positive acceleration of the car. Notice that at point *T*, the car momentarily stops rolling backward and begins moving forward.

BTW

It is a good idea to get to know these three groupings of graphs because we repeatedly see constant position, constant velocity, and constant accelerated motion in physics.

The equation of this line is $v = v_0 + at$. The *x-t* graph is the most complicated. Since the slope of this graph equals the velocity of the car and the car's velocity is changing, the slope of this line will also be changing.

This produces a parabolic curve. The equation of this parabola is $x = x_0 + v_0 t + \frac{1}{2}at^2$ where x_0 is the starting position of the car, v_0 is the initial velocity of the car, and a is the acceleration of the car. Note that at point *T* the slope of the *x-t* graph is zero, indicating that the car momentarily stops before changing direction.

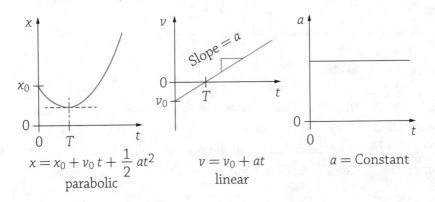

$$x = x_0 + v_0 t + \frac{1}{2}at^2$$
parabolic

$$v = v_0 + at$$
linear

$$a = \text{Constant}$$

Kinematics Equations

Motion diagrams give us a simple picture of motion. Graphs give us a more detailed picture of position, velocity, and acceleration. Our final representation will be algebraic. Our graphical representation of motion has given us two useful equations:

$$x = x_0 + v_0 t + \frac{1}{2} a t^2$$

$$v = v_0 + at$$

If we solve the bottom equation for t and substitute it into the top equation, we get a third equation:

$$v^2 = v_0^2 + 2a(x - x_0)$$

These equations can be used for any object that has a constant acceleration.

BTW

There are six variables in these equations. To get organized, I have my students create a table to keep them all straight. Any variable you don't know, label with a question mark. Indicate the variable you are trying to find.

EXAMPLE

▶ Let's go back to our imaginary car rolling slowly backwards and the driver applying the gas. The car is rolling backwards at 4 m/s when the driver applies the gas so that the car accelerates forward at a constant rate of 2 m/s².

▶ How long does it take for the car to stop rolling backward? Organize our data in a table:

x_0	0
x	?
v_0	−4 m/s
v	0
a	2 m/s²
t	What we want to find

▶ Since we are not given a starting position, assume it to be zero. We do not know the final position and are looking for t. Our starting velocity is negative because it is backwards, and the ending velocity is zero because that is when we stop rolling backwards. Which equation do we use? Use the equation that has the variable that we are looking for and where we know all the other variables in the equation:

$$v = v_0 + at$$

$$0 = -4\,\text{m/s} + (2\,\text{m/s}^2)t$$

$$t = 2\,\text{s}$$

▶ How far has the car gone after 10 s? Organize your data in a table:

x_0	0
x	What we want to find
v_0	−4 m/s
v	?
a	2 m/s²
t	10 s

▶ All of the initial condition variables stay the same, but now we are looking for the position at the 10 s mark where we do not know the velocity. Choose your equation and solve:

$$x = x_0 + v_0 t + \frac{1}{2}at^2$$

$$x = 0 + (-4\,\text{m/s})(10\,\text{s}) + \frac{1}{2}(2\,\text{m/s}^2)(10\,\text{s})^2$$

$$x = 60\,\text{m}$$

▶ What is the velocity of the car when it reaches 100 m? Organize your data in a table:

x_0	0
x	100 m
v_0	−4 m/s
v	What we want to find
a	2 m/s^2
t	?

▶ All of the initial condition variables are still the same, but now we are looking for the velocity at the 100 m position and we don't know the time it takes to get there. Choose your equation and solve:

$$v^2 = v_0^2 + 2a(x - x_0)$$

$$v^2 = (-4\,\text{m/s})^2 + 2(2\,\text{m/s}^2)(100\,\text{m} - 0)$$

$$v^2 = 416\,\text{m}^2/\text{s}^2$$

$$v = 20.4\,\text{m/s}$$

Here is another example for us to practice on.

You are driving at 15 m/s when a cute puppy runs out in front of your car 10 m ahead. You slam on your brakes. What acceleration do you need to stop in time? Organize your data and pick an equation to use:

x_0	0
x	10 m
v_0	15 m/s
v	0 (We have to stop so we don't hurt the puppy!)
a	What we want to find
t	What we want to find

$$v^2 = v_0^2 + 2a(x - x_0)$$

$$0 = (15\,\text{m/s})^2 + 2a(10\,\text{m} - 0)$$

$$a = -11.3\,\text{m/s}^2$$

▶ Do you know why the acceleration is negative? It's because the car is slowing down.

▶ How long does it take you to stop? Our data is already organized, as presented in the earlier table. We just need to pick an equation to use:

$$v = v_0 + at$$

$$0 = 15\,\text{m/s} + (-11.3\,\text{m/s}^2)t$$

$$t = 1.3\,\text{s}$$

Freefall

When you drop a baseball, it accelerates downward toward the floor. We call this the **acceleration caused by gravity** and it is designated by the symbol g. On Earth the acceleration from gravity is $g = 9.8$ m/s² downward toward the ground. This means that every second the baseball goes 9.8 m/s faster than the second before as long as we are ignoring the effects of air resistance.

 IRL On other planets, the acceleration caused by gravity will be different. On Mars, $g = 3.8$ m/s². On the Moon, $g = 1.6$ m/s². This means that objects fall slower on the Moon and Mars than on Earth.

EXAMPLE

▶ Let's put all of this together. You throw a baseball upward at 39.2 m/s. Sketch the y-t, v-t, and a-t graphs for the ball.

▶ Notice that our position graph shown here is for the y-direction this time.

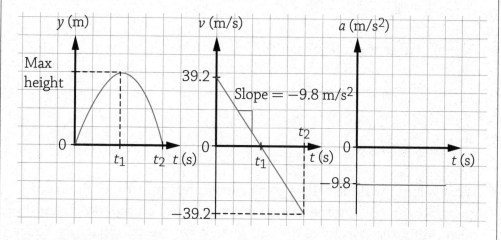

▶ Since the acceleration is downward and constant, the acceleration is a negative horizontal line and the velocity will begin at 39.2 m/s and have a slope of -9.8 m/s². The position graph is a parabola pointing downward because of the negative acceleration. Note that t_1, the time to reach maximum height, is half of t_2, the time to return to ground. Also note that the velocity is zero at the maximum height at t_1.

▶ Now let's use our equations. How long does it take the ball to stop going upward?

y_0	0
y	What we want to find
v_0	39.2 m/s
v	0 (at the top)
a	−9.8 m/s²
t	What we want to find

$$v = v_0 + at$$
$$0 = 39.2 \, \text{m/s} + (-9.8 \, \text{m/s}^2)t$$
$$t = 4 \, \text{s}$$

▶ How high does the ball go?

$$v^2 = v_0^2 + 2a(y - y_0)$$
$$0 = (39.2 \, \text{m/s})^2 + 2(-9.8 \, \text{m/s}^2)(y - 0)$$
$$y = 78.4 \, \text{m}$$

▶ What velocity will the ball be going when it returns to you? From our graph, we can see that the motion is symmetrical on the way up and down. Thus, the velocity of the ball when it returns is −39.2 m/s. You can prove this with equations as well.

Two-Dimensional (2D) Motion

So far, we have only considered motion in one dimension. What happens if the direction of motion changes?

The following figure shows a girl walking around in the city as she goes from her apartment, to a coffee shop, and then to work.

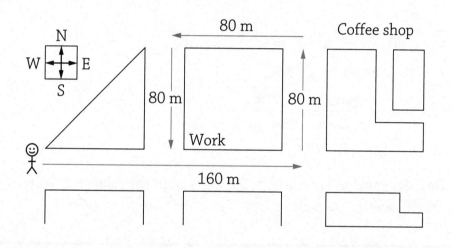

The distance she walked to work is 400 m, but her displacement is 80 m east from the starting position to work. From the starting position to the coffee shop her distance is 240 m, and her displacement is 179 m at an angle of 27° north of east.

Remember that displacement is a vector pointing from the starting point to the coffee shop. So, make a triangle, use the Pythagorean theorem, and use the trigonometry functions to find the magnitude and direction! See the following figure.

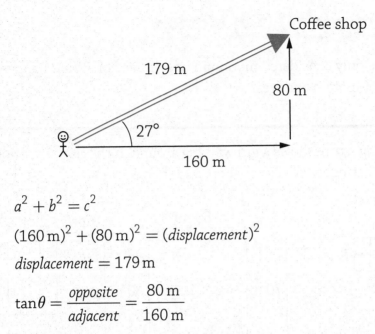

$$a^2 + b^2 = c^2$$

$$(160\,\text{m})^2 + (80\,\text{m})^2 = (displacement)^2$$

$$displacement = 179\,\text{m}$$

$$\tan\theta = \frac{opposite}{adjacent} = \frac{80\,\text{m}}{160\,\text{m}}$$

$$\theta = 27°$$

Circular motion is a common 2D motion. In the following figure, you see the Earth orbiting the Sun.

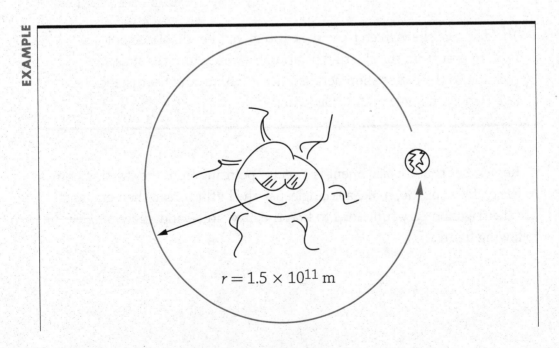

$r = 1.5 \times 10^{11}\,\text{m}$

▶ How fast is the Earth traveling around the sun? Velocity is $v = \dfrac{\Delta x}{\Delta t}$. The distance around a circle is the **circumference**: $C = 2\pi r$.

$$v = \frac{2\pi r}{\Delta t} = \frac{2\pi(1.5 \times 10^{11}\,\text{m})}{(365\,\text{days})(24\,\text{hours/day})(3{,}600\,\text{s/day})} = 30{,}000\,\text{m/s}$$

▶ This is the instantaneous velocity because the Earth is constantly changing direction and therefore accelerating.

Circular motion will be covered in more detail in Chapter 5.

 IRL   The Earth's orbit is actually a little elliptical, like all the planets. Pluto and comets have much more elliptical orbits.

Every time you throw an object through the air, it moves forward but also arcs toward the ground. This is a type of 2D motion called **projectile motion**. Consider the following figure as we go through a specific example.

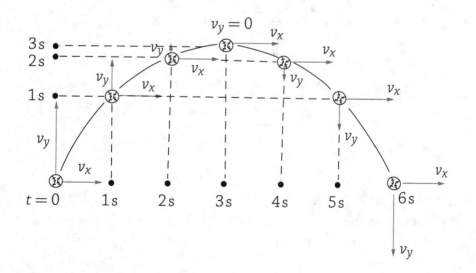

Projectile motion is two separate motions occurring at the same time. In the y-direction the ball accelerates downward at 9.8 m/s^2 due to gravity. But in the x-direction, there aren't any forces to speed up or slow down the ball so it travels at a constant speed. The figure shows a baseball thrown up and to the right. Notice how the motion diagram in the x-direction shows a constant velocity motion. The y-direction motion diagram shows the ball slowing down on the way up and speeding up on the way down. When we overlap these two motions, the path of the ball forms a parabolic arc. As long as air resistance isn't too great, everything that launches through the air follows a parabolic arc: footballs, Olympic gymnasts, and cars flying off cliffs.

Let's look at an example of 2D projectile motion.

EXAMPLE

▶ While filming an action movie, a director wants a car to be driven horizontally off a 20-m-high cliff at 30 m/s, as seen in the following figure. How long will it take the remotely controlled car to hit the ground, and how far will it travel?

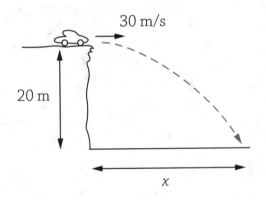

▶ Some of our data is x-direction information and some is y-direction information. We need to keep this information separate. Sort out the data in a table, keeping the "y-stuff" separate from the "x-stuff."

x_0	0	y_0	20 m
x	What we want to find	y	0
v_0	30 m/s	v_0	0
v	30 m/s	v	?
a	0	a	−9.8 m/s^2
t	What we want to find	t	What we want to find

▶ Notice in our table that only the time is shared in both columns because it is a scalar without direction. In the x-direction, the acceleration is zero and the x-velocity remains the same, leaving us with only this equation to use: $x = x_0 + v_0 t$. Both the time and final position are unknown, leaving the equation unsolvable. We will have to start with y-direction information instead. Setting up our y-equation, we can solve for the time to impact:

$$y = y_0 + v_0 t + \frac{1}{2}at^2$$

$$0 = 20\,\text{m} + 0 + \frac{1}{2}(-9.8\,\text{m/s}^2)t^2$$

$$t = 2.0\,\text{s}$$

▶ Now that we have the time, we can go back to the x-information and find out how far the car traveled horizontally before landing:

$$x = x_0 + v_0 t + \frac{1}{2}at^2$$

$$x = 0 + (30\,\text{m/s})(2.0\,\text{s}) + 0$$

$$x = 60\,\text{m}$$

▶ How fast will the car be going when it hits the ground below the cliff? The x-velocity stays the same, but the y-velocity is increasing because of gravity:

$$v_{y\text{-direction}} = 0 + (-9.8\,\text{m/s}^2)(2.0\,\text{s})$$

$$v = -19.6\,\text{m/s}$$

▶ To find the car's total resultant velocity, we need to add the x- and y-velocities using geometry:

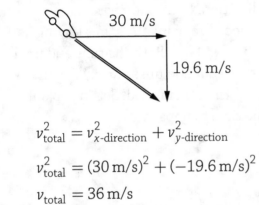

30 m/s

19.6 m/s

$$v_{\text{total}}^2 = v_{x\text{-direction}}^2 + v_{y\text{-direction}}^2$$

$$v_{\text{total}}^2 = (30\,\text{m/s})^2 + (-19.6\,\text{m/s})^2$$

$$v_{\text{total}} = 36\,\text{m/s}$$

▶ Yikes ... That's about 80 mph ... good thing the car was remotely controlled by the stunt coordinator!

REVIEW QUESTIONS

Let's review what we have learned about kinematics by answering the following questions.

1. Which are vectors and which are scalars: position, displacement, distance, velocity, speed, acceleration.

2. A boy walks 40 m east and then 30 m south in 100 s. What is the boy's distance and displacement, velocity, and speed for the journey?

3. Sketch a motion diagram of a car starting at rest, speeding up, and then traveling at a constant velocity.

4. Describe what the object is doing in this motion diagram.

$t = 0$ 2 s   4 s   6 s   8 s   10 s   12 s   14 s

5. Convert the information in this motion diagram into a position-time graph.

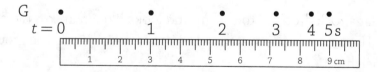

6. Calculate the speed of the object in the motion diagram in Question 5 from $t = 0$ to 2 s.

7. Describe the motion of the object in this position-time graph. What is the velocity of the object?

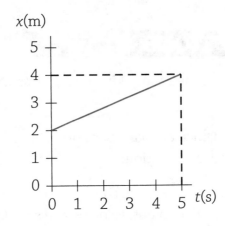

8. Describe the motion of the object in this velocity-time graph. How far has the object traveled in 5 s? What is the acceleration of the object?

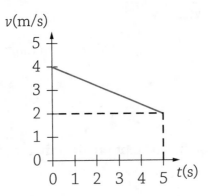

9. Sketch what a rock dropped to the ground looks like on an x-t, v-t, and a-t graph. Use a coordinate system where up is positive and the ground is the zero height.

10. A villain pushes a puppy off a building. Our hero grabs a parachute and jumps off the building a second after the puppy in order to catch it and save it. Will the hero be able to catch and save the puppy?

11. A girl throws a football straight up at 19.6 m/s. How high does it go, and how long does it take for the football to return to her?

12. A car is traveling 10 m/s when the driver steps on the gas and accelerates at 4 m/s² for 8 s. How far has the car traveled after 8 s? How fast is the car going at 8 s?

13. A horse on a carousel travels at 4.5 m/s and takes 6.8 s to make one complete trip around. What is the radius of the carousel?

14. A rock is dropped from a cliff at the same time an identical rock is launched horizontally from the cliff. Which rock hits the ground below first?

15. Describe the *x*- and *y*-velocities of an object thrown upward at an angle through the air.

Flashcard App

 **Forces**

MUST KNOW

⚡ Forces only exist when two objects interact with each other, *and* this interaction always creates an equal and opposite force-pair on the two objects.

⚡ Objects can't exert a force on themselves. Therefore, an object that's sitting still won't move, *and* an object that's already moving won't stop unless a different object exerts a force on it.

⚡ Forces from outside the object can accelerate the object.

I n the previous chapter, we developed several ways to describe the motion of things but never really talked about *why* objects move the way they do. In this chapter, our goal is to learn the "why" of motion. The question is: What makes certain objects sit still and others move about? In order to grasp this, we will explore the concept of a force. A force is a push or a pull that tries to change the motion of things. Forces are vectors, which means they have a magnitude and direction. A push to the right is different from a push to the left. Therefore, we will have to use a little geometry and trigonometry to understand force vectors. The unit for force is $kg \cdot m/s^2$ which we call a **newton** (N). I'll introduce you to some forces you encounter in daily life, and we will build a model for what forces do and how they can change the motion of objects.

Like many great scientists, Newton did a lot of great and also dumb things, but he is best known for his three laws of motion. They are actually the connection between forces and motion. Newton's laws are deceptively simple and far reaching. I think it is easier to understand Newton's ideas by first understanding that forces are an interaction between two different objects before we investigate what forces do to those objects. For this reason, I'm going to introduce Newton's Laws to you, starting with the Third Law.

Newton's Third Law

Forces only occur when one object interacts with another object. When an object exerts a force on another, the second object always replies with a duplicate force in the opposite direction. Forces are always produced in an equal magnitude, but opposite in direction pairs.

Stand up. Grab your shoelaces. Try to pull your feet off the ground. It's impossible. Why? Your hand is pulling your feet up, but your feet are pulling your hands down with an equal but opposite force. Since your hands and feet are connected by your body, you can't lift yourself up. This is why a single object cannot exert a net force on itself. We say that the forces are internal to the object and thus cancel out. OK, but what if your shoes are not connected to your feet? Now we have two separate objects. You pull the shoes up off the ground with a force and the shoes pull you down with the same force. Since you are bigger than the shoes, they move and you don't feel much. Now try to pull a 30-lb child up off the ground. You can probably still lift the child upward, but you start to notice the downward force that the child is exerting on you, and it becomes much harder on you to lift the object.

Here are some Third Law examples:

▶ A hammer is used to drive a nail. The two interacting objects are the hammer and the nail. The force from the hammer pushes the nail into the board. What does the force from the nail on the hammer do? It pushes upward and stops the hammer from moving downward.

▶ How does a person jump? You have to push down on the floor so that the floor pushes you upward with the same force in the opposite direction. Don't believe? Climb onto a bed and have a friend video you jumping upward. Now watch in slow motion. You'll see that you push the bed downward and that the bed pushes you upward.

▶ When you walk you will place your foot on the ground and push backward. This produces a force-pair between your foot and the ground. Your foot pushes the ground backward, and the ground pushes your foot forward. The force from the ground makes you move forward. Now, go to an ice-skating rink and try to walk forward. It is much harder, because the ice is slippery, making it harder for you and the ground to exert a force-pair on each other.

▶ A rocket pushes gas out the back end. In response, the gas pushes upward and launches the rocket into space.

Let's say that you get into a fight with a friend over who is better at physics. They get mad and hit you in the face with their fist. You have the satisfaction of knowing that your face exerted just as big a force back on their fist. The force from your face will cause their hand to slow down and possibly break their knuckle. Unfortunately, the force from their fist will make your head move backward and probably do more serious damage, because knuckle bones are stronger than face bones.

IRL In action movie fights it is not uncommon to see the "hero" head-butt the "bad guy." According to the Third Law, is this a good fighting strategy? No! The force the hero applies to the bad guy is the same as the bad guy applies to the hero. The hero is likely to get a concussion!

Newton's First Law

This law is commonly referred to as the law of **inertia**. Inertia is the idea that objects don't want to change what they are doing. An object will naturally do one of two things:

1. If already sitting still (at rest), an object will remain sitting still. (Velocity = Constant of Zero).

2. If already moving, an object will continue to move in a straight line at a uniform rate. (Velocity = Constant).

   Notice how rest is just a special case of constant velocity where the constant is zero.

So, it appears that the natural state of motion for our universe is constant velocity. Why? A force is required to change the motion of an object. Remember from Newton's Third Law that objects can't exert a net force on themselves. In order to change the motion of an object, we need an outside force from another object to change the motion of anything.

Here are some First Law examples:

▶ A book on a table won't start moving by itself.

▶ A bowling ball keeps rolling (until it hits something).

▶ In space, constant velocity motion is even easier to see. Pull up a video of astronauts in space and see for yourself. When an astronaut lets go of their pen it just floats there without moving. If the pen accidently has a velocity when the astronaut lets go of it, the pen will keep moving in a straight line until it runs into something.

▶ Why do we wear seat belts in a car (or other vehicle)?

▶ Because, once a car is in motion, we're also in motion. If a car comes to a sudden stop, our bodies want to stay in motion. Seat belts prevent us from continuing to move (and spread the force through the torso).

Here is the bottom line for Newton's First Law: If there are no outside forces acting on an object, it will be in a state of constant velocity—at rest or moving in a straight line at a constant speed. This is also true if there are outside forces acting on the object, but they cancel out. In physics terms, the sum of the forces is zero or the net force is zero ($\sum F = 0$).

 IRL Baby car seats face backwards because a baby's neck is not strong enough to stop their head in a forward collision. A baby seat also has side pads for support and is inclined so the baby's neck doesn't have to do much work in an accident and won't get injured.

Newton's Second Law: $\sum F = F_{net} = ma$

When there is a net external force acting on an object, it will accelerate. More precisely: The vector sum of the external forces on an object will cause the object to accelerate in the same direction as the net force. Objects with less mass are more easily accelerated. Objects with more mass are harder to accelerate due to their higher inertia. The equation that represents this is $\sum F = F_{net} = ma$. Notice that Newton's First Law is a special case of the Second Law when acceleration is zero!

Here are some Second Law examples:

EXAMPLES

▶ A small child sits in a swing. The parent gives the child a push to accelerate them.

▶ An older child sits in a swing. It takes more force to accelerate the more massive child.

▶ A car sits at a stop light. The light turns green. The driver steps on the gas. This causes a force forward that accelerates the car forward. A greater force will accelerate the car more quickly.

▶ A car is traveling forward toward a stop sign. The driver applies the brakes, which causes a force in the opposite direction the car is going in order to accelerate the car backwards and stop the car.

Here is the bottom line for the Second Law:

■ If there is a net force on an object, it will accelerate in the direction of that net force.

■ The larger the mass of the object, the harder it is to accelerate.

■ The equation $\sum F = F_{net} = ma$ only relates force to mass and acceleration. If you want to know the velocity or displacement of an object, you will need to use what we learned in the last chapter on kinematics.

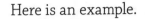

Here is an example.

> A 50-kg person is standing still when a friend pulls them to the right with a force of 40 N. A second friend pulls them to the left with a force of 140 N. Does the person accelerate? If so, what is the acceleration and direction?
>
> There is a net force of 100 N acting on the person to the left; therefore, they will accelerate to the left at 2 m/s².
>
> $$\sum F = F_{net} = ma$$
>
> $$40N - 140N = (50kg)a$$
>
> $$a = -2m/s^2$$

Common Forces We Encounter in Everyday Life

Newton's Laws seem simple enough until we realize that there are lots of forces in the world and usually more than one acts on an object at the same time. The equation $\sum F = ma$ tells us to "sum up" all the forces acting on the object. So, first we have to know what kind of forces there are.

Here's a list of common forces we encounter and a description of their behavior.

Force	Symbol	Equation/Description
Gravity	F_G	$$F_G = G\frac{m_1 m_2}{r^2}$$
		This is the general equation for the gravitational force between any two masses (m_1 and m_2) in kilograms separated by a distance (r) in meters. G is the universal gravitational constant: $G = 6.67 \times 10^{-11} \, Nm^2/kg^2$.
		Gravity always attracts the two masses toward each other. We use this equation to find the gravity forces between astronomical bodies like planets, moons, and stars.

Force	Symbol	Equation/Description
Weight	F_g	$$F_g = mg$$ This is a simplified equation for gravity that is only good to find the force of gravity on an object near the surface of a planet. Gravity always acts downward toward the center of the planet. g is the acceleration caused by gravity. (g is also called the gravitational field at the surface of the planet.) On Earth, $g = 9.8\text{m/s}^2$, but on the Moon $g = 1.6\text{m/s}^2$.
Normal	F_N	The normal force is a contact force between objects. The normal force (F_N) always pushes perpendicular away from the surface that the object is in contact with. Objects must touch for there to be a normal force. For example, when you stand on the floor, the floor exerts a normal force upward on your feet perpendicular to the floor. If you jump off the floor, the normal force disappears because you are no longer in contact with the floor.
Friction	F_f	$$F_f \leq \mu F_N$$ Friction is another contact force between objects that try to slide past each other. The force of friction (F_f) is always parallel to the surface. Friction acts in the opposite direction to the sliding or intended sliding direction of an object across a surface. For example, if you slide across the floor in your socks to the right, friction on you from the floor will be to the left. μ is the coefficient of friction between the two surfaces in contact and depends on how rough or smooth the surfaces are. The coefficient of friction does not have any units. F_N is the normal force.
Tension	F_T	The tension force tries to pull an object back into its normal shape when it is being tugged or stretched. The tension force (F_T) pulls back through objects like strings and ropes towards their center. F_T is distributed throughout the string and pulls inward from both ends of the string, rope, cable, etc. Imagine two people in a tug-of-war with a rope between them. The rope is under tension. The rope will apply an inward force on both of the people pulling on the rope.
Spring, or Elastic	F_s	$$F_s = k\Delta x$$ The spring force (F_s) always acts opposite in direction to the way the spring or rubber band is stretched or compressed (Δx) in meters. Spring force depends on how stretchy the spring is. This is measured by the spring constant (k), which has units of newton/meter. If you stretch a spring, the spring will pull inward from both ends. If you compress a spring, the spring will push outward from both ends. Springs are discussed in more detail in Chapter 5.

Force	Symbol	Equation/Description
Buoyancy	F_b	$$F_b = \rho V g$$ Objects that are surrounded by a fluid will experience a buoyancy force due to the pressure of the fluid on the object. The buoyancy force depends on the density (ρ) of the fluid and how much volume (V) of fluid is displaced by the object. Buoyancy force always pushes an object upward in the opposite direction of gravitational acceleration (g). This force is discussed in detail in Chapter 6.
Electric	F_E	$$F_E = \left| k\frac{q_1 q_2}{r^2} \right|$$ The electric force acts between two charged objects (q_1 and q_2) in coulombs separated by a distance (r) in meters. k is the Coulomb's law constant: $k = 9.0 \times 10^9 \text{Nm}^2/\text{C}^2$. Electric charge comes in two types: positive or negative. Opposite charges attract. Like charges repel. The electric force is discussed in more detail in Chapter 10.
Magnetic	F_M	$$F_M = qvB\sin\theta$$ A magnetic force occurs on a moving charge (q) in coulombs moving at a velocity (v) in meters per second through a magnetic field (B) measured in units of tesla. Magnetic forces are discussed in more detail in Chapter 13.

"The Procedure" for Setting Up and Solving $\sum F = ma$ Problems

Take a look back at the forces table just presented. A lot of different forces can act in different directions on an object. So, the real difficulty when working with forces is building the "$\sum F$" or net force side of the $\sum F = ma$ equation. What we need is a procedure to help us assemble Newton's Second Law: $\sum F = ma$.

Here is the procedure I use with my students to help them keep everything organized. If you follow these steps, you'll be led through a process that facilitates setting up Newton's Second Law correctly.

1. Draw a **free body diagram** (FBD). This is also called a **force diagram**.

2. Determine if the object is accelerating.

3. Set up a coordinate system parallel and perpendicular to the acceleration.

4. Build $\sum F = ma$ for the two directions of your coordinate system.

5. Solve!

Use these steps every time, even if you don't think you need them! Let's discuss each of these steps in detail one at a time.

Step 1 Draw an FBD

A free body diagram is a simple picture that shows all the forces acting on the object and helps you visualize what the forces and the object are doing. Here is how to draw an FDB.

- First, draw a *simple* picture of the object. When I say simple, I mean simple. This is not artwork. You will see that in the free body diagram that follows, I am drawing boxes to represent the object.

- Then draw an arrow for each force vector that acts on the object. These force arrows should

  - Be attached/connected to the object and point away from the object in the direction of the force.

  - Have a length that represents their appropriate magnitude.

    - Small forces should have short vectors.

    - Large forces should have long vectors.

    - Equal forces should have arrows of the same length.

You can also mark vectors equal with hatch marks on the vector, like you do in geometry to show that angles or sides of a triangle are equal.

- *Never, never, ever* draw any other vector arrows attached to the object because it just confuses the drawing.

  - Never draw displacement, velocity, or acceleration vectors attached to your FBD.

  - You may draw these vectors off away from your FBD as reference, but never attached to it.

Here are three examples of free body diagrams. Before we look at them, let's remind ourselves of the following:

▶ Normal force, F_N, is always perpendicular to the surface.

▶ Friction, F_f, is always parallel to the surface and opposite to the sliding direction.

▶ Gravity, $F_g = m_g$, is always downward toward the center of the Earth.

▶ Tension, F_g, always acts along the rope or string that is being pulled on.

▶ Draw bigger forces with longer arrows.

▶ If you don't know the name of a force, just give it a name like I did on the rocket FBD. The name of the force is called "thrust" but I just called it F.

A child is pulling a box at a constant speed across a rough surface that has friction.

$v =$ Constant

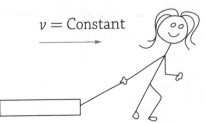

FBD for the box

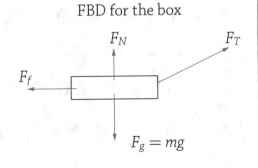

A cow on a skateboard is accelerating down a hill with negligible friction because of the wheels on the skateboard.

FBD of cow on skateboard rolling down a hill

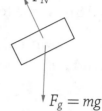

A rocket, with engines firing, is accelerating off the launch pad.

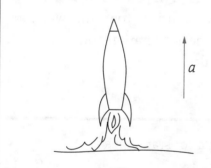

FBD of rocket

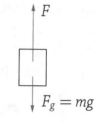

Step 2 Determine If the Object Is Accelerating

If the object is at rest or moving at a constant velocity, the acceleration is zero and the forces on the object must cancel out. Objects that are not accelerating are said to be in **equilibrium**. When in equilibrium, $\sum F = 0$.

If the object is accelerating, it is very important to determine which

BTW

A few key phrases can tell us if there is acceleration or not. Look for these phrases:
- *Rest (or velocity) equals zero.*
- *Constant velocity (or velocity) is a constant.*
- *Acceleration is zero (or not accelerating).*

Each statement means that a = 0. Therefore, either v = 0 or v = constant, and that means that the object is in equilibrium and all the forces cancel out.

$$\sum F = 0.$$

direction it is accelerating in because accelerating objects must have a net force in the direction of the acceleration. Always ask yourself: Is it accelerating in the x-direction, y-direction, or some other direction? For example:

- A cannonball shot through the air will be accelerating downward at $a = g = 9.8 \, \text{m/s}^2$ due to gravity.

- A ball rolling down a hill accelerates down the hill parallel to the inclined surface.

Step 3 Set Up a Coordinate System Parallel and Perpendicular to the Acceleration

Set up the coordinate system so that one axis is in the direction of the acceleration, if there is any, and the other perpendicular to it. This way you will always get $\sum F = ma$ in the direction of the acceleration and $\sum F = 0$ in the other direction. This will simplify your calculations!

▶ Draw an FBD for the box pulled by girl at a constant speed

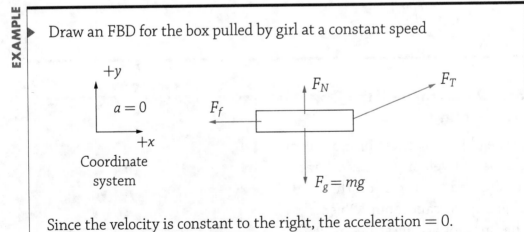

Since the velocity is constant to the right, the acceleration $= 0$. Therefore $\sum F = 0$ in every direction. Picking a standard x-y coordinate system is best in this example.

Notice that all the forces are in the x-y direction except the tension force.

▶ Draw an FBD of cow on skateboard rolling down a hill

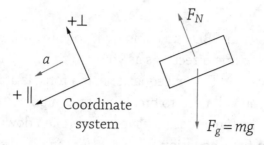

Because the cow accelerates down the hill, it is best to choose an inclined coordinate system. This gives us:

$\Sigma F_{\parallel} = ma_{\parallel}$ in the parallel direction

$\Sigma F_{\perp} = 0$ for the perpendicular axis

The skateboarding cow is not accelerating perpendicular to the hill, so all the forces must cancel out in that direction.

▶ Draw an FBD of the rocket

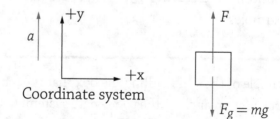

The rocket accelerates up, so we will use a standard x-y system. Thus:

$\Sigma F_y = ma_y$ and $\Sigma F_x = 0$

Step 4 Build $\Sigma F = ma$ for the Two Directions of Your Coordinate System

Look at each force in your FBD and compare it to your coordinate system. Are your forces in the same directions as your coordinates? If so, you can go ahead and build your $\Sigma F = ma$ equations. If a force is at an angle to the coordinate system, you will need to break it into components before building $\Sigma F = ma$. After all the force vectors have been broken down, all that is left to do is build the net force equation, $F_{net} = \Sigma F = ma$, and solve!

Step 5 Solve

The examples we just worked through didn't have any numerical values. They were all symbolic. I did this on purpose so that you could see how to set the problems up. This is an important skill to learn because it allows you to see the "physics" in the equations and the relationship between the variables. Always work out physics problems with symbols first and only put numbers in, if there are any, at the very end.

　　Steps 4 and 5, building and solving $\Sigma F = ma$, are demonstrated in each of the following examples.

▶ Notice in our example of the child pulling the box that the tension from her pulling on the rope is at an angle to the coordinate system. It must be broken into x- and y-components. I labeled them F_{Tx} and F_{Ty} in the drawing.

▶ Draw an FBD for the box pulled by the child at a constant speed:

$+y$

$+x$

$a = 0$

F_f

F_N　$F_{T_y} = F_T \sin \theta$

F_T

θ　$F_{T_x} = F_T \cos \theta$

$F_g = mg$

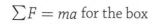

$\sum F = ma$ for the box

x	y
$\sum F_x = 0$	$\sum F_y = 0$
$F_{T_x} - F_f = 0$	$F_N + F_y - mg = 0$
$F_{T_x} = F_f$	$F_N + F_y = mg$
$\boxed{F_f = F_T = \cos\theta}$	$\boxed{F_N + F_T \sin\theta = mg}$

▶ In our example of the cow on a skateboard, Here the weight of the cow must be broken into vectors parallel and perpendicular to the incline.

Draw an FBD of cow on skateboard rolling down a hill

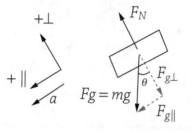

$\sum F = ma$ for cow

<u>Perpendicular</u>

$$F_N - F_{g\perp} = 0$$
$$F_N = F_{g\perp}$$
$$\boxed{F_N = mg\cos\theta}$$

<u>Parallel</u>

$$\sum F_\| = ma_\|$$
$$F_{g\|} = ma_\|$$
$$mg\sin\theta = ma_\|$$
$$\boxed{g\sin\theta = a_\|}$$

▶ In our rocket example, all the forces are in the y-direction and nothing more needs to be done.

FBD of rocket:

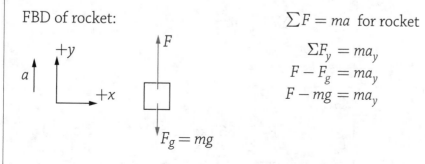

$\sum F = ma$ for rocket

$$\sum F_y = ma_y$$
$$F - F_g = ma_y$$
$$F - mg = ma_y$$

This has been a lot of technical information and may seem intimidating, but, like any skill, it can be learned with practice. So, let's look at some specific examples. In our first example, we have an object that is not accelerating.

▶ A string holds up a 4.0-kg weight. Calculate the tension in the string.

▶ The weight is not accelerating; therefore, $\sum F = 0$, and the tension force upward must cancel the downward force of gravity. See the following figure.

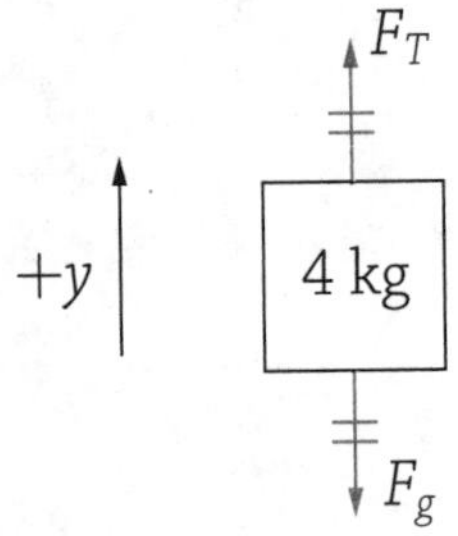

$$\sum F_y = 0$$
$$F_T - F_g = 0$$
$$F_T - mg = 0$$
$$F_T = mg$$
$$F_T = (4.0\,\text{kg})(9.8\,\text{m/s}^2)$$
$$\boxed{F_T = 39\,\text{N}}$$

F_T (up), 4 kg box, F_g (down), $+y$

Now let's look at an example where the object is accelerating vertically.

▶ An 860-kg elevator is accelerating downward at 1.5 m/s². What is the tension in the cable that holds up the elevator?

▶ The elevator is accelerating downward; therefore, the tension upward must be less than the force of gravity. We will use a y-direction coordinate system. See the following figure.

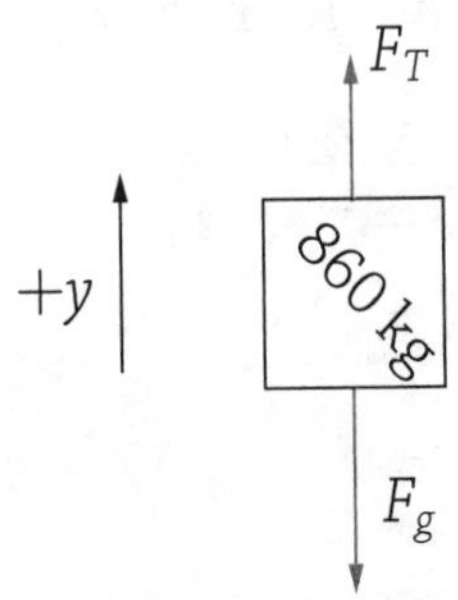

$$\sum F_y = ma$$
$$F_T - F_g = ma$$
$$F_T - mg = ma$$
$$F_T = mg$$
$$F_T = m(a + g)$$
$$F_T = (860\,\text{kg})(1.5\,\text{m/s}^2 + 9.8\,\text{m/s}^2)$$
$$\boxed{F_T = 9,718\,\text{N}}$$

F_T (up), 860 kg box, F_g (down), $+y$

Now consider an object that is accelerating horizontally across a floor.

▶ A 6.4-kg box moving across a floor at 3.8 m/s slides to a stop in 1.9 m. Calculate the force of friction.

▶ Since the problem didn't say which way the box was sliding, let's choose to the right. The box is slowing down. This means the box will be accelerating to the left. We'll use an x-direction coordinate system. See the following figure. This problem will take a little longer to solve because we have to calculate the acceleration. Notice how the acceleration is negative because it is in the opposite direction of the velocity.

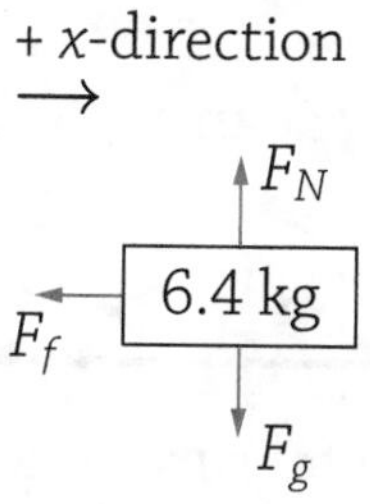

+ x-direction
→

Finding acceleration:

$x_0 = 0$
$x = 1.9\,\text{m}$
$v_0 = 3.8\,\text{m/s}$
$v = 0$
$a = $
$t = ?$

Find F_f:

$$v^2 = v_0^2 + 2a(x - x_0)$$
$$0 = (3.8\,\text{m/s})^2 + 2a(1.9\,\text{m})$$
$$a = -3.8\,\text{m/s}$$

substitute

$$\Sigma F_x = ma$$
$$F_f = ma$$
$$F_f = (6.4\,\text{kg})(-3.8\,\text{m/s}^2)$$
$$F_f = -24\text{N}$$

▶ Now let's calculate the coefficient of friction between the floor and the box.

▶ The coefficient of friction is the ratio of the friction force divided by the normal force. It is always positive.

Finding the coefficient of friction:

$$F_f = \mu F_N$$

Remember that $F_N = mg$ in this situation.

$$\mu = \frac{F_f}{F_N}$$

$$\mu = \frac{F_f}{mg} = \frac{24N}{(6.4\,kg)(9.8\,m/s^2)}$$

We drop the negative sign on friction because coefficient of friction is always positive.

$$\mu = 0.38$$

Here is an example where one of the forces is acting at an angle. We will need to break the force into its components.

▶ A child pulls a 35-kg sled to the left along a horizontal surface covered by snow with a force of 22 N at an angle of 30° above the horizontal. See the following figure. Assume the friction is negligible. Calculate the acceleration of the sled and the normal force from the ground pushing up on the sled.

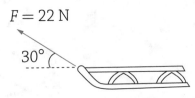

$F = 22\ N$

$30°$

▶ To answer this question, let's choose an x-y coordinate system with left being positive because the sled will accelerate to the left.
See the following figure. We have to set up and solve $\sum F = ma$ for both the x- and y-directions.

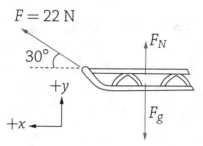

$$F = 22\,\text{N}$$

$$\sum F_x = ma$$
$$F\cos 30° = ma$$
$$(22\,\text{N})\cos 30° = (35\,\text{kg})a$$
$$a = \frac{(22\,\text{N})\cos 30°}{35\,\text{kg}}$$
$$\boxed{a = 0.54\,\text{m/s}^2}$$

$$\sum F_y = 0$$
$$F\sin 30° + F_N = F_y$$
$$F\sin 30° + F_N = mg$$
$$F_N = mg - F\sin 30°$$
$$F_N = (35\,\text{kg})(9.8\,\text{m/s}^2)$$
$$\quad - (22\,\text{N})\sin 30°$$
$$\boxed{F_N = 332\,\text{N}}$$

For our last example, let's look at a situation where our object is sitting at an angle. This will require us to use an angled coordinate system.

▶ A 1,600-kg truck is sitting stationary on a 20° inclined hill (see the following figure). How large is the friction force that holds the truck in place, and how large is the normal force from the hill on the truck?

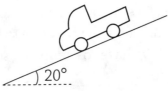

Even though the truck is not accelerating, it's going to tilt the coordinate system because we're looking for friction and normal force, which are parallel and perpendicular to the incline. (See the next figure.) This means we need to break gravity into parallel and perpendicular components.

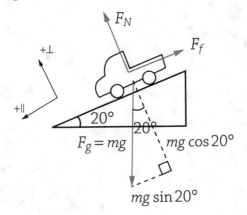

$$\Sigma F_\perp = 0$$
$$F_N - mg\cos 20° = 0$$
$$F_N = mg\cos 20°$$
$$F_N = (1{,}600\,\text{kg})$$
$$(9.8\,\text{m/s}^2)\cos 20°$$
$$\boxed{F_N = 14{,}700\,\text{N}}$$

$$\Sigma F_\| = 0$$
$$F_f - mg\sin 20° = 0$$
$$F_f = mg\sin 20°$$
$$F_f = (1{,}600\,\text{kg})$$
$$(9.8\,\text{m/s}^2)\sin 20°$$
$$\boxed{F_f = 5{,}400\,\text{N}}$$

REVIEW QUESTIONS

Let's practice Newton's Laws and forces by answering the following questions.

1. You and a friend are on an ice-skating rink. You give your friend a push forward. What happens to you?
   a. You will move forward in the same direction as your friend.
   b. You will stay where you are.
   c. You will move backward in the opposite direction of your friend.

2. A father and his young daughter are ice skating together. The girl says, "Push me, Daddy!" So the father stands behind the girl and launches her forward with a big push. Use Newton's Second and Third Law to explain what, if anything, happens to the father due to this push.

3. A car is traveling at a constant speed along a straight road. Are any forces acting on the car? If so, what are they? Is a net force acting on the car? Explain your reasoning.

4. Give an example of a Newton's Third Law force-pair. Explain what each force is doing to the other object.

5. You are in your car at a stop light when you are rear-ended by a careless driver. How does the headrest keep you safe?

6. A magician has a glass on a table. She says she is going to pull the white tablecloth out from under the glass. To add more drama, she fills the glass with red wine. How does filling the glass with wine help her with her stunt?

7. You are a passenger in a car when the driver makes a hard left turn. Use the concept of inertia to explain why you feel pushed against the passenger door.

8. Which way does the force of gravity always point?

9. Explain when there is a normal force and friction force. Also explain in which direction they point.

10. A football is flying through the air. Why does the football accelerate downward? Why does the football continue to move forward at a constant rate?

11. You are playing with a 0.20-kg yo-yo. The yo-yo moves up and down. (See the following figure.) At some points it is stationary, at others it is accelerating upward, and at other times it is accelerating downward. When is the tension in the string greatest and least? Calculate the tension in the string when the yo-yo is accelerating upward toward your hand at 3.0 m/s².

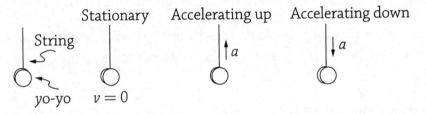

12. A 30-kg child running at 4.5 m/s slides across a slippery floor and comes to a stop in 2.2 s. (See the following figure.) Calculate the force of friction and the coefficient of friction between the child's socks and the floor.

13. A skier slides down a 40° mountain slope on super-slippery, no-friction snow. (See the following figure.) What is the skier's acceleration?

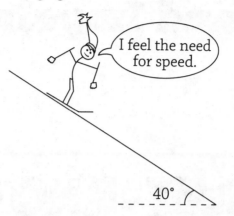

14. A father is holding his 26-kg child at a 30° angle in a swing (see the following figure). What horizontal force must the father push to hold the child in this position?

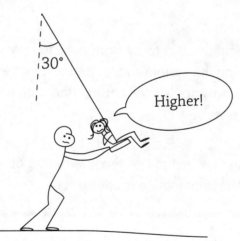

Flashcard App

3 Momentum and Impulse

MUST ⚡ KNOW

⚡ Momentum is a vector description of mass in motion.

⚡ An impulse is a force from outside a system that acts on the system over a period of time. A net external impulse will change the momentum of the system.

⚡ When we include all the interacting objects in a system, momentum is conserved.

omentum is a concept that pops right out of Newton's Second Law when you write it in a different form. Let's see how:

Start with Newton's Second Law:

$$\sum F = ma$$

$$\downarrow$$

Substitute in the definition of acceleration, $a = \dfrac{\Delta v}{\Delta t}$, to get:

$$\sum F = m\frac{\Delta v}{\Delta t}$$

$$\downarrow$$

Multiply both sides by Δt to get rid of the fraction:

$$\sum F \Delta t = m\Delta v$$

$$\downarrow$$

Substitute in the definition of Δv to get:

$$\sum F \Delta t = m(v_f - v_i)$$

$$\downarrow$$

Distribute the mass:

$$\sum F \Delta t = mv_f - mv_i$$

And we are done!

In this chapter, we will analyze this new equation and see how it gives us a different picture of the world and how it operates. You will learn what **momentum**, **impulse**, and a **system** are. Most importantly, we'll get our first look at one of the all-important conservation laws of physics. So, let's get started!

Momentum

The right-hand side of our new equation has the term, mv, which is called momentum. Think of momentum as a vector way to represent mass in motion. Momentum has units of kg m/s and since it is a vector, the direction the object is going in makes a difference. There is a special symbol for momentum: $p = mv$. It is a little odd that the symbol is the letter p. You might guess it should be m but that letter is already used for mass. Since we have run out of letters, scientists chose the symbol that was derived from a Latin word.

Question: Does a car traveling north at 60 mph have the same momentum as an identical car traveling south at the same speed? No! Momentum is a vector. Both cars would have the same magnitude of momentum, but the two cars have opposite directions. One would have a positive momentum and the other would have a negative momentum. Another question: Which has a greater magnitude of momentum: a bus or a car? Well. . . that depends. How fast are they going? If both are moving at the same speed, then the bus, which has a larger mass, would have more momentum. Yet another question: Can a mosquito have the same momentum as a bus? Sure! There are two possibilities. If both are stationary, then both have a momentum of zero. If the bus is moving, the mosquito would have to be flying extremely fast to have the same momentum as the bus because the bus has a much larger mass:

$$(mv)_{mosquito} = (mv)_{bus}$$

So, what have we learned so far? Momentum is a vector, which means the direction of the motion makes a difference. The momentum of objects is zero when they are not moving. Momentum is the combination of mass times velocity, which means you can have a large momentum if you have a large enough mass, or a large enough velocity, or both. Now look back at the right side of our new equation: $mv_f - mv_i$. This represents the change in momentum and can be rewritten as $mv_f - mv_i = p_f - p_i = \Delta p$.

EXAMPLE

▶ The driver of a 1,100-kg car hits the brakes and slows the vehicle down from 40 m/s to 25 m/s in 4.6 s. Let's find three different momentum-related quantities:

▶ What is the initial momentum, p_i, of the car?

$$mv_i = (1{,}100\,\text{kg})(40\,\text{m/s}) = 44{,}000\,\text{kg}\cdot\text{m/s}$$

▶ What is the final momentum, p_f, of the car?

$$mv_f = (1{,}100\,\text{kg})(25\,\text{m/s}) = 28{,}000\,\text{kg}\cdot\text{m/s}$$

▶ What is the car's change in momentum, Δp?

$$\Delta p = p_f - p_i = (28{,}000\,\text{kg}\cdot\text{m/s}) - (44{,}000\,\text{kg}\cdot\text{m/s}) = -16{,}000\,\text{kg}\cdot\text{m/s}$$

▶ Δp is negative because it is in the opposite direction the car was originally moving.

Let's look at another example.

EXAMPLE

▶ A 0.145-kg baseball is thrown from the pitcher at 45 m/s and is hit directly back to the pitcher at the same speed. What is the change in momentum of the baseball? Don't jump to a conclusion! Remember that momentum is a vector.

▶ The answer is not zero. Watch out for the signs on the velocity. I assumed that the initial velocity was +45 m/s. When the ball is hit back toward the pitch, the velocity will be negative.

$$\Delta p = m(v_f - v_i) = 0.145\,\text{kg}\big((-45\,\text{m/s}) - (+45\,\text{m/s})\big)$$
$$= -13\,\text{kg}\cdot\text{m/s}$$

BTW

A common mistake when calculating momentum is to forget about the direction in which the object is traveling. A car traveling to the right along the positive x-axis will have a positive momentum. When the car turns around and travels in the negative direction, it will also have a negative momentum because the velocity is now negative.

Impulse

The left side of our new equation, $F\Delta t = \Delta p$, is called **impulse**. Impulse is an external force that acts over a period of time to change the momentum of an object. Impulse is a vector with units of newton times seconds (N · s). Note that the change in momentum of the object will equal the impulse and be in the same direction as the impulse as well.

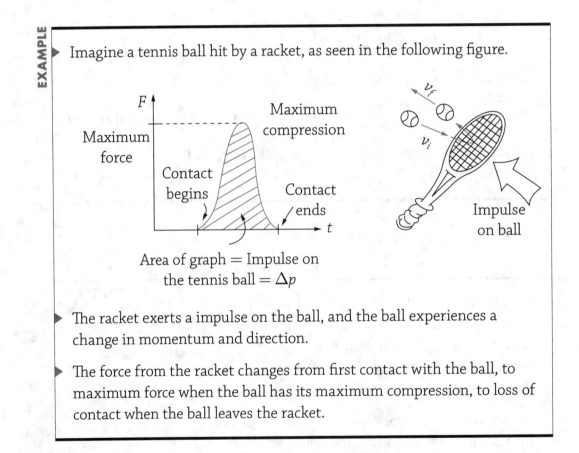

Imagine a tennis ball hit by a racket, as seen in the following figure.

Area of graph = Impulse on the tennis ball = Δp

▶ The racket exerts a impulse on the ball, and the ball experiences a change in momentum and direction.

▶ The force from the racket changes from first contact with the ball, to maximum force when the ball has its maximum compression, to loss of contact when the ball leaves the racket.

This is typical of all collisions. The area under the force curve equals the impulse.

▶ Safety engineers work hard to prevent injuries in accidents. Imagine a 1,000-kg car traveling 20 m/s colliding with a tree and coming to a complete stop.

▶ The car experiences a change in momentum of $\Delta p = m(v_f - v_i) = 1{,}000\,\text{kg}(0 - 20\,\text{m/s}) = -20{,}000\,\text{kg} \cdot \text{m/s}$. This means there must have been an impulse of $F\Delta t = -20{,}000\,\text{kg} \cdot \text{m/s}$. (The negative sign means the impulse was in the opposite direction of the initial velocity.)

▶ Now consider how much time it took to stop the car. If the time is longer, the force will be smaller, but if the time between the beginning and end of contact is small, the force can be very large. Notice in the following figure that the shorter collision time produces a much larger force. In the graph, the long collision time is approximately three times longer. This means the maximum force will be only about a third of the short collision time crash. Engineers use seat belts and airbags to increase the time it takes a person to slow down during a car crash. This reduces the forces on passengers and saves lives in accidents.

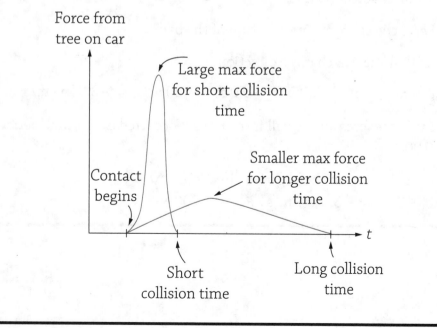

This is an important finding. This is why it hurts to fall on concrete but is fun to fall down on a bed. The bed gives and extends the collision time, making the force more enjoyable. Other examples of this phenomena are running shoes, football pads, and boxing gloves. All are designed to extend the time of impact and minimize the force on the athlete. However, there are times when you want the force of impact to be large, like when you use a hammer. The head of a hammer is made of solid steel with little flexibility. This ensures that when you strike a nail you get maximum force to drive the nail into a board. Hammers wouldn't be very useful if they were made of something squishy.

 IRL   The safety equipment in a car is designed to extend the time of the collision and reduce the forces on the passengers. Seat belts stretch. Airbags cushion. Dashboards bend. Even the front and rear of the car are designed to crumple up to reduce injuries and deaths in accidents.

EXAMPLE

▶ A tennis racket hits a 0.059-kg tennis ball traveling at 20 m/s back the other direction at 30 m/s.

▶ What is the change in momentum of the ball?

▶ Note that the ball changes direction:

$$\Delta p = m(v_f - v_i) = 0.059 \,\text{kg}(-30 \,\text{m/s} - 20 \,\text{m/s}) = -3.0 \,\text{kg} \cdot \text{m/s}$$

▶ What is the force on the ball if the impact occurred over a time period of 0.0025 s?

▶ The force will be in the negative direction: $F\Delta t = \Delta p$,
$F(0.0025 \,\text{s}) = -3.0 \,\text{kg} \cdot \text{m/s}$, $F = -1{,}200 \,\text{N}$

Systems: What They Are and Why They Are Important

A **system** is a collection of objects that are being considered together for physics purposes. By choosing the system strategically, we make the physics easier to figure out or show important details that are otherwise missed. For momentum, we want to choose a system that includes all the objects that are interacting and are creating net forces on each other.

For example: A car collides with a truck. The two exert a force on each other, so we would want to include both in our system. When we do this the force between the two vehicles becomes an internal force to the system and no longer affects the overall motion of the system. This is very important. The internal parts of our system, the car and truck, will have changes to their motion, but the overall motion of the system as a whole won't change at all. This leads us to the idea of conservation of momentum.

Conservation of Momentum

Consider two pool balls, the cue ball and the eight-ball. The eight-ball is initially at rest while the cue ball is traveling toward it. See the following figure.

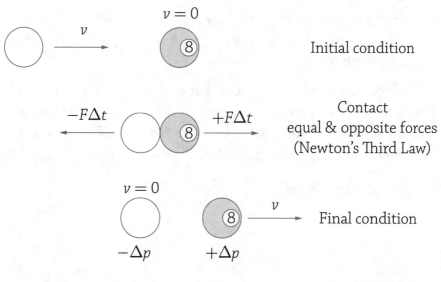

When the two collide, they exert Newton's Third Law forces on each other that are equal and opposite in direction, and the time of contact between the balls is also identical. Therefore, both balls experience an equal but opposite in direction impulse, which means they receive equal but opposite changes in momentum. Notice that the eight-ball gets an increase in momentum and the cue ball has a negative change in momentum. By combining the two balls into a common system, we find that the system as a whole has no change in momentum at all. This is called **conservation of momentum**: $p_i = p_f$. Whenever we include all of the interacting objects into the system so that there aren't any outside net forces, the momentum will be conserved. The internal parts of the system exchange momentum, but the overall momentum of the system remains unchanged. Brilliant!

So here are the rules of conservation of momentum:

1. Choose a system so that there aren't any net external forces to produce external impulses that will change the momentum of the system. For example: When two cars collide, we need to include both cars in the system. You don't need to include the Earth because the gravity on the cars is canceled by the normal force from the road on the cars. We also are assuming that forces like friction and air resistance are small enough that they don't produce a significant external impulse and can be ignored.

2. You must keep track of all the objects both before and after the interaction. For example: When two cars collide, we need to track their velocity both before and after the collision.

3. All objects must be free to move. For example: When two cars collide, both are free to move. But when a car hits a tree, momentum will not be conserved because the tree is connected to the Earth, which exerts forces on the tree preventing it from moving.

Let's consider an example.

EXAMPLE

A stationary 1,000-kg cannon fires an 80-kg clown at a speed of 75 m/s. See the following figure. What is the recoil speed of the cannon?

Before

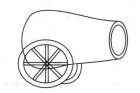

After

v_{cannon}

v_{clown}

▶ The two interacting objects are the cannon and the clown, so we must include them both in our system. The clown is free to move forward, and let's assume the gun can move backward as well. Therefore, momentum should be conserved for the system. Next, I have set up the conservation of momentum equation. Notice that the initial momentum is zero.

$$p_i = p_f$$

$$\left(mv_{cannon} + mv_{clown}\right)_i = \left(mv_{cannon} + mv_{clown}\right)_f$$

▶ Note that the initial velocity of both the cannon and clown are zero, so they disappear from the equation.

$$0 = \left(mv_{cannon} + mv_{clown}\right)_f$$

$$0 = \left((1{,}000\,\text{kg})v_{cannon} + (80\,\text{kg})(75\,\text{m/s})\right)_f$$

$$v_{cannon} = -6.0\,\text{m/s}$$

▶ Why is the velocity of the cannon negative? The force that pushes the clown forward pushes the gun backward.

▶ Now think about yourself running forward. You push the Earth backward with friction between your foot and the ground. Meanwhile, the Earth is pushing you forward with an equal and opposite friction forward. This is just like the clown and the cannon, except the Earth is so incredibly massive that its recoil velocity is too small to notice.

Let's look at another example.

EXAMPLE

▶ Two train cars collide as shown in the following figure. What is the final velocity of the train car on the right? (Note that I am choosing left as the positive direction for velocity in this problem so we don't have negative numbers.)

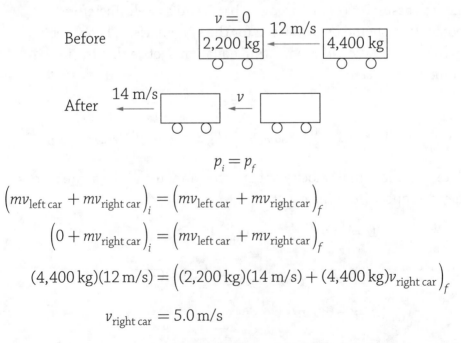

$$p_i = p_f$$

$$\left(mv_{\text{left car}} + mv_{\text{right car}}\right)_i = \left(mv_{\text{left car}} + mv_{\text{right car}}\right)_f$$

$$\left(0 + mv_{\text{right car}}\right)_i = \left(mv_{\text{left car}} + mv_{\text{right car}}\right)_f$$

$$(4{,}400\,\text{kg})(12\,\text{m/s}) = \left((2{,}200\,\text{kg})(14\,\text{m/s}) + (4{,}400\,\text{kg})v_{\text{right car}}\right)_f$$

$$v_{\text{right car}} = 5.0\,\text{m/s}$$

Center of Mass

The center of mass is the point on an object where the object can be balanced and is marked by the symbol ◓. Think of the center of mass as a single point that describes the location of all the mass of our system. In the figure, the center of mass for our two pool balls would be halfway between the cue ball and the eight-ball because they have the same mass.

If the objects aren't the same mass, the center of mass will be closer to the heavier object, like for our railroad cars.

Remember from Chapter 2 that objects in motion stay in motion and if they are at rest, they will stay at rest unless there is a net outside force acting on the object. We called this Newton's First Law. Now consider what we've done when we define a system of interacting objects. We specifically choose a system where there isn't a net outside force acting on it. This means that our collection of objects will obey Newton's First Law! That means the parts of our system may be moving all about, but the motion of the center of mass of our system won't change.

EXAMPLE

For example, a large asteroid is sitting at rest in space. A company wants to mine the asteroid for natural resources. They plan to blow the asteroid up and bring the smaller pieces back to Earth and sell them. When the asteroid blows up, it breaks into many pieces that go all over the place, but the center of mass of the system stays at rest where it started. See the following figure.

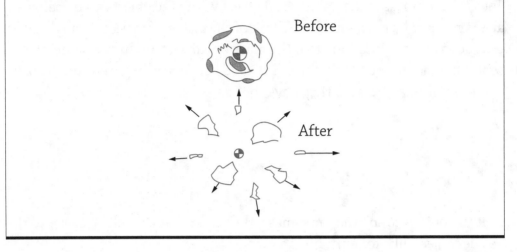

IRL One of Earth's great disasters happened 65 million years ago when a huge asteroid hit the Earth, causing the extinction of 70% of all life on Earth, including all the dinosaurs. To prevent this from happening in the future, NASA tracks near-Earth asteroids and comets to see if they are likely to hit us. You might think that a good way to save the Earth from an approaching asteroid is to blow it up with nuclear weapons. That's actually not the best idea. Remember that the center of mass of all the asteroid rubble after it is nuked will still be on track to hit us. We would have to pulverize the asteroid into small enough pieces so that it would all burn up in the atmosphere. That would be very hard to do. A much easier plan is to hit the asteroid with nukes while it is far enough away and just nudge it off course so that it misses us completely.

Let's look a couple of examples demonstrating center of mass.

A person stands on one side of a boat, as seen in the following figure. When the person walks to the left side of the boat, what happens to the boat and the center of mass of the person-boat system?

As the person walks left, the boat moves to the right so that the center of mass stays in the same spot. See the following figure.

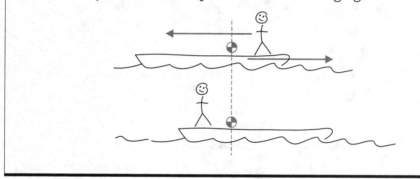

Now consider an example with two astronauts floating in space.

Two astronauts are in space connected by a tether, as shown in the following figure. The astronaut on the right has a motor that reels in the tether. Where is the center of mass of the two astronauts? Describe the motion of both astronauts and the motion of the center of mass.

The center of mass will be between the two astronauts. See the next figure. The motor will pull both astronauts inward toward the center of mass. Since there are no outside forces, the two will meet at the center of mass, and the center of mass will not move.

Types of Conservation of Momentum: Collisions and Kinetic Energy

Physics describes collisions or interactions between objects by what happens to the **kinetic energy** in the collision. Kinetic energy is the energy of motion and will be discussed in more detail in the next chapter. For now, think of kinetic energy as a scalar way to describe mass in motion. Momentum is a vector way to describe mass in motion. There are four types of collisions:

■ **Explosions**   These are also called separation collisions. In these collisions, the system gains kinetic energy from some other type of energy source. For example, when a bomb explodes, chemical energy is converted into kinetic energy. Another example would be two people standing next to each other and they push each other apart. They had no motion before the interaction but after the interaction both are moving. A key feature of these collisions is that the system explodes apart. Hence the name.

■ **Elastic**   These are collisions where the objects bounce off each other and the kinetic energy of the system remains exactly the same. Collisions between atoms are elastic, which will be explained in more detail in Chapters 7 and 8.

Collisions between atoms and subatomic particles like protons and electrons are elastic because they don't actually touch each other when they collide. They repel each other with the electric force and "bounce" off each other before they connect.

■ **Inelastic**   These are also bouncing-type collisions but some of the kinetic energy is lost to sound, heat, and the deformation of the objects colliding. An example of an inelastic collision would be a car crash because these interactions produce sound and heat and damage the vehicles. Another inelastic collision is bouncing a basketball on the floor. If you drop the basketball, it will not bounce back to the same height because some of the kinetic energy is transformed into other types of energy in each collision. Most collisions are inelastic.

It is very important to remember that momentum is conserved in all of these different types of collisions!

■ **Completely inelastic**   This is a special case of inelastic collisions where the interacting objects stick together after the collision. An example would be a football player grabbing and tackling another player. In completely inelastic collisions, the maximum amount of kinetic energy is lost while still obeying conservation of momentum.

Here is an example of a completely inelastic collision where the two interacting objects stick together after the collision.

EXAMPLE

▶ A 90-kg football player has the ball and is running toward the goal at 6 m/s when he is met head on by a 140-kg linebacker running at 4 m/s. Assume the linebacker tackles and holds onto the other player. Will there be a touchdown?

$$p_i = p_f$$

$$\left(mv_{runner} + mv_{linebacker}\right)_i = (m_{runner} + m_{linebacker})v_{both\ players_f}$$

$$(90\,kg)(6\,m/s) + (140\,kg)(-4\,m/s) = \left((90\,kg + 140\,kg)v_{both\ players}\right)_f$$

$$v_{both\ players_f} = -0.087\,m/s$$

▶ The final velocity is in the direction of the linebacker, so no touchdown.

2D Collisions

Let's say you are out having fun with your friends at the bowling alley. You are trying to pick up a single pin spare. The roll is a little to the side, and the bowling ball hits the pin off center and the pin shoots off at an angle as shown in the following figure. What happens to the ball?

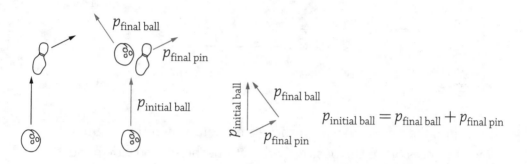

Remember that momentum is a vector. The original momentum of the ball was in the y-direction with no x-momentum at all. Therefore, the final momentum of the ball-pin system must also have only y-momentum and no x-momentum after the collision. Therefore, the ball must move off to the left with enough negative x-momentum to cancel out the positive x-momentum of the pin. The combined y-momentum of the ball and pin must equal the original y-momentum of the ball before the collision. See the figure again for reference.

IRL Police forensics investigators can use conservation of momentum to reconstruct car accidents to calculate the speed and direction of cars before and after the collision to determine if a driver was speeding or otherwise at fault.

Let's take a look at an example.

A clown shot from a cannon collides with and bounces off a second clown, as shown in the following figure. Sketch the direction the second clown takes after the collision and explain your reasoning.

$p_{\text{final clown 1}}$

$p_{\text{initial clown 1}}$

The original momentum was in the x-direction. When the original clown bounces off in the positive y-direction, the second clown must move in the negative y-direction. The x-momentum of the two clowns must equal the original x-momentum of the clown shot from the cannon.

$p_{\text{final clown 1}}$

$p_{\text{initial clown 1}}$

$p_{\text{final clown 2}}$

REVIEW QUESTIONS

Let's practice our momentum skills by answering the following questions.

1. Calculate the initial momentum, final momentum, and change in momentum of a 1,600-kg car that goes from 20 m/s to 45 m/s.

2. What is the change in momentum of a 0.056-kg golf ball that hits the floor at 15 m/s and bounces back up at 12 m/s? If the collision took 0.03 s to occur, calculate the force from the floor on the golf ball.

3. How do you catch a water balloon without popping it?

4. What would hurt worse: Hitting a brick wall and stopping, or hitting a brick wall and bouncing off backward?

5. An astronaut is trapped out in space floating away from his spaceship. All he has is a wrench. How will he get safely back to the spaceship?

6. Two carts of mass m and $2m$ have a spring between each other, as seen in the following figure. A string holds the two carts together and compresses the spring. The string is cut, and the carts fly apart with the right cart moving away at a velocity of v_R. What is the velocity of the left cart in terms of v_R? What type of collision is this? Explain your answer.

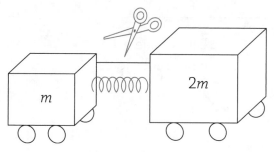

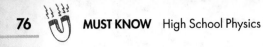
7. A 1,900-kg truck moving 15 m/s north collides with a 1,000-kg car traveling south at 30 m/s. The cars stick together after the collision. What is the final speed of the mangled wreckage?

8. A 20-kg child running at 4.5 m/s with an exercise ball collides with her stationary grandfather whose mass is 80 kg. The grandfather gets pushed forward at 1.0 m/s after the collision. What is the velocity of the granddaughter after the collision?

9. A pool player has only one more shot to make to win a game of 8-ball. The player lines up the cue ball and calls her shot to be the top right pocket, as shown in the following figure. She hits the cue ball and pockets the 8-ball in the correct pocket as shown. Does she win the game? Explain your answer.

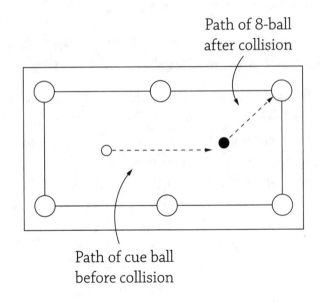

Path of 8-ball after collision

Path of cue ball before collision

10. A truck is traveling east while a car is traveling north, as seen in the following figure. They collide in the intersection at the point marked with an *x* in the figure. Assuming the cars stick together after the collision, in which of the directions will the two vehicles go off together? Explain your reasoning. Hint: Remember that momentum is a vector quantity.

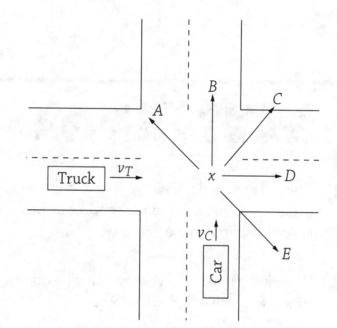

4 Energy and Work

 MUST KNOW

 There are many different types of energy, and all of them involve the movement and/or location of the object carrying the energy.

Work is the transfer of energy from one object/system to another through a force.

When you include all the objects that are interacting into one single system, we find that the energy of the system can be transferred around between objects and transmuted from one form into another form, but the total energy of the system is always conserved.

nergy is the most all-encompassing idea in all of science. Energy weaves its way through physics, biology, chemistry, and environmental science, from the largest scales of astronomy, to the smallest scales of nuclear physics, down to the fabric of space itself. If you learn only one concept in all of science, energy is what you should concentrate on. For all its greatness, energy is actually a very hard thing to define. The most common definition is that energy is the capacity to do **work** or make things move. Unfortunately, this is not a very satisfying definition. It turns out the more basic the concept, the harder it is to define, so let's describe energy in a way that you get a better feel for what it is.

Energy is a scalar idea. Remember that a scalar is a magnitude— only without any direction. This makes it very different than velocity, acceleration, forces, and momentum, which are all vector ideas and require the use of trigonometry. Being a scalar, energy does not have a direction and the math is much easier, in general, to solve.

Energy is not a physical thing. You can't touch it or feel it. Instead, energy is carried by something, either because of its movement or its location. For example, a car has energy because of its movement. A rock on the top of a mountain has more energy than one in the valley because of its location. A pot of boiling water has energy because of the vibrations of the atoms inside. Nuclear fuel in a reactor has energy because of the arrangement of the atoms inside the fuel. So, energy is related to the movement and location of objects.

Energy cannot be directly measured; it has to be calculated. Energy is always conserved, and it all has to be accounted for. An object doesn't just magically gain energy or lose energy. Energy has to move into, or out of, the object from someplace else for its total energy account to change. Energy can transform from one form into another and move from one object to another, but the total amount stays the same. Energy is like the accounting system of the universe.

The units for energy are $kg \cdot m^2/s^2$ which is called a **joule** (J). All forms of energy have the same unit and can be mathematically added to determine the total energy of the system.

Our goal in this chapter is to understand and be able to calculate the value of common forms of energy to discover how energy moves from place

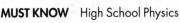

to place and transforms from one form into a different form and how energy is conserved.

Kinetic Energy

Kinetic energy is the energy of organized motion. That means all the atoms of the object are moving together. This motion comes in two forms: **translational** and **rotational**. The person on the left in the following figure is translating (changing location), the one in the middle is rotating (spinning about a point), and the last is translating and rotating at the same time. (In this chapter we will concentrate on translational kinetic energy and will discuss rotational kinetic energy in Chapter 5.) The equation for translational kinetic energy is $K = \frac{1}{2}mv^2$ where m is the mass of the object in kg and v is the velocity of the object in m/s.

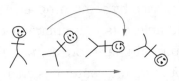

| Translation kinetic energy only | Rotation kinetic energy only | Both translation & rotational kinetic energy |

IRL What kind of kinetic energy does the Earth have? The Earth is moving through space and also rotating, so it has both translational and rotational kinetic energy.

EXAMPLE

► Imagine a 1,000-kg car traveling in the positive x-direction at 40 m/s and an identical car traveling in the opposite direction at −40 m/s. Do the two cars have the same kinetic energy?

▶ Notice that the velocity is squared, so it does not matter which direction the vehicle is going; kinetic energy is always positive or zero if it isn't moving.

$$K_{car} = \frac{1}{2}mv^2 = \frac{1}{2}(1,000 \text{ kg})(40 \text{ m/s})^2 = 800,000 \text{ J}$$

▶ What happens to the energy of the car if it doubles from 40 m/s to 80 m/s? Kinetic energy is proportional to the velocity squared: $K \propto v^2$. Therefore, when the velocity doubles, the kinetic energy would quadruple.

Potential Energy

Potential energy is a stored form of energy with the capability of being released to cause motion. All forms of potential energy are due to the relative positions of the interacting parts of the system. An object by itself cannot have potential energy. Only when two or more objects are interacting do we have potential energy. Think of a rubber band. By itself, it has no potential energy. It's only when another object comes along and stretches it that it has stored energy ready to be released. For this reason, we always talk of the potential energy of the system of interacting objects.

Gravitational Potential Energy (U_g)

This form of stored energy is due to the gravity between two objects.

▶ Imagine a glass sitting on a table as seen in the following figure.

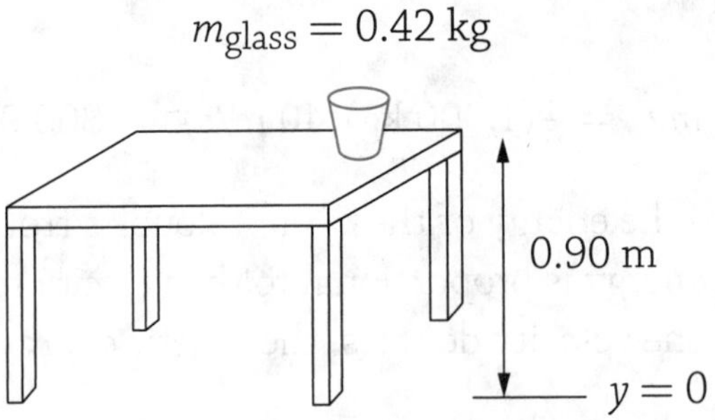

$$m_{glass} = 0.42 \text{ kg}$$

0.90 m

$y = 0$

▶ The Earth pulls the glass downward and the glass pulls up on the Earth. The Earth-glass system has potential energy ready to be converted into motion if only given the chance to fall. The equation of gravitational potential energy is $U_g = mgy$ where m is the mass of the object in kg, $g = 9.8 \text{ m/s}^2$, and y is the vertical location of the object relative to a reference point. Calculate the gravitational potential energy of the glass relative to the floor.

$$U_g = mgy = (0.42 \text{ kg})(9.8 \text{ m/s}^2)(0.90 \text{ m}) = 3.7 \text{ J}$$

▶ What is the gravitational potential energy of the glass if it fell to the floor? Zero. Now imagine that this table is on the second floor of an apartment building and we walk the glass down to the first floor. The glass now has negative gravitational potential energy. That's no big deal—negative potential energy just means we are below the zero point we selected. You get to pick where the zero height is, so we usually try to avoid negative potential energy by picking the lowest point of the object's motion to be the zero point.

▶ Not everyone will choose the same zero point, so we usually write our equation in terms of change in energy ($\Delta U_g = mg\Delta y$) to indicate that no matter what reference you pick, the change in energy will be the same.

Let's look at another example.

EXAMPLE

▶ Imagine that you have a safe that you need to move to your third-floor apartment. (See the following figure.) You can move the safe one of four ways: A) Take it in the front door and up the stairs. B) Move it below the window and use ropes to pull it up to the apartment. C) Build a ramp and push it up through the window. D) Rent a helicopter and drop it in through the window.

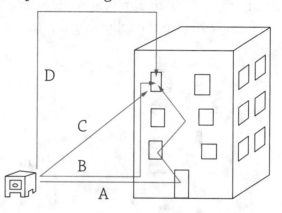

▶ Which of these paths will create the biggest change in gravitational potential energy for the safe from beginning to end?

▶ They are all the same! Each path has the exact same change in height in our equation, $\Delta U_g = mg\Delta y$.

This is an important feature of energy; the pathway it takes to get from one place to another does not matter. The change in energy will remain the same.

Spring/Elastic Potential Energy

This form of potential energy is due to a spring or other elastic object that is being deformed by another object to create built-up energy as the spring tries to return to its normal shape. The equation for spring potential energy is $\Delta U_s = \frac{1}{2}k(\Delta x)^2$, where Δx is the distance the spring has been stretched or compressed in meters. k is the spring constant in units of N/m. The spring constant is a number that tells us how "springy" the spring is. Literally it tells us how many newtons of force it takes to stretch or compress the spring 1 meter.

EXAMPLE

▶ A spring attached to the ceiling has an original length of 0.30 m and a spring constant of 45 N/m. (See the following figure.)

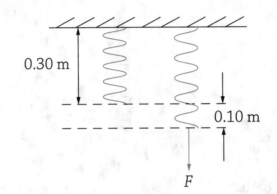

0.30 m

0.10 m

F

▶ The spring is stretched downward 0.10 m by a force. How much potential energy does the spring have?

$$\Delta U_s = \frac{1}{2}k(\Delta x)^2 = \frac{1}{2}(45 \text{ N/m})(0.10 \text{ m})^2 = 0.23 \text{ J}$$

▶ What length would the spring have in order to give it four times the energy as when it was stretched 0.10 m? To generate four times the potential energy, we need to stretch the spring twice as far: $\Delta U_s \propto x^2$. That means the final spring length needs to be the original length plus an extra stretched length of 0.20 m, for a total length of 0.50 m.

Mechanical Energy and Other Forms of Potential Energy

Kinetic energy, gravitational potential, and spring potential energies are grouped together under the umbrella of what is called **mechanical energy** because they all deal with physical objects. There are many other forms of energy besides mechanical that will be explored later in the book: internal and thermal energy in Chapters 8 and 9, electric potential energy in Chapter 10, wave energy in Chapters 15 and 16, photon energy in Chapter 19, and nuclear/mass energy in Chapter 20. For now, we will concentrate on mechanical energy.

Work

In the English language work refers to a job, a way to make money, or something that is hard or difficult to do. In physics, the word **work** means something completely different. Work is the transfer of energy from one object/system to another through a force.

EXAMPLE

▶ A parent picks up their child from the floor, as seen in the following figure.

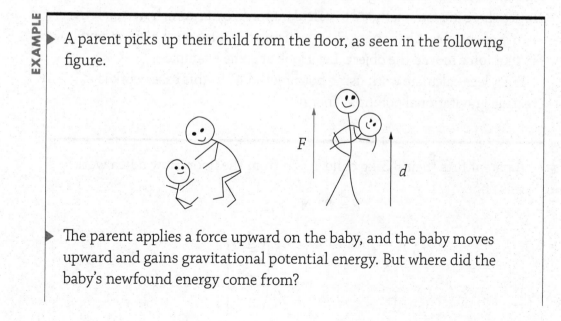

▶ The parent applies a force upward on the baby, and the baby moves upward and gains gravitational potential energy. But where did the baby's newfound energy come from?

▶ It was transferred from the parent to the child. The parent's energy decreased and the baby's energy increased. This exchange is called work. We say the parent did work *on* the baby because that is the way the energy moved, from the parent to the child.

IRL When you lift and move objects around all day, you are doing work, which transfers energy from you to these other objects. To replenish your energy, you eat chemical potential energy (food), which your body converts into thermal energy (heat to keep you warm), electrical energy (nervous system and brain), and kinetic energy (pumping blood and moving your limbs). In the United States, our food labels list calories, which is another unit of energy. Food labels in European countries list the energy content of food directly in joules. The next time you travel, check out the food labels and keep your energy levels up.

Here is how we calculate work: $W = \Delta E = Fd\cos\theta$, where ΔE is the change in energy of the system in joules, F is the force in newtons causing the change in energy, d is the displacement in meters that the force moves the object, and θ is the angle between the force and displacement vectors. Notice there are two methods to calculate work: by finding how much the energy of the object has changed or by using the force on the object and how far that force moved the object. Let's look at some examples:

First, let's calculate work using both methods. In this case, work is creating gravitational potential energy.

EXAMPLE

▶ A parent lifts their 8.3-kg child 1.3 m from the floor. How much work was done?

$$F = 81 \text{ N}$$

$$d = 1.3 \text{ m}$$

▶ Thinking in terms of energy:

$$W = \Delta E = \Delta U_g = mg\Delta y$$
$$= (8.3 \text{ kg})(9.8 \text{ m/s}^2)(1.3 \text{ m}) = 110 \text{ J}$$

▶ Thinking in terms of forces: The force upward to lift the child will be equal to the child's weight.

▶ Since the lifting force is upward and the displacement is also upward, the angle θ will be zero. $W = Fd\cos\theta = F_g d\cos 0 = mgy = 110 \text{ J}$.

▶ Notice we get the same answer if we calculate work using the change in energy or forces and displacement. Both methods will always agree.

> **BTW**
>
> *When calculating work, using the change in energy is almost always easier because energy is a scalar concept; therefore, the math will be easier to do. So, always think about the situation in terms of energy first when trying to calculate work.*

Sometimes there are forces on the object and the object may even be moving but no work is being done.

EXAMPLE

▶ A 1,650-kg sports car is driving along at a constant velocity of 50 m/s. How much work is done on the car while it moves along for 1,000 m? (See the following figure.)

50 m/s

▶ Thinking in terms of energy: The car is traveling at a constant rate. Its kinetic energy is not changing. Nor is there any indication that its potential energy is changing. Therefore, the work is zero.

▶ Thinking in terms of forces: In a free body diagram of the car we see that there is a gravity force downward being canceled by a normal force upward and a friction force propelling the car forward being canceled by an air drag force backward.

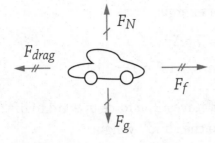

▶ The forces must cancel because the car is not accelerating. Since there is no net force on the car, the work must be zero:

$W = Fd\cos\theta = (0)d\cos\theta = 0.$

Be careful when the force is at an angle to the direction of the motion! Only the part of the force in the direction of the motion does any work. In this next example, kinetic energy is being produced.

▶ A 42-kg wagon is pulled 3.1 m to the right by a force at an angle as shown in the following figure. How much work is done?

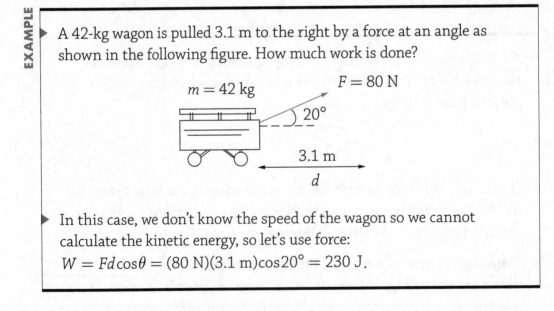

▶ In this case, we don't know the speed of the wagon so we cannot calculate the kinetic energy, so let's use force:

$W = Fd\cos\theta = (80 \text{ N})(3.1 \text{ m})\cos 20° = 230 \text{ J}.$

Here is an example that involves negative work.

A 1,300-kg car traveling 36 m/s in the positive x-direction stops. How much work was done just to stop the car?

Negative work is required to stop the car:

$$W = \Delta E = \Delta K = \frac{1}{2}mv_f^2 - \frac{1}{2}mv_i^2 = 0 - \frac{1}{2}(1,300 \text{ kg})(36 \text{ m/s})^2$$
$$= -840,000 \text{ J.}$$

A net force was needed to stop the car. What direction was this force?

Well, the car was traveling in the positive x-direction. So, the force must have been in the negative x-direction. The force was in the opposite direction of the displacement. This means that the angle θ between the force and displacement was 180°. $\cos 180° = -1$. Anytime a force is in the opposite direction of motion, negative work will be done.

Here is one last example.

A person stands with a 5.8-kg book bag on their back. How much work does the person do on the book bag?

This one is easy when we think in terms of energy. Neither the kinetic nor the potential energy of the book bag changes when the person just stands still. Work = 0.

Now the person with the book bag gets on an escalator and rides upward 4.3 m to the second floor and then gets off and stops. How much work is done on the book bag this time? (See the following figure.)

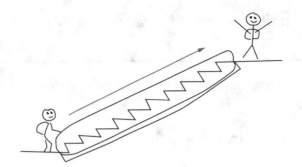

▶ Well, the book bag begins and ends at rest, so $\Delta K = 0$. But there is a change in gravitational potential energy,

$$W = \Delta U_g = mg\Delta y = (5.8 \text{ kg})(9.8 \text{ m/s}^2)(4.3 \text{ m}) = 240 \text{ J}.$$

Let's review. Work is the transfer of energy into or out of a system. Positive work transfers energy into the system. Negative work transfers energy out of the system. There has to be a force acting on the object, and the object must move for there to be work. A force in the direction the object is going does positive work because it adds kinetic energy to the object. A force in the opposite direction to the object's motion does negative work, because the object slows down and loses kinetic energy. Forces perpendicular to the motion do not change the energy of the object because they neither speed up nor slow down the object; therefore, the kinetic energy does not change. This is what happens to a planet in a circular orbit around a star.

We can also find work using a graph.

EXAMPLE

▶ Let's go back to our parent lifting their child and graph the force they apply as a function of distance. (See the following figure.)

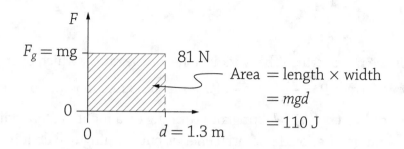

▶ Notice how the area of the graph equals the same as our value for work calculated earlier!

The area under a force displacement graph will always give you the amount of work done on the object. This is very useful when the force acting on the object is not constant. For example, take the force required to stretch a spring $F_s = k\Delta x$.

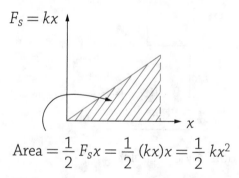

$$\text{Area} = \frac{1}{2} F_s x = \frac{1}{2} (kx)x = \frac{1}{2} kx^2$$

The farther you stretch the spring, the more force is required, as seen in the previous graph. The area of this graph is a triangle. Calculating the area tells us that work is equal to $\frac{1}{2}kx^2$. This makes sense because work equals the change in energy and is equal to the potential energy of the stretched spring $W = \Delta E = \Delta U_s = \frac{1}{2}k(\Delta x)^2$.

So now we have three ways to determine the work: $W = \Delta E = Fd \cos \theta =$ Area of the F-d graph.

Power

Power is the rate at which the work is done, or the rate at which energy is being transferred: $P = \dfrac{W}{\Delta t} = \dfrac{\Delta E}{\Delta t}$. Power is measured in joules per second, which we call **watts** (W). Walking and running up a flight of steps will both increase your gravitational potential energy, but running will do it in a shorter period of time and generate more power.

EXAMPLE

▶ Imagine a 2,000-kg elevator being pulled upward by a cable at a constant rate a total distance of 12 m from the second floor to the fifth floor in 16 s.

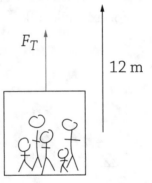

F_T

12 m

What is the work done by the cable?

▶ $W = \Delta E = \Delta U_g = mg\Delta y = (2{,}000 \text{ kg})(9.8 \text{ m/s}^2)(12 \text{ m}) = 240{,}000 \text{ J}$

▶ What was the power generated by the force?

▶ $P = \dfrac{W}{\Delta t} = \dfrac{\Delta E}{\Delta t} = \dfrac{240{,}000 \text{ J}}{16 \text{ s}} = 15{,}000 \text{ W}$

▶ What was the tension in the cable?

▶ $W = Fd\cos\theta$

$240{,}000 \text{ J} = F_T d\cos\theta = F_T(12 \text{ m})\cos\theta$

$F_T = 20{,}000 \text{ N}$

Conservation of Energy

There are many types of energy: mechanical energy (translational kinetic, rotational kinetic, gravitational potential, and spring/elastic potential) and others that will be explored later in the book. Work transfers energy from object to object. It's time to put this all together in one simple equation:

$$E_1 + W = E_2$$

This equation is the general form of conservation of energy. It simply says that a system starts with an initial amount of energy, E_1, and ends with a final amount of energy, E_2. When there is work, W, done on the object between the initial and final states, the energy of the system will be changed by that amount. On the other hand, if no work is done to change the energy of the system, $W = 0$, then the energy of the system remains unchanged, $E_1 = E_2$.

EXAMPLE

▶ Imagine a roller coaster that starts at rest at the top of the first hill as seen in the following figure. Assume there is no air resistance or friction to do work on the roller coaster car. Look what happens to the energy of the Earth-car system.

▶ At point A, the energy is all gravitational potential because the car is at rest 80 m above the ground. At point B, the car is at its lowest point and has only kinetic energy. Since no work is being done, the kinetic energy at point B must be equal to the starting potential energy at point A.

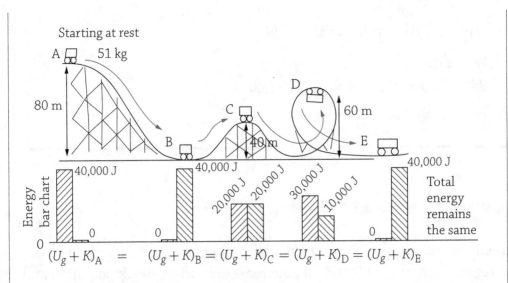

The energy bar chart in the figure gives us a visualization of this transformation of energy at the different points of the car's journey. At each point along the trip, the car's kinetic and potential energies add up to the same amount of total energy.

At point D, the car will have more potential energy than at point C, because point D is higher than point C. Therefore, the car must be moving faster at point C than at point D. At point E, the car is back to the same speed it had at point B because all the energy is kinetic again.

BTW

Notice how the roller coaster car is going in loops and changing direction. This would be an awful problem to analyze with forces or kinematics because they are vector concepts. We would have to account for all the directional changes. On the other hand, energy is a scalar quantity, so we don't care about how the object gets from one location to another. The roller coaster car has the same velocities at points B and E because the car has the same kinetic energy at those locations, regardless of the path taken to get there. This makes energy a particularly useful way to evaluate the world!

IRL Real roller coasters experience wind resistance and friction. These forces transfer energy out of the roller coaster car. This is why the first hill on a roller coaster is the highest and the car can never reach the same height again unless energy is added to the car somewhere in the middle of the ride.

Let's look at another situation where energy is conserved even though the energy is changing from one type to another.

EXAMPLE

▶ Imagine a 5.0-kg block is compressed 0.40 m against a spring with a force constant of 1,000 N/m, as seen in the following figure at point A. When the block is released, it slides along a frictionless horizontal track to point B and then ascends the ramp to a maximum height h and stops at point C.

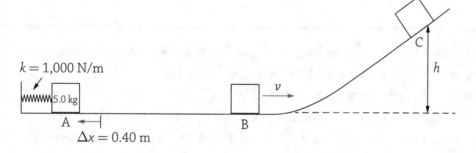

▶ Let's analyze the energy of this situation. First, consider the types of energy the block-Earth-spring system has at each point. At point A, the system energy is all spring potential. At point B, the system energy is all kinetic, and at point C, all the energy is converted into gravitational potential.

▶ The energy bar chart below the following figure shows this transformation from the original spring potential energy to kinetic energy for the box at point B and finally to gravitational potential energy for the box at point C. Notice that the energy totals at each point are the same. Now we can calculate the velocity of the block at point B and the maximum height of the block at point C.

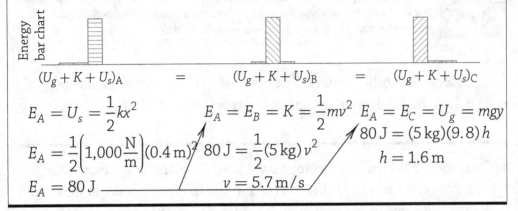

$(U_g + K + U_s)_A$ $=$ $(U_g + K + U_s)_B$ $=$ $(U_g + K + U_s)_C$

$$E_A = U_s = \frac{1}{2}kx^2 \qquad E_A = E_B = K = \frac{1}{2}mv^2 \quad E_A = E_C = U_g = mgy$$

$$E_A = \frac{1}{2}\left(1{,}000\frac{N}{m}\right)(0.4\,m)^2 \quad 80\,J = \frac{1}{2}(5\,kg)\,v^2 \quad \begin{array}{l} 80\,J = (5\,kg)(9.8)\,h \\ h = 1.6\,m \end{array}$$

$$E_A = 80\,J \qquad\qquad\qquad v = 5.7\,m/s$$

Now consider the same example, but this time let's add some friction. This will make the situation more challenging because friction will do work on the block.

EXAMPLE

This time there is a 3.0-m-long rough patch with a coefficient of friction $\mu = 0.24$ between points A and B, as seen in the following figure. (Remember the equation for friction from Chapter 2: $F_f = \mu F_N$.) Notice that negative work is done on the block this time because the friction force is directed to the left, while the block displaces to the right. Thus, the angle $W = \Delta E = Fd\cos\theta$ is 180° and causes the work, $Fd\cos\theta$, to be negative.

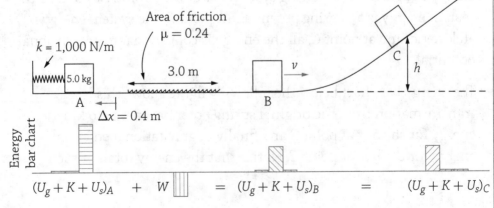

$(U_g + K + U_s)_A$ $+$ W $=$ $(U_g + K + U_s)_B$ $=$ $(U_g + K + U_s)_C$

▶ Notice how the energy bar chart has changed. There is now negative work that decreases the energy at points B and C—this reduces both the velocity at B and the height at C. The new velocity at B and height at point C are calculated in the next figure.

$$E_A + W = E_B$$
$$U_{s_A} + F_f d \cos\theta = K_B$$

How I converted F_f into μmg:

$$F_N = F_g = mg$$

$$\left(\frac{1}{2}kx^2\right)_A + \mu mg\, d \cos\theta = \left(\frac{1}{2}mv^2\right)_B$$

$$F_f = \mu F_N = \mu mg$$

$$\frac{1}{2}\left(1,000\,\frac{N}{m}\right)(0.4\,m)^2 + (0.24)(5\,kg)(9.8\,m/s^2)(3.0\,m)\cos 180° = \frac{1}{2}(5\,kg)\,v^2$$

$$80\,J - 35\,J = 2.5\,v^2$$
$$v = 4.2\,m/s$$

▶

$$E_A + W = E_C$$
$$80\,J - 35\,J = mgh = (5\,kg)(9.8\,m/s^2)h$$
$$h = 0.92\,m$$

Energy is a very powerful concept in physics. In this chapter, I introduced you to the basics of energy. How energy moves from one object to another through work, how energy and transform from one type to another, and how energy is conserved. In this chapter, we concentrated on mechanical energy; kinetic energy, gravitational potential energy, and spring potential energy. But energy is much bigger than just mechanics. The ideas of energy are a guiding principle throughout all of physics and we will visit energy again and again in later chapters.

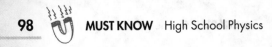
REVIEW QUESTIONS

Show your energy skills by answering the following questions.

1. A pole vaulter runs, plants their pole, bends the pole, and barely clears the bar as seen in the following figure. Describe the energies at each location indicated.

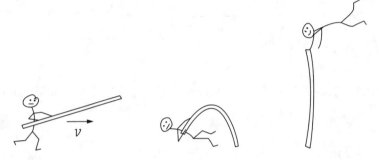

2. Identical twins go from the first floor to the second floor of a hotel. One walks up the stairs while the other runs up. How do their work and power compare?

3. How much power does it take to accelerate a 1,600-kg car from rest to 30 m/s in 3.4 s?

4. Jack and Jill run up a hill to fetch a pail of water. (See the following figure.) Jack and Jill are competitive, so they race. Jack has a mass of 80 kg and runs up the 20-m-tall hill in 10 s. Jill, who has a mass of 60 kg, takes a shortcut and runs to the top in 7 s. Which one does the most work? Which one generated the most power?

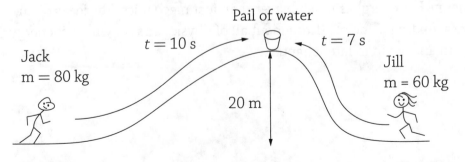

Pail of water

Jack
m = 80 kg

$t = 10$ s

$t = 7$ s

Jill
m = 60 kg

20 m

5. Jane is having fun with a lion on top of a 16-m-tall cliff, as seen in the following figure. Tarzan feels left out. He wants to swing over to Jane, but his cliff is only 12 m tall. How fast will he have to run in order to get to Jane and the lion?

6. A roller coaster starts at rest at point *A* and passes through points *B* and *C*, as seen in the following figure. How fast is the roller coaster going at point *B* and at point *C*?

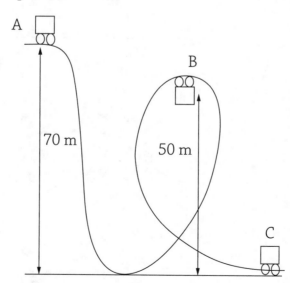

7. Two baseballs are thrown from the edge of a cliff at the same velocity. One is thrown straight up and the other is thrown horizontally, as seen in the following figure. Which ball is traveling the fastest when it lands at the bottom of the cliff?

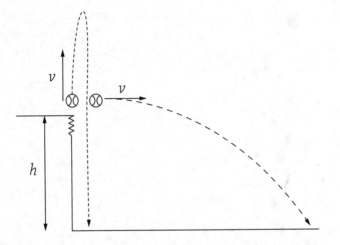

8. A roller coaster car starts from rest at point A and launches at an angle at point B. (See the following figure.) How high will the car go compared to the height it started from? Higher, lower, or the same height?

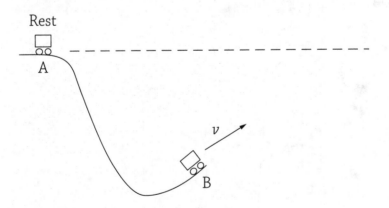

9. A fire traps a 75-kg man on the roof of a 12-m-tall building. (See the following figure.) The fire department puts out a trampoline with a spring constant of 3,900 N/m for the man to jump onto. What is the maximum distance the trampoline stretches while saving the man's life?

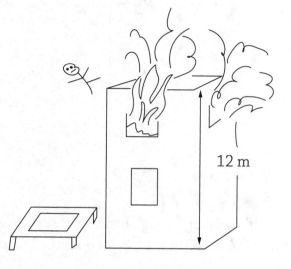

12 m

10. An 80-kg skydiver falls 100 m before pulling his parachute. (See the following figure.) Calculate the speed of the skydiver just before he opens the parachute if there is no air resistance. Recalculate the speed of the skydiver when there is 200 N of air resistance.

100 m

11. A spring with a force constant of 650 N/m is compressed 4.6 m and used to launch a 60-kg clown at the circus, as shown in the following figure. How high will the clown go?

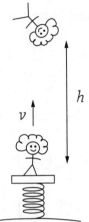

12. A 3.6-kg box is sliding along a rough horizontal surface with a coefficient of friction of $\mu = 0.4$ at 6.0 m/s. How far does the box slide before coming to rest?

13. Three identical crates slide down 3-m-tall ramps of different shapes, as seen in the following figures. Assume there is no friction between the crate and the ramp. Rank the speeds of the crates at the bottom of the ramps from greatest to least.

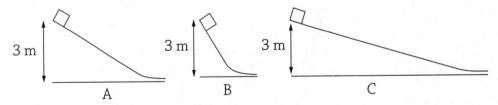

14. A quarterback throws a 22 m/s pass along an arc to a receiver. At the top of the arc, the ball is traveling 18 m/s. What speed will the ball be traveling when it is caught by the receiver? How high above the release point did the football get?

Flashcard App

Three Special Situations: Oscillations, Circular Motion, and Rotational Motion

103

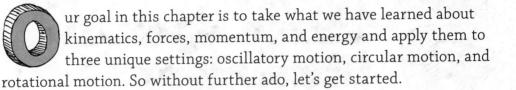

ur goal in this chapter is to take what we have learned about kinematics, forces, momentum, and energy and apply them to three unique settings: oscillatory motion, circular motion, and rotational motion. So without further ado, let's get started.

Equilibrium and Stability

Place a marble on a table. As long as the table is level, the marble will stay right where you placed it because it is in equilibrium. The forces acting on the marble—gravity and normal force—cancel out. (See the following figure.) Now take two curved bowls, one right side up and one upside-down, as seen in the figure. Place a marble right in the center of each bowl. Once again, the marbles are in equilibrium, but the marble on the upside-down bowl is in a very precarious position. If it gets nudged a little to one side or the other, it will roll off, never to return. This is called an **unstable equilibrium position**. On the other hand, the right-side-up bowl is in a **stable equilibrium position** because, if the marble moves away from this spot, a **restoring force** will push the marble back toward equilibrium.

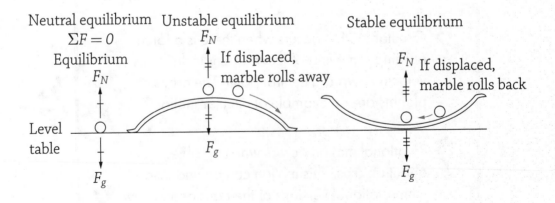

Oscillatory Motion

Whenever a system has a stable equilibrium position with a restoring force, there will be a repeated motion called **oscillatory motion**. Once displaced, the marble in a bowl will move back and forth about the equilibrium position. This is a very common motion in nature. Some examples include a child on a swing, waves on the ocean, a tree swaying back and forth in the breeze, a plucked guitar string, and the ground vibrations during an earthquake.

A Mass Attached to a Spring

All oscillatory motion is basically the same. We will look at a very simple example—a mass attached to a spring—and it will tell us the basics of all oscillating systems. The following figure shows a spring attached to a wall on one end and a mass on the other. Assume the whole contraption sits on a smooth table with no friction.

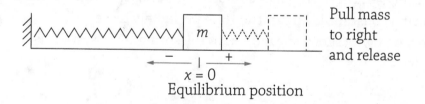

Equilibrium position

When we pull the mass to the right, the spring will pull the mass back to the equilibrium position. The mass will shoot past the equilibrium position to the left until the spring finally stops the mass and then pushes it back toward equilibrium again. This motion will repeat back and forth forever if there is no friction.

Equilibrium — force is cancel out

The maximum displacement of the mass from equilibrium is called the **amplitude** (A) of the motion, which is measured in meters. The time it takes to complete one oscillation is called the **time period** (T), measured in seconds. The number of oscillations that are completed

per second $\left(f = \dfrac{oscillations}{\Delta t} \right)$ is called the **frequency** (f),

measured in hertz (Hz). Time period and frequency are

the reciprocals of each other: $T = \dfrac{1}{f}$. The time period of

a spring/mass system depends on the spring constant

and the mass attached to the spring: $T_s = 2\pi\sqrt{\dfrac{m}{k}}$,

BTW

We will use these same terms (frequency, time period, and amplitude) with waves in Chapters 15 and 16.

where m is the mass in kg and k is the spring constant in N/m. If the mass is bigger, the spring moves slower, making the time period larger. If the spring constant is larger, the time period decreases.

The next figure shows the motion of the spring-mass system as it oscillates.

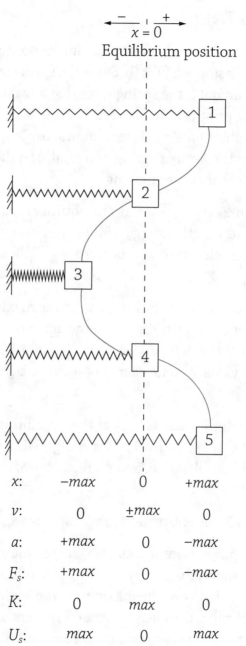

Let's analyze the motion:

- At location 1, the mass is at its maximum positive position and has momentarily stopped ($v = 0$). Since the spring is stretched to its maximum amount, the spring force is at a maximum toward the equilibrium position. Therefore, the acceleration is also at a maximum toward equilibrium. At this maximum position, all of the system energy is in the form of spring potential, and there is no kinetic energy because the mass is not moving.

- At location 2, the mass is at the equilibrium position where the velocity and kinetic energy will be at a maximum. The spring force and acceleration will be zero, as well as the spring potential energy.

- At location 3, the mass has stopped at the maximum negative displacement. The restoring force and acceleration will now be at a maximum to the right, and the spring potential energy is back to its maximum value. Kinetic energy is again zero because the velocity is zero.

- At location 4, the mass is back at the equilibrium position again, where the velocity, as well as the kinetic energy, will be at a maximum. The spring force, acceleration, and spring potential energy will all be zero again.

- At location 5, the entire motion repeats over again.

As the spring/mass system oscillates, energy must be conserved. The following graph shows the energy as a function of position. Notice that as the mass moves between the maximum amplitude on the right and left of the equilibrium, the energy converts from potential to kinetic and back again, but the total energy remains constant in accordance with conservation of energy.

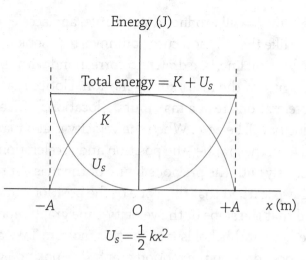

When the position is plotted as a function of time, it forms a cosine wave with the equation $A\cos(2\pi ft) = A\cos\left(\dfrac{2\pi t}{T}\right)$. See the following figure.

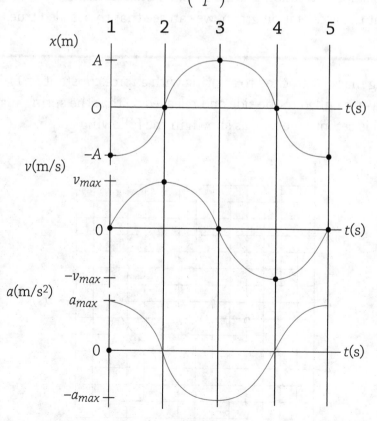

A plot of the velocity will produce a sine curve, and acceleration will be a cosine curve like the position curve but inverted. Look closely at the previous graph. Vertical lines are drawn to correspond with the locations of our previous figure of the mass/spring motion. Notice that when the position and acceleration are at a maximum—locations 1, 3, and 5 on the graph—the velocity will be zero. When the velocity is at a maximum—locations 2 and 4 on the graph—the position and acceleration are zero. This matches up precisely with the previous figure of the mass/spring motion.

In Chapter 1, we learned that the slope of the position-time graph equals the velocity and that the slope of the velocity-time graph represents the acceleration. Let's see if this holds true for this motion. Take a look at the vertical line for location 2 on the motion graph. The position graph has its steepest positive slope, which means the velocity should be at its maximum positive value. Looking at the velocity graph, this is true. The velocity graph has a zero slope at location 2, which means the acceleration should be zero. Looking at the acceleration graph, we can see that this is also true.

▶ A 10-kg mass is attached to a spring with a force constant of 11 N/m. The mass is pulled to the side and released so that the spring-mass system is set into motion, as shown in the following graph.

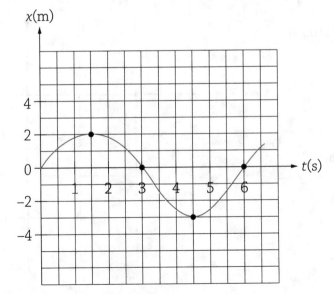

Let's see what we can determine from this graph.

▶ What is the amplitude of the motion? Remember that the amplitude is from the equilibrium to the maximum position. Therefore, the amplitude is 2 m.

▶ What is the time period of the motion from the graph? Time period is the time to complete one full oscillation. From our graph, we can see that this is 6 s.

▶ We can calculate the time period using the equation

$$T_s = 2\pi\sqrt{\frac{m}{k}} = 2\pi\sqrt{\frac{(10\,\text{kg})}{(11\,\text{N/m})}} = 6.0\,\text{s}.$$ Perfect! This matches up with the graph!

▶ What is the frequency of the motion? Frequency is the inverse of the time period, $T = \dfrac{1}{f} = \dfrac{1}{6\,\text{s}} = 0.17\,\text{s}.$

▶ We can write the equation for the position of the mass at any time. This is a sin wave: $x = A\sin(2\pi f t) = A\sin\left(\dfrac{2\pi t}{T}\right) = (2\,\text{m})\sin\left(\dfrac{2\pi t}{6\,\text{s}}\right).$

▶ At what times is the mass at equilibrium? Equilibrium is when the position is zero at 0 s, 3 s, and 6 s.

▶ At what times is the mass moving at its greatest velocity? Maximum velocity occurs at the equilibrium at 0 s, 3 s, and 6 s.

▶ At what times is the spring force at its greatest value? Maximum force from the spring is when it is stretched/compressed the farthest at 1.5 s and 4.5 s.

▶ At what time is the spring potential energy at its greatest value? Maximum spring potential energy is at the maximum position $(U_s = \dfrac{1}{2}kx^2)$ at 1.5 s and 4.5 s.

▸ At what times is the kinetic energy at its greatest value? Maximum kinetic energy is when the velocity is at its greatest, which is when the mass is at the equilibrium position ($K = \frac{1}{2}mv^2$) at 0 s, 3 s, and 6 s.

▸ Calculate the total energy of the spring-mass system. Total energy = spring potential + kinetic energy. But the maximum displacement of the kinetic energy is zero. So, all we need to do is calculate the potential energy at the amplitude to find the total energy of the system: $U_{smax} = \frac{1}{2}kx_{max}^2 = \frac{1}{2}(11\,\text{N/m})(2\,\text{m})^2 = 22\,\text{J}$.

Pendulums

Another excellent example of oscillatory motion are **pendulums**. They consist of a mass on the end of a string and swing back and forth, as seen in the following figure.

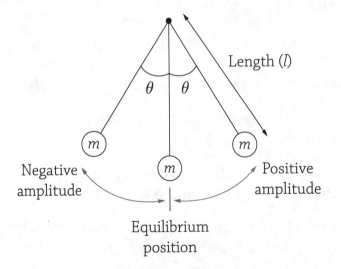

Pendulums will have the same motion, force, and energy relationships as the spring-mass system. The equation to find the time period of a pendulum is $T_p = 2\pi\sqrt{\dfrac{l}{g}}$, where l is the length of the pendulum in meters, g is the acceleration caused by gravity, and T_p is the time period in seconds.

Circular Motion

What happens to a bowling ball when a net force pushes it? It accelerates! Look at the next figure.

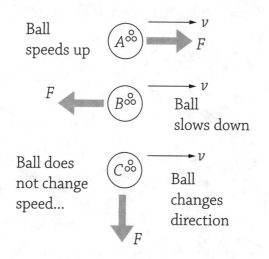

In case A, the ball speeds up because the force acts in the same direction as the velocity. In case B, the ball will slow down because the force accelerates the ball in the opposite direction of the velocity. But what happens in case C? The force is perpendicular to the bowling ball's velocity, so the ball won't speed up or slow down. Instead, the perpendicular force makes it change directions. The bowling ball will turn into an arcing path. If the perpendicular force continues, the ball will turn through a complete circle at a constant speed. This is called **uniform circular motion**.

Acceleration, Forces, and Velocity in Circular Motion

The acceleration toward the center of the circular pathway is called

centripetal acceleration: $a_{center} = \dfrac{v^2}{r}$, where v is the velocity of the object

and r is the radius of the curved path in meters. The centripetal acceleration is perpendicular to the velocity of the object, as seen in the following figure of a car going around curves and a planet going around the sun. The velocity is always tangent (parallel) to the path and in the direction the object is going. The centripetal acceleration always points toward the center of the circular path.

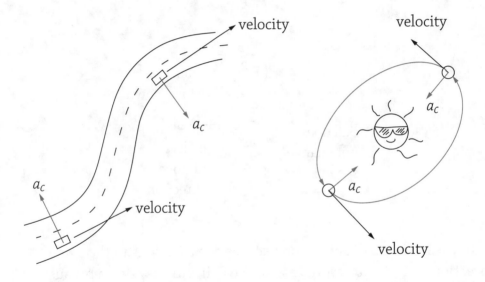

Sometimes people have a hard time believing there is a force pushing toward the center of a curved path. When you are a passenger in a car and the driver makes a hard-left turn, most people assume that they are "thrown" toward the right-side door by some magical force. But your body simply wants to continue forward in a straight line because you have mass and inertia (Newton's First Law). The door and seatbelt supply the force to make you move in a circular path with the car. If you don't believe this, remove the passenger door and don't wear your seatbelt and let the driver turn hard-left. You will not turn left with the car! You will continue on a straight path because there is no force toward the center of the circle causing you to turn left!

> **IRL** I used to teach with two colleagues who liked to play golf. During one game, a rabbit jumped out in front of the golf cart. On instinct, the driver jerked the wheel and the golf cart shot left in a tight turn. Unfortunately, golf carts don't have doors, and the passenger kept going straight, flew from the vehicle, and broke her hip. On the bright side, her hip healed quickly, the rabbit was saved, and I got a real-world circular motion "physics in action" story out of it.

The next figure shows an object going around in a circle clockwise.

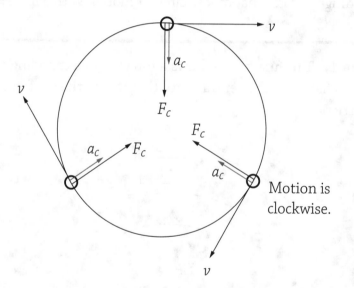

Motion is clockwise.

Notice that the velocity is always tangent to the path and that both the force and the acceleration are toward the center of the circle. We can calculate all three as follows:

- **Velocity of object moving in a circle**

$$v_{circle} = \frac{\Delta x}{\Delta t} = \frac{circumference}{time\ period} = \frac{2\pi r}{T},$$ where r is the radius of the circle

and T is the time to complete one circle.

- **Centripetal acceleration toward the center of the circular path**

$$a_{center} = \frac{v^2}{r}$$

- **Force toward the center** $\sum F_{center} = ma_{center}$

This force is often called the **centripetal force**, but this can cause confusion. This is not a special new force. Instead, it is a normal everyday force or combination of forces that work together to make an object move in a circle. This is the most critical step in understanding circular motion, so let's practice finding the force that causes circular motion with a few examples.

EXAMPLES

Example 1   This first figure shows a top view of a car going around a turn on a flat road. The central force that causes the car to move in a circle is friction.

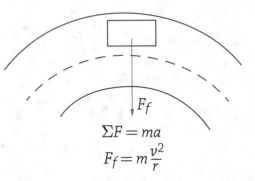

$$\Sigma F = ma$$
$$F_f = m\frac{v^2}{r}$$

Example 2   An amusement park ride called The Rotor is a big rotating drum in which the passengers stand against the inside wall of the cylinder. As the ride rotates, the passengers are stuck to the wall while the floor drops out. Passengers are held up by friction with the wall. The central force is the normal force from the wall of the cylinder.

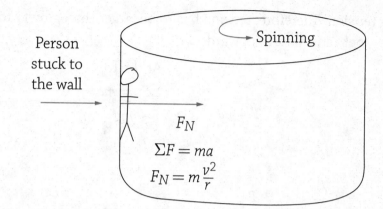

Person
stuck to
the wall

Spinning

F_N

$\Sigma F = ma$

$F_N = m\dfrac{v^2}{r}$

▶ **Example 3** A child attaches a baseball to the end of a string and rotates it around in a horizontal circle. The string drops down at an angle θ. The forces in the y-direction must cancel. The horizontal component of the tension is the central force that causes the ball to move in a circle.

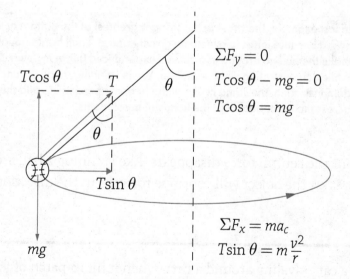

$T\cos\theta$ T θ

θ

$T\sin\theta$

mg

$\Sigma F_y = 0$

$T\cos\theta - mg = 0$

$T\cos\theta = mg$

$\Sigma F_x = ma_c$

$T\sin\theta = m\dfrac{v^2}{r}$

▶ **Example 4** The same child now rotates the baseball in a vertical circle. At the top of the circular motion, gravity and the tension in the string work together to produce the central force downward toward the center of the circle. At the bottom of the circle, gravity is fighting tension. Therefore, tension has to be larger than gravity to produce a net central force upward.

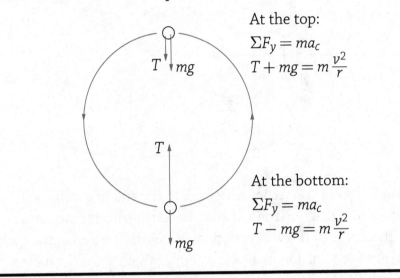

At the top:

$$\Sigma F_y = ma_c$$

$$T + mg = m\frac{v^2}{r}$$

At the bottom:

$$\Sigma F_y = ma_c$$

$$T - mg = m\frac{v^2}{r}$$

IRL   In last example, the stress in the string is greatest at the bottom of the swing because it has to overcome gravity and still pull hard enough to make the ball go around in a circle. I experienced this in my own classroom while demonstrating circular motion. While spinning a ball around in a demonstration, the string broke at the bottom of the swing and the ball shot across the room right over the heads of my students.

If at any point the central force disappears, like spinning a ball around and the string breaks, the object will continue forward in the direction of the velocity.

EXAMPLE

▶ Imagine a car traveling around a turn when it hits a patch of ice. Remember from the concept of inertia that an object in motion will continue in straight-line motion at a constant velocity unless an outside force changes the motion. When the car hits the ice, friction can no longer push the car around in the circular path, and the car slides in a straight line off the road. See the following figure.

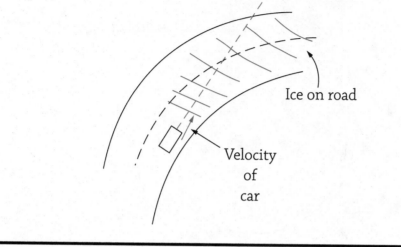

Path car will travel

Ice on road

Velocity
of
car

Circular Orbits

Imagine throwing a ball horizontally, as seen in the next figure. Because of gravity, the ball always falls toward the Earth.

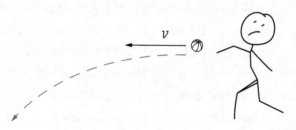

v

But imagine that you had a "cyborg power arm" strong enough to throw the ball extremely fast. As we zoom our view out from the Earth in the figure and exaggerate your height just a bit, we see that the faster you throw the ball, the farther it goes before hitting the Earth. In addition, the Earth is curved, making it harder for the ball to reach the surface.

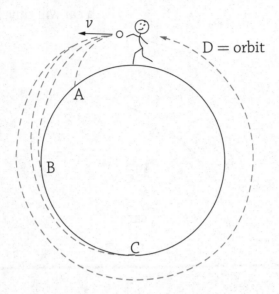

Along path *A*, the curvature of the Earth hardly matters. But you can see that launching with higher velocities, the ball will travel a quarter (*B*) or even halfway around the planet (*C*). If you really put some effort into it, you could throw the ball so fast, it would fall but never reach the surface of the planet and would fall forever in a circle (*D*). After you launch the ball, about 90 minutes later you could actually turn around and catch the ball. This is a state of constant **freefall** called **orbit**. Objects in freefall experience **weightlessness**, which is a state where the object does not appear to be affected by gravity. You would experience weightlessness in orbit, but also if you are in an elevator and then cut the supporting cable! (Well . . . at least until you hit the ground floor.)

IRL   Obviously, a person cannot throw a ball into orbit because our arms are much too weak and the drag from the atmosphere would slow the ball down. But this is essentially what NASA and private companies do when they put something into orbit. They simply launch the object out of the atmosphere and then get it going fast enough so that it falls but never reaches the planet's surface.

With gravity as the central force, we can derive several useful equations:

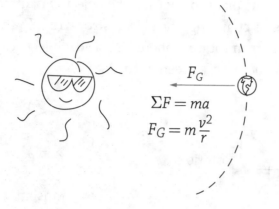

$$\Sigma F = ma$$

$$F_G = m\frac{v^2}{r}$$

$$\sum F_{center} = ma_{center}$$

$$F_G = m_{planet}\frac{v^2}{r}$$

$$G\frac{m_{sun}m_{planet}}{r^2} = m_{planet}\frac{v^2}{r}$$

$$G\frac{m_{sun}}{r} = v^2$$

$$v_{orbit} = \sqrt{\frac{Gm_{sun}}{r}}$$

The mass of the planet and one of the radius terms on each side of the equation cancel out.

This equation shows us that the velocity of an orbit only depends on the mass of the object being orbited and the radius of the orbit. If we plug in the velocity of objects moving in circles, $v_{circle} = \dfrac{2\pi r}{T}$, and solve, we get an equation for the time period of the orbit called Kepler's Third Law:

$$v_{orbit} = \frac{2\pi r}{T} = \sqrt{\frac{Gm_{sun}}{r}}$$

$$T_{orbit}^2 = \left(\frac{4\pi^2}{Gm_{sun}}\right)r_{orbit}^3$$

Rotating versus Translating

When an object moves from one place to another, the object is **translating**.
When an object's orientation is changing, the object is **rotating**. In the
following figure, I have the person rotating about her head. Objects can
rotate and translate at the same time, as seen in the figure.

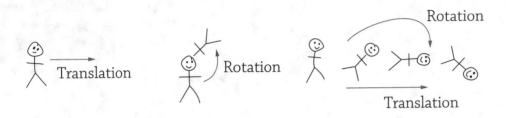

In nature, objects naturally rotate about their center of mass.

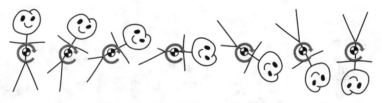

Rotation about center of mass

To make an object rotate about a point other than the center of mass
requires an outside force to hold the object in place about the pivot point.

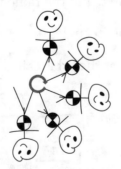

Rotation about point (the person's foot)

 **IRL** Engineers like to design rotating parts so that they spin about their center of mass, if possible, because it doesn't require an extra force to hold the part in place and there will be less stress on the parts. Objects that don't have their center of mass properly located on the rotational axis will vibrate and shake. Just like that wobbly ceiling fan that drives you crazy at home!

There Is a Rotational Equivalent for Every Translational Concept

The nice thing about rotational motion is that it's not really anything new. Everything we have learned about motion, Newton's Laws, momentum, and energy, has a rotational analog. Here is a table of the translational idea followed by its corresponding rotational counterpart.

Concept	Translation	Translational Units	Rotation	Rotational Units
Displacement	$\Delta x = x_f - x_i$	meters: m	$\Delta \theta = \theta_i - \theta_f$	radians: rad
Velocity	$v = \dfrac{\Delta x}{\Delta t}$	meters/second: m/s	$\omega = \dfrac{\Delta \theta}{\Delta t}$	radians/second: rad/s
Acceleration	$a = \dfrac{\Delta v}{\Delta t}$	meters/second squared: m/s²	$\alpha = \dfrac{\Delta \omega}{\Delta t}$	radians/second squared: rad/s²
Agent of change	Force: $\sum F = ma$	newton: N	Torque: $\tau = rF\sin\theta$ $\tau = r_\perp F$ $\tau = rF_\perp$ $\sum \tau = I\alpha$	newton·meter: Nm
Inertia: (resistance to change)	Mass: m	kilogram: kg	Rotational inertia $I = \sum mr^2$	kilogram·meter squared: kg·m²
Momentum	$p = mv$	kilogram·meters/ second: kg·m/s	$L = I\omega$ $L = r_\perp mv$	kilogram·meter squared/second squared: kg·m²/s²
Energy of motion	$K_{trans} = \dfrac{1}{2}mv^2$	joules: J	$K_{rotational} = \dfrac{1}{2}I\omega^2$	joules: J

Rotational Kinematics

Rotational motion is just like one-dimensional translational motion. Instead of keeping track of the object's location and how fast it is moving, rotational kinematics is concerned with the object's orientation and how fast it is spinning. The mechanical link between translation and rotation comes from realizing that the distance covered moving along the circumference of a circle divided by the radius of the circle equals the angle in radians:

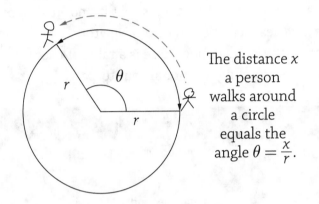

The distance x a person walks around a circle equals the angle $\theta = \dfrac{x}{r}$.

I know that the unit of radians scares some people, but trust me, we don't need to make this hard! We have already learned all of this in Chapter 1. Here is a comparison between the translational and rotational equations of motion. They are all exactly the same. Only the symbols have changed. Even the θ-ω-α graphs are identical to the x-v-a graphs of motion.

1D Translational Equations	Rotational Equations
	$\theta = \dfrac{x}{r}$
	Remember:
	2π radians = 1 revolution
$v = \dfrac{\Delta x}{\Delta t}$	$\omega = \dfrac{\Delta \theta}{\Delta t}$
	$\omega = \dfrac{v}{r}$

1D Translational Equations	Rotational Equations
$a = \dfrac{\Delta v}{\Delta t}$	$\alpha = \dfrac{\Delta \omega}{\Delta t}$
	$\alpha = \dfrac{a}{r}$
$x = x_0 + v_0 t + \dfrac{1}{2} a t^2$	$\theta = \theta_0 + \omega_0 t + \dfrac{1}{2} \alpha t^2$
$v = v_0 + at$	$\omega = \omega_0 + \alpha t$
$v^2 = v_0^2 + 2a(x - x_0)$	$\omega^2 = \omega_0^2 + 2\alpha(\theta - \theta_0)$

Let's look at an example of rotational kinematics.

EXAMPLE

▶ A tire rotating at 4.0 rad/s rolls to a stop in 10 s. The tire has a radius of 0.45 m. Let's calculate how many rotations the tire made, how far the tire rolled, and the acceleration of the tire.

▶ First, calculate the rotational acceleration of the tire. Remember to keep track of your variables just like in translational motion.

θ_0	0
θ	What we want to find
ω_0	4 rad/s
ω	0
α	What we want to find
t	10 s

$$\omega = \omega_0 + \alpha t$$
$$0 = 4\,\text{rad/s} + \alpha(10\,\text{s})$$
$$\alpha = -0.4\,\text{rad/s}^2$$

▶ Now let's figure out how many revolutions the tire makes before stopping. We will solve for the rotation in units of radians and then must convert the answer into units of revolutions.

$$\theta = \theta_0 + \omega_0 t + \frac{1}{2}\alpha t^2$$

$$\theta = 0 + (4\,\text{rad/s})(10\,\text{s}) + \frac{1}{2}(-0.4\,\text{rad/s}^2)(10\,\text{s})^2 = 20\,\text{radians}$$

$$\theta = 20\,\text{rad} \times \left(\frac{1\,\text{revolution}}{2\pi\,\text{radians}}\right) = 3.2\,\text{revolultions}$$

▶ Next, let's calculate how far the tire rolls before stopping. Convert rotational displacement to translational displacement using our connector equation. Be sure to use radians:

$$\theta = \frac{x}{r}$$
$$x = \theta r = (20\,\text{rad})(0.45\,\text{m}) = 9\,\text{m}$$

▶ Can you sketch the θ-ω-α graphs of motion for the tire? This is a constant acceleration situation. Remember that the slope of the displacement graph equals the rotational velocity and that the slope of the rotational velocity equals the acceleration. (See the following graphs.)

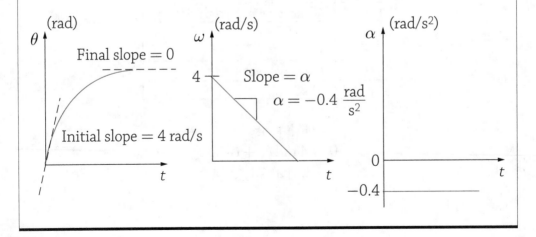

Rotational Inertia

For translational motion, the shape of the object and where the mass is located do not matter one bit. Mass is mass no matter what, but inertia is a bit different in rotation.

Do this experiment. Stand up and grab two weights or books. Put your feet together and hold the weights against your chest and twist back and forth, as seen in the following figure. Now hold the weights out and twist back and forth. Notice how it does not feel the same. It is harder to change your motion when the weights are held out farther from the rotational axis. This is because when rotating, not all of the mass is being moved the same distance. The mass farther from the pivot point has to move farther than the mass along the axis.

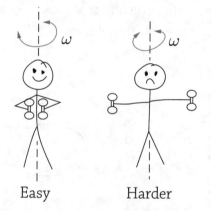

Easy   Harder

Translational inertia is simply the mass, m, but rotational inertia, I, depends on where that mass is located. Our equation for rotational inertia is $I = \sum mr^2$, where m is the mass in kg and r is the radius of the mass from the rotational axis. Look at the following figure. Both objects have the same mass and are rotating about their centers. Object A has a smaller rotational inertia than object B because its mass is in closer to the rotational axis.

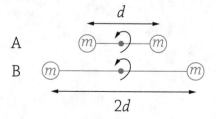

Calculating the rotational inertia for both:

$$I_A = \sum mr^2 = m\left(\frac{d}{2}\right)^2 + m\left(\frac{d}{2}\right)^2 = \frac{md^2}{2}$$
$$I_B = \sum mr^2 = md^2 + md^2 = 2md^2$$

The masses for B are twice as far from the rotational axis, but the rotational inertia for B is four times that of A because the radius is squared in our equation.

The rotational inertia for six common objects is shown in the next figure. All of these shapes have a different equation to calculate their rotational inertia. Notice that the farther the object's mass is located from the rotational axis, the larger the rotational inertia is.

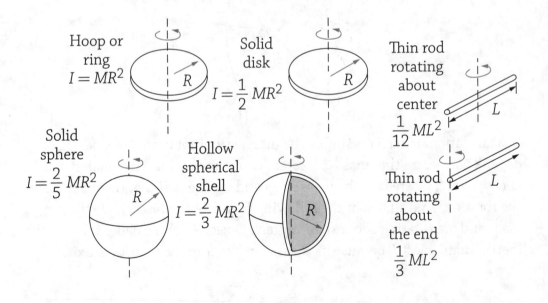

Look at the next example and let's see if you get it.

The following figure shows three objects of identical mass and height. All are rotating about their centers, as shown.

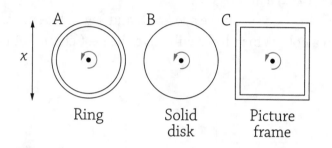

Let's rank the rotational inertia of the objects from greatest to least: $C > A > B$. The picture frame has some mass all the way out at the corners, which is farther from the rotational point than any of the other shapes. The solid disk is the only one that has mass near the rotational axis.

Which of these objects would be the easiest to start spinning? The smaller the rotational inertia, the easier it is to change the rotational motion; therefore, B—the solid disk—will be the easiest to start spinning.

Which of the objects would be the hardest to stop spinning? The higher the rotational inertia, the harder it is to change the rotational motion; therefore, C—the picture frame—will be the hardest to stop once it is spinning.

Newton's First Law in Rotational Form

Newton's First Law: An object at rest (not spinning) will stay at rest, and an object that is spinning will continue spinning at a constant rate unless acted on by an outside **torque**. That all seems fairly straightforward except for the new term: torque. A torque is a force that acts at a radius from the pivot point that tries to turn an object.

In the following figure you see a wrench and several identical forces trying to turn a bolt. Which force will have the best chance of turning the bolt?

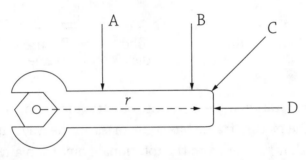

The forces at the far-right end will produce more torque because they have a larger radius, but D won't do anything because it is pushing in the wrong direction. This tells us that torque depends on how far the force is relative to the rotational axis and the direction of the force. The equation for torque is $\tau = rF\sin\theta$ where r is the distance from the rotational point to the application point of the force, F, and θ is the angle between the radius and the force. Torque, τ, is measured in newton-meters.

Consider the forces B, C, and D in the figure. All are applied at the same radius, and all have the same magnitude, but because of their angle, the torques are different. B produces the most torque because it is perpendicular to the radius. D produces no torque to rotate the bolt because it is parallel to the radius. Force C has a component parallel and perpendicular to the radius.

Only the perpendicular component of the force will produce any torque. So, we can say that $\tau = rF_{\perp}$. Another way to look at force C is to visualize the force to be shifted along its line of action until it is perpendicular to the rotational point. (See the following figure.) This minimum radius, or perpendicular radius, is called the **lever arm**. So, we can say that $\tau = r_\perp F$. This gives us two different ways to consider the torque: The actual radius times the perpendicular component of the force, or the actual force times the perpendicular radius (lever arm).

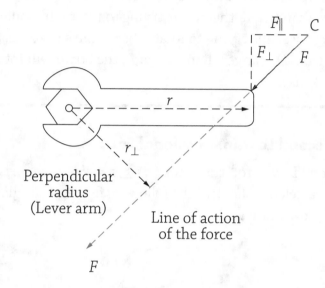

Let's practice with an example.

A square board has a pivot point in the middle, as shown in the following figure. Four identical forces are acting on the board as shown.

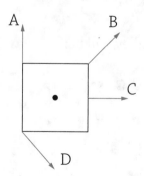

▶ Let's rank the magnitude of the torques from greatest to least created by the four forces: $D > A > B = C = 0$. The line of action for both B and C passes through the pivot point, making their torque zero. A has a shorter lever arm than D.

▶ Which way will the board rotate? Forces B and C don't supply any torque, so we can ignore them. D is trying to rotate the square counterclockwise. A is trying to rotate the square clockwise. Since the torque from D is larger than that of A, the board will rotate counterclockwise.

Newton's Second Law in Rotational Form

Newton's Second Law in rotational form is $\sum \tau = I\alpha$. If there is a net torque, the object will accelerate. Look at the following figure of a pulley and a string that is pulling downward with a force.

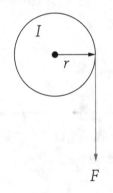

The acceleration of the pulley will be:

$$\sum \tau = I\alpha$$
$$rF = I\alpha$$
$$\alpha = \frac{I}{rF}$$

EXAMPLE

▶ A 5.0-kg pulley in the shape of a uniform disk with a radius of 0.20 m has a string draped over the top as seen in the following figure. The right side of the string is pulled with a force of 100 N. The left side of the string is pulled with a force of 40 N. What is the acceleration of the pulley?

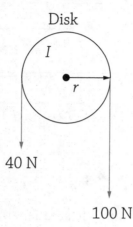

Disk

I

r

40 N

100 N

▶ The two strings are producing torques in opposite directions. This means that one will be positive and the other will need to be negative since they are "fighting each other":

$$\sum \tau = I_{Disk}\, \alpha$$

$$(rF)_{clockwise} - (rF)_{counter-clockwise} = \frac{1}{2} M_{pulley} r^2 \alpha$$

$$\frac{2(F_{clockwise} - F_{counter-clockwise})}{M_{pulley} r} = \alpha$$

$$\frac{2(100\,\text{N} - 40\,\text{N})}{(5.0\,\text{kg})(0.20\,\text{m})} = I\alpha$$

$$\alpha = 120\,\text{rad/s}^2$$

If there is no net torque, the object won't rotate. The next figure shows a child of mass m on a see-saw a distance d from the pivot point. A bigger child of mass $2m$ wants to play. The two children don't want the see-saw to rotate. Where should the bigger child sit?

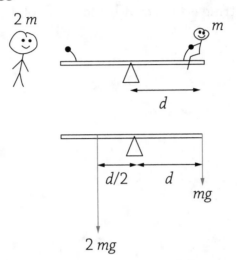

The torques have to cancel out. The smaller child is producing a clockwise torque. Since the larger child has twice the downward force of gravity, they will need to sit on the left side $d/2$ from the pivot to produce an equal counterclockwise torque.

EXAMPLE

A large crane is used to lift cars off a ship and place them on the dock, as seen in the following figure. A 5,000-kg counterweight is used to balance the crane. Where should the counterweight be placed to cancel out the torque of the car?

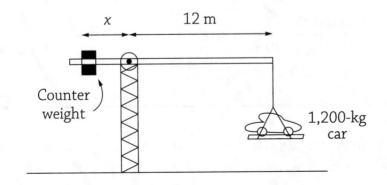

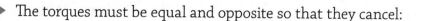

▶ The torques must be equal and opposite so that they cancel:

$$\sum \tau = 0$$
$$(rF)_{clockwise} = (rF)_{counter-clockwise}$$
$$(12\,\text{m})(mg)_{car} = x(mg)_{counter\ weight}$$
$$(12\,\text{m})m_{car} = xm_{counter\ weight}$$
$$(12\,\text{m})(1{,}200\,\text{kg}) = x(5{,}000\,\text{kg})$$
$$x = 2.88\,\text{m}$$

Newton's Third Law in Rotational Form

This is an easy one. Newton's Third Law "rotationalized" simply states that torques always come in pairs. When one object exerts a torque on another object, an equal (but opposite in direction) torque is exerted back on the original object. You can see this in a helicopter. The fuselage exerts a torque on the rotor blades, causing them to spin. The rotor blades exert an opposite torque on the fuselage, which will cause the fuselage to spin in the opposite direction. Since the fuselage has a larger rotational inertia, it will spin at a slower rate. This is a problem for the pilot and would make the vehicle dangerous to fly. To stop the fuselage rotation, a design engineer can do one of two things: 1) put a tail rotor on the helicopter to stop the fuselage from spinning or 2) put two rotor blades on the vehicle so that the torques from the two blades cancel out without the need for a tail rotor.

Rotational Kinetic Energy

Here is a quick review of what we learned in Chapter 4. There are many types of energy. Energy can change forms from one type to another, but the total energy of a system stays constant unless work is done on the system to either transfer energy into or out of the system. Our general equation for conservation of energy is $E_1 + W = E_2$. Kinetic energy is the energy of motion. There are translational ($K_{translational} = \frac{1}{2}mv^2$) and rotational kinetic ($K_{rotational} = \frac{1}{2}I\omega^2$) energies. Remember that energy is in units of joules (J).

Consider a solid disk at rest at the top of a ramp with enough friction to make the disk roll without slipping, as seen in the following figure. The disk starts with only gravitational potential energy. At the bottom, the disk has both translational and rotational kinetic energies. But how is it split between the two?

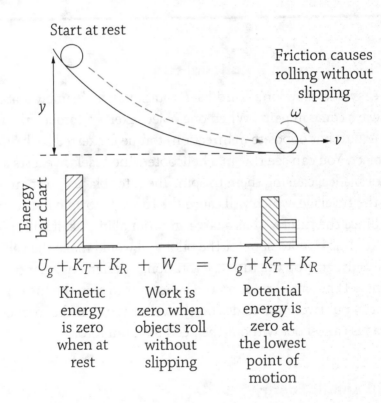

Start with conservation of energy:

$$E_1 + W = E_2$$
$$mg\Delta y = \frac{1}{2}mv^2 + \frac{1}{2}I\omega^2$$

Remember that $I_{disk} = \dfrac{1}{2}mr^2$, and $\omega = \dfrac{v}{r}$.

$$mg\Delta y = \frac{1}{2}mv^2 + \frac{1}{2}\left(\frac{1}{2}mr^2\right)\left(\frac{v}{r}\right)^2$$

$$mg\Delta y = \underbrace{\left(\frac{1}{2}mv^2\right)}_{translational\ energy} + \underbrace{\left(\frac{1}{4}mv^2\right)}_{rotational\ energy}$$

Notice that the translational kinetic energy is twice as big as the rotational kinetic energy term. The final velocity is:

$$v = \sqrt{\frac{4}{3}g\Delta y}$$

What would happen if we sprayed the ramp with oil so the solid disk slips down the ramp without rolling? Will the disk be going faster or slower at the bottom? (See the following figure.)

> **BTW**
>
> When rolling without slipping, friction does no work because the friction is static and does not work through a distance. On the other hand, when friction is kinetic, the object must be sliding. Sliding friction will convert mechanical energy into thermal energy. This will have the effect of heating the object up and slowing the object down.

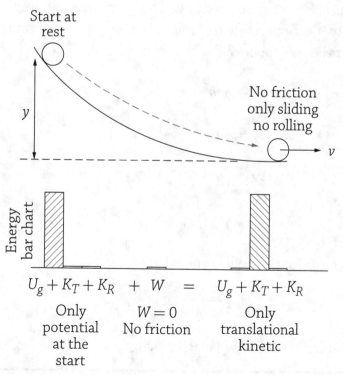

Notice that the energy bar chart changes. The disk only has translational kinetic energy at the bottom of the ramp, so it must be going faster. Using conservation of energy, we can calculate that its final speed will be $v = \sqrt{2g\Delta y}$.

▶ Solid and hollow spheres of the same radius and mass are placed at the top of identical ramps and race to the bottom. (See the following figure.) Which one wins the race? Explain your reasoning.

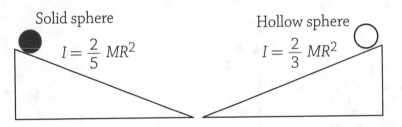

Solid sphere $I = \dfrac{2}{5}MR^2$ Hollow sphere $I = \dfrac{2}{3}MR^2$

▶ The solid sphere wins. The easy conceptual explanation is that the solid sphere has less inertia, so gravity will have an easier time accelerating it. The harder equation way to explain is to set up conservation of energy for the two spheres. For the solid sphere:

$$mg\Delta y = \frac{1}{2}mv^2 + \frac{1}{2}I\omega^2$$

$$mg\Delta y = \frac{1}{2}mv^2 + \frac{1}{2}\left(\frac{2}{5}mr^2\right)\left(\frac{v}{r}\right)^2$$

$$mg\Delta y = \frac{1}{2}mv^2 + \frac{1}{5}mv^2 = \frac{7}{10}mv^2$$

$$v_{solid\ sphere} = \sqrt{\frac{10}{7}g\Delta y}$$

▶ Repeating this for the hollow sphere gives us:

$$v_{hollow\ sphere} = \sqrt{\frac{6}{5}g\Delta y}$$

▶ The solid sphere wins: 10/7 is larger than 6/5.

Now let's consider a tricky rotational example.

EXAMPLE

▶ A ball starts from rest at the top of a curved ramp, rolls without slipping, and launches straight upward at the bottom of the ramp as seen in the following figure. Will the ball go higher than it started from, the same height it started from, or lower than it started from? Justify your answer.

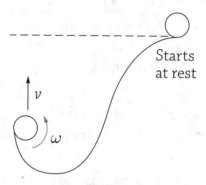

Starts at rest

▶ Careful! When the ball leaves the ramp, it is still spinning. This traps some of the energy in rotational kinetic energy that cannot convert back into gravitational potential energy as the ball ascends. Only the translational kinetic energy converts back into gravitational potential energy. Therefore, the ball will not reach the height it started from.

Rotational Momentum and Conservation of Momentum

Translational momentum is a vector description of mass in motion. **Rotational momentum**, also called **angular momentum**, is a vector description of mass in rotational motion. There are two ways to calculate angular momentum. For an object spinning about an axis, use $L = I\omega$, where I is the rotational inertia and ω is the rotational velocity. For objects moving around a rotational point, use $L = r_\perp mv$, where m is the mass of the object, v is the velocity of the object, and $r_\perp$ is the minimum perpendicular distance the object's path is from the rotational axis. Rotational momentum has units of $\text{kg} \cdot \text{m}^2/\text{s}$.

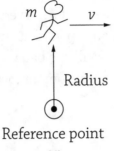

Let's look at a few situations. In the next figure, we see a child running toward and then jumping on a stationary merry-go-round.

Rotational axis

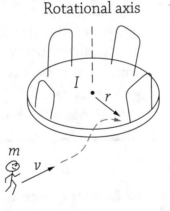

This looks like a classic conservation of momentum problem. We have interacting objects—a child and a merry-go-round—that we can include in our system. Unfortunately, linear momentum won't be conserved in this case because the merry-go-round is not free to move. It is attached to the ground. But angular momentum will be conserved because both the child and the merry-go-round are free to move about the rotational axis of the merry-go-round. The initial angular momentum of the child will equal the final combined angular momentum of the child/merry-go-round system:

$$L_i = L_f$$

$$(r_\perp mv)_{initial\ for\ child} = (r_\perp mv)_{final\ for\ child} + (I\omega)_{final\ for\ merry\text{-}go\text{-}round}$$

In this next figure, we have a planet in an eliptical orbit.

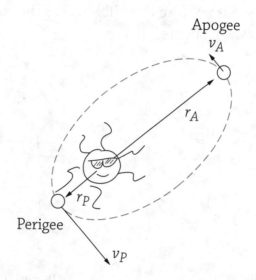

The angular momentum of the planet will be conserved during the orbit. This is because the force of gravity from the star passes through the orbital center of the motion, which makes it parallel to the radius. Therefore, gravity does not produce a torque on the planet. When the planet passes through apogee, the farthest point from the star, the radius is large. When passing through perigee, the closest point of orbit, the radius is much smaller. Conservation of angular momentum tells us that the planet must be moving much faster at perigee and slow at apogee because of the radius difference:

$$L_i = L_f$$

$$(r_\perp mv)_{apogee} = (r_\perp mv)_{perigee}$$

$$(r_\perp v)_{apogee} = (r_\perp v)_{perigee}$$

IRL This is why we don't see comets for very long. They zip through the solar system at high speed during perigee and then disappear for decades because they travel so slow at their apogee.

We see an excellent example of conservation of angular momentum every time we watch ice skating. Usually at the end of the routine, the skater begins a spin that speeds up to an incredible rate as she pulls in her arms and legs. (See the following figure.)

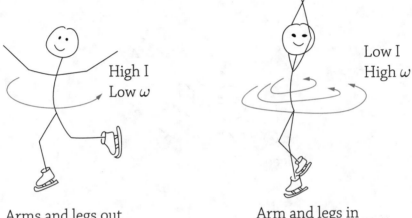

Arms and legs out
High rotational inertia
Low-speed spin

Arm and legs in
Low rotational inertia
High-speed spin

Why does this happen? Because she is decreasing her rotational inertia, and as a consequence her rotational velocity must increase to keep the rotational momentum constant.

$$L_i = L_f$$
$$(I\omega)_i = (I\omega)_f$$

A skater who reduces her rotational inertia to one-third of its original value will triple her spin rate.

In the following example, we will consider different situations where translational and rotational momentum will be conserved and will not be conserved.

▶ We throw a baseball at a long board that is nailed to the wall so that it is free to rotate, as shown in the following figure. Which conservation of momentum laws apply to this interaction?

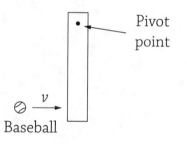

▶ The board is not free to translate. It is only free to rotate about the nail. So only angular momentum about the nail is conserved.

▶ What if we take our baseball and board to space, put the board outside our spaceship, and throw the baseball? Which conservation of momentum laws apply to this interaction?

▶ In space, the board is free to translate and rotate. Both translational and rotational momentum are conserved.

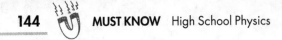

One last example for rotational momentum.

A gymnast wants to do a flip during a tumbling routine. What can he do to speed up his rotation rate once he has left the ground?

The gymnast needs to pull his arms and legs in to reduce his rotational inertia. This will speed up his rotation rate.

REVIEW QUESTIONS

Wow! We have covered a ton of material. Let's practice our new skills by answering the following questions.

1. You have two objects. The first is a mass on the end of a spring to form a spring-mass oscillating system. The second is a mass on the end of a string to form a swinging pendulum. When set into motion, both oscillate with a time period of 1.0 s. If you quadruple the mass of each device, will the time periods still be the same? Explain.

Questions 2 to 7: The following figure shows the motion of a pendulum.

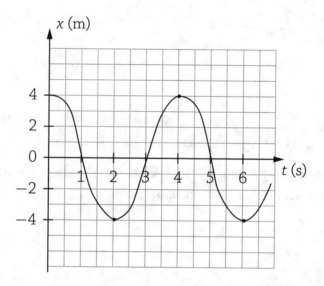

2. Determine the amplitude, time period, and frequency.

3. Write the equation of motion for the pendulum.

4. What times is the pendulum at its maximum gravitational potential energy?

5. What times is the pendulum at its maximum kinetic energy?

6. What times is the pendulum at the equilibrium position?

7. What times is the pendulum at rest?

8. Which planet is moving faster: Earth or Mars? Explain your reasoning.

9. Which planet has a longer year: Earth or Mars? Explain your reasoning.

10. A car is driving with a constant speed over the top of a hill as shown in the following figure. Add arrows to represent the forces acting on the car while it is in this position. (Assume that friction and drag are small.)

11. Rank the forces acting on the car from greatest to least. Explain your reasoning.

12. Starting with $\sum F_{center} = ma_{center}$, write the symbolic equation for the car using the forces acting on the car.

Questions 13 to 16: A toy plane attached to a string flies around in a circle. The following figure shows a top view of the plane in motion.

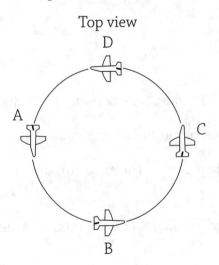

Top view

13. Draw a vector to indicate the direction of the net force acting on the plane at location A.

14. Draw a vector to indicate the direction of the plane's acceleration at location B.

15. Draw a vector to indicate the direction of the plane's velocity at location C.

16. Sketch the path the plane would take if the string pulling the plane around in the circle breaks at location D.

Questions 17 to 20: Starting from rest, a 0.12-m-radius ball rolls down a slope for 3.2 s with a rotational acceleration of 1.1 rad/s^2.

17. What is the final velocity of the ball?

18. How many revolutions did the ball make on its trip down the slope?

19. Calculate how far the ball traveled.

20. Sketch the θ-ω-α graphs of motion for the ball.

21. Which is easier to do, rotate a baseball bat about its center of mass or about its end? Explain.

22. You have a flat tire and your phone is dead. You are having trouble getting the lug nut off the rim so that you can change the tire. Explain how you could use a long pipe to help get the lug nut off.

23. The following figure shows a force pulling on a wrench. Calculate the torque from the force.

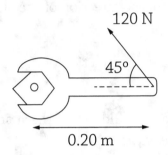

24. A construction worker takes a 55-kg board of length 6.0 m and places it so that 2.0 m are extending out beyond the edge of a building as shown in the following figure. The worker has a mass of 40 kg. How far out beyond the edge of the building can the worker walk before the board begins to rotate? Assume that the edge of the building is the pivot point.

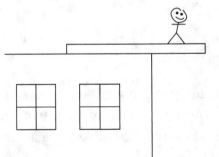

25. A block of ice slides down a ramp without friction and a sphere rolls down without slipping, as seen in the following figure. Assume the ramps are the same height and that the ice and sphere have the same mass. Which wins in a race to the bottom? Justify your answer with an energy bar chart.

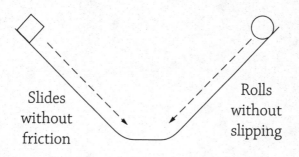

Slides without friction

Rolls without slipping

26. A disk and a hoop rolling at the same rate encounter a ramp. Both roll up without slipping to a stop. (See the following figure.) Which rolls higher? Assume that the disk and hoop have the same radius and mass.

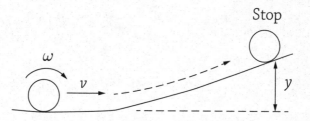

Stop

ω

v

y

27. A planet in an elliptical orbit has an apogee radius that is four times larger than its perigee radius. How many times faster is the planet traveling at perigee compared to apogee?

28. A child walks from the center of a rotating merry-go-round to the outer edge. When she does this, she doubles the rotational inertia of the child/merry-go-round system. What happens to the rotational velocity of the merry-go-round?

Flashcard App

PART TWO

The Physics of Nonsolid Behavior

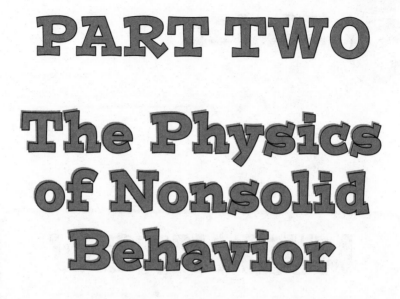

Solids have a ridged, defined shape, but not all objects are solid. Fluids can flow and move, and gases don't have a definite volume. In this section, we'll investigate the physics of systems of atoms that don't have a rigid structure or form.

 **Fluids**

 Fluids and gases have common properties and behaviors.

 Gravity increases the pressure the deeper you go into a fluid. This causes a buoyancy force upward on objects in the fluid.

 Moving fluids must obey conservation of mass and conservation of energy.

So far, we have learned the behavior of solid ridged objects. But what happens when the atoms that make up the object don't hold their shape? Our goal in this chapter is to apply what we know about forces, motion, and conservation to systems of atoms that flow. We will find some surprising results, so let's get started.

Common Properties of Gases and Liquids

Gases and liquids don't seem very similar. Take a look at the following table.

Property	Gases	Liquids
Molecular behavior	Gas molecules are not connected to each other and are far apart compared to their size. Each molecule flies around free. The molecules only interact with each other when they collide.	Liquid molecules are very close together and exert weak bonding on each other. The bonding is strong enough to hold the fluid together but is not strong enough to hold a definite shape.
Volume and shape	Gases do not have a defined surface. They expand to fill the volume of their container. The container provides a boundary surface for the gas.	Liquids have a definite volume and a defined surface. A container confines the liquid, but the liquid has its own surface on top.
Compressibility	Due to the empty space between the molecules, the volume of a gas can easily be changed. (Gases are said to be **compressible**.)	Since the molecules are close together, a liquid cannot really be compressed. Due to the bonding between the molecules, liquids can't be expanded. (Liquids are **incompressible**.)

For all their differences, gases and liquids actually have a great deal in common:

- Both gases and liquids **flow**.

- Since they flow, neither gases nor liquids have a definite shape like a solid object does.

- Both gases and liquids exert **pressure** on the containers that confine them.

Due to their similarities, gases and liquids have common properties and are lumped into a group called **fluids**. A fluid is any material or substance that flows. Thus, both gases and liquids are fluids. (Since gases have the unique ability to expand and contract, they have some special properties that liquids do not possess. These distinct properties of gases will be discussed in Chapter 7.)

Volume and Mass Density

Volume is the amount of space a fluid takes up. In physics, we measure volume in m^3.

Mass-to-volume ratio is called **mass density**, or just density for short: $\rho = \dfrac{m}{V}$. The units for density are kg/m^3. The following table gives the densities for some common fluids.

Liquids	Density (kg/m³)	Gases	Density (kg/m³)
Ethyl Alcohol	790	Helium	0.179
Oil (Average)	900	Steam (Water 100°C)	0.598
Water (at 4°C)	1,000	Air	1.29
Mercury	13,600	Carbon Dioxide	1.98

The density of gases is much smaller than liquids. Remember that the density of a substance is the same no matter how much of the material you have. A small glass of water and a swimming pool of water both have the same density.

Pressure

First of all, **pressure** is not a force. We love saying things like: "When I dove to the bottom of the swimming pool, the pressure hurt my ears." Technically, it is the water that exerts a force on your eardrums that hurts your ear. So, what exactly is pressure?

- Pressure is a ratio of the force to area: $P = \dfrac{F}{A}$

- The units are newton/meter^2 which we call a **pascal**, or Pa for short. 1 pascal $= 1$ Pa $= 1$N $/ 1$m^2.

- Pressure is everywhere in a fluid.

- Pressure is a scalar quantity, which means it does not have a specific direction. At any one point in a fluid, the pressure is the same in all directions. If you take a pressure gauge (a device that measures pressure) and submerge it in a fluid, the gauge will measure the exact same pressure no matter which direction you point the device.

- There are two causes for pressure in a fluid: the thermal motion of the atoms, and gravity.

Thermal Motion of Molecules Creates Pressure

Remember that fluids are made up of lots and lots and lots of molecules. Each of these molecules is vibrating around in a random fashion due to thermal energy. The hotter the fluid, the faster the vibrations of the molecules will be. (Thermal energy will be discussed in more detail in Chapters 8 and 9.) These vibrating molecules collide with anything the fluid comes in contact with. The following figure shows the molecules colliding with the wall of a container. Each collision imparts a small perpendicular force on the wall. Thus, the net force caused by fluid pressure will always be perpendicular to the surface the fluid is in contact with. When you dive into a swimming pool, the force from the water is pushing perpendicular to every part of your body at the same time.

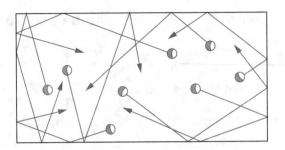

The forces caused by fluid pressure can be enormous over large areas. The next time you fly on a jet, remember that is it the force caused by air pressure on the wings that holds up you and the 400,000-kg passenger jet.

Gravity Creates Additional Pressure in Fluids

The following figure shows a stationary liquid in a glass. Since the liquid is stationary, it must be in static equilibrium, and the forces at any point in the fluid must sum up to zero. Each point must support the weight of all the fluid above itself; otherwise, the liquid above would accelerate downward due to its weight. Point B is deeper in the liquid than point A. Therefore, at point B the counteracting force must be greater than at point A. This means that the pressure at point B must be greater than at point A simply because it has to support more fluid above itself in the Earth's gravitational field.

BTW

Because the density of gases is very small, the change in pressure with vertical height is small. On Earth you would have to go to the top of a 30-story building just to decrease atmospheric air pressure 1%. But if you have ever flown in a plane, you certainly notice the change in pressure on your ears as the plane takes off or lands. A good rule of thumb is to ignore pressure differences in gases unless the change in vertical height is 100 m or more.

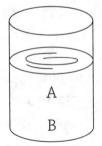

Pressure Submerged in a Fluid

The thermal and gravitational effects work together to create the atmospheric pressure we have here on Earth. Near the Earth's surface, the average atmospheric pressure is approximately $p_{atmosphere} = 100{,}000$ Pa $= 1$ atmosphere $= 1$ atm.

The pressure inside a fluid is given by the equation $P = P_0 + \rho g h$. The next figure shows a glass of water. P_0 is the pressure of the atmosphere above the water. h is the depth below the surface of the water in meters. ρ is the density of the water in units of kg/m³. g is the acceleration caused by gravity, which, of course, is 9.8 m/s² on Earth. Notice something very peculiar? The only physical dimension that matters is h, how deep you are in the fluid. That means the pressure would be the same 10 cm deep in a glass of water as it is 10 cm deep in a swimming pool.

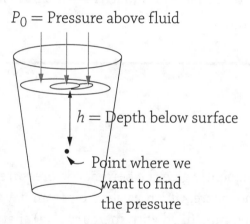

$P_0 =$ Pressure above fluid

$h =$ Depth below surface

Point where we want to find the pressure

It also means that along a horizontal line in a fluid, the pressure is exactly the same. The pressures at P_1 and P_2 in the following figure must be exactly the same.

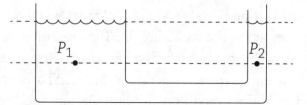

Let's practice using the static pressure equation in the next example.

▶ The following figure shows a scuba diver in a lake. What is the water pressure on the scuba diver?

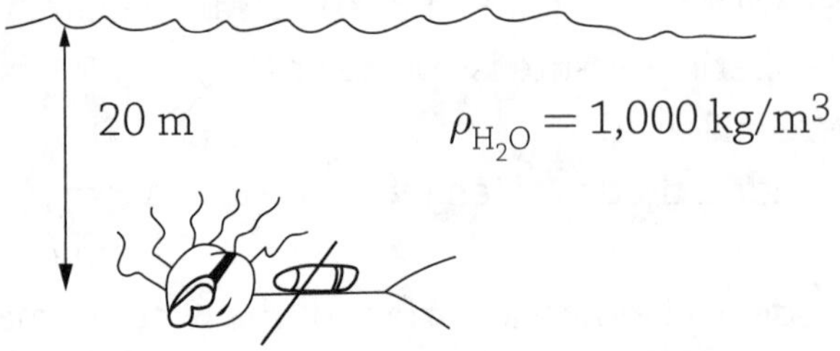

20 m $\rho_{H_2O} = 1{,}000 \text{ kg/m}^3$

▶ To solve this problem, we need to use the static pressure equation.

$$P = P_0 + \rho g h$$

$$P = 100{,}000 \text{ } Pa + (1{,}000 \text{ kg/m}^3)(9.8 \text{ m/s}^2)(20 \text{ m})$$

$$P = 2.96 \times 10^5 \text{ } Pa$$

Barometers

Weather reports frequently give the barometric pressure. This is the pressure in the atmosphere, which varies a bit from day to day. The original barometer was a long tube filled with mercury (Hg) that was inverted into an open container also filled with mercury. (See the following figure.) Some, but not all, of the mercury will flow out. This leaves a vacuum gap at the top of the test tube. Why doesn't the mercury all flow out? The pressure at point $A = 100{,}000$ Pa. The pressure at point B must equal the pressure at point A because they are at the same height. The pressure at point B is $P_B = P_0 + \rho g h$. But, $P_0 = 0$ because there is a vacuum on top. Therefore, the mercury can't flow out of the tube because there isn't any pressure in the vacuum at the top of the tube to push it out. Or, put another way, the mercury stays in the tube because the atmospheric pressure at point *A* pushes, or holds, the mercury up in the tube.

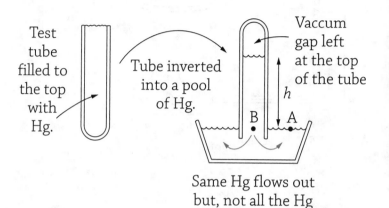

IRL You can try this at home. Take a clear glass with you the next time you take a bath. Submerge the glass to fill it completely. Invert the glass beneath the surface, keeping it completely filled, and pull the glass upward. The water will stay in the inverted glass as long as you don't break the surface and let air bubble in through the open bottom. The water is held up inside the glass due to the atmospheric pressure pushing down on the water outside the glass.

Pascal's Principle

An external pressure applied to any point of a confined fluid increases the pressure everywhere in the fluid. In the following figure a force is added to the piston on the left, adding extra pressure to the fluid on the left side. This extra pressure is transmitted to all parts of the fluid. The pressure goes up everywhere in the fluid by the same amount. This is called Pascal's principle.

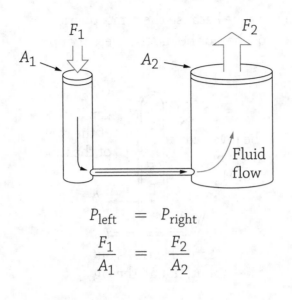

$$P_{\text{left}} = P_{\text{right}}$$

$$\frac{F_1}{A_1} = \frac{F_2}{A_2}$$

 IRL You use Pascal's principle every time you squeeze a toothpaste tube. It doesn't matter where you squeeze. The pressure is distributed undiminished throughout the entire tube, which forces the toothpaste out.

We put this property to good use in devices called hydraulics (liquid devices) and pneumatics (gaseous devices). In the previous figure, you see two different-sized cylinders of fluid attached by a connecting tube. When force F_1 pushes down on the left cylinder's piston, fluid flows into the right cylinder moving the piston up. The extra pressure applied to the left cylinder is transmitted to the right cylinder. This creates a force F_2 that pushes upward. Now look closer. Since the area A_2 is larger than area A_1, the only way for the pressure ratios F/A to remain equal is for the force F_2 to be larger than F_1. Thus, this device multiplies our force output by a ratio $= A_2/A_1$.

Buoyancy Force

Consider the block immersed in the container of fluid, as shown in the following figure.

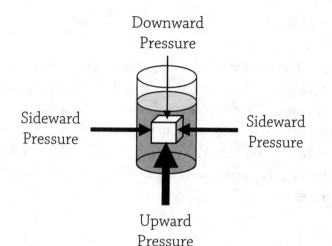

We know:

- Pressure in a fluid increases with depth $P = P_0 + \rho g h$.

- The bottom of the block is deeper in the fluid than the top.

- Therefore, the upward pressure on the bottom of the block must be greater than the downward pressure on the top of the block.

- The pressure on the sides of the block must be identical and thus cancel out.

- This means that every object surrounded by a fluid must receive an upward force due to the difference in pressures on the top and bottom of the object.

We call this force **buoyancy**: $F_b = (\rho V_{displaced\,fluid})g$, where ρ is the density of the fluid in units of kg/m³ and V is the volume of the fluid that has been displaced by the object. Volume has units of m³. For an object to experience a buoyancy force, it must displace fluid from a region of space. Buoyancy is always directed upward opposite to gravity.

BTW

A long time ago in a country far, far away, the Greek mathematician/scientist Archimedes, while taking a bath, recognized that the buoyancy force must be equal to the weight of the fluid displaced by the object: $F_b = (m_{displaced\,fluid})g =$ *weight of displaced fluid. Apparently, Archimedes was so excited by this discovery that he ran through the streets of Syracuse naked yelling, "Eureka!"*

Sink or Float?

When do objects float and when do they sink? The simple answer is that when the weight of the object is greater than buoyancy, the object sinks. When the weight of the object is less than the maximum buoyancy force, the object will float. Another way to answer is to compare the density of the object versus the density of the fluid. When the density of the object is greater, the object will sink to the bottom. When the density of the fluid is greater, the fluid will sink below the object and force the object to float.

> **IRL** You are surrounded by air. Therefore, you are submerged in the fluid atmosphere and receive a buoyant force from the atmosphere upward of about 2 N (half a pound). Not nearly enough to make you float. The average density of humans is 985 kg/m³, so we sink in air ($\rho_{air} = 1.29$ kg/m³) and only just barely float in water ($\rho_{water} = 1,000$ kg/m³).

Ships are made of steel, and steel is way denser than water! How do ships float? The structure of the ship is made of steel, but the shape of the hull displaces a large volume of water with mostly empty space inside the ship, as seen in the following figure. This reduces the average density of the ship to a value less than that of water.

BTW

Caution: One common mistake is to assume that the volume of the object is always equal to the volume of the displaced fluid. Objects that float do not displace the fluid equal to their volume. Only the submerged portion of the object displaces fluid. (See the following figure.)

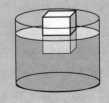

▶ Three blocks sit at the bottom of glasses of water, as seen in the next figures.

▶ Which of these objects receives the largest buoyancy force?

▶ The answer is C because the fluid is the same in each case, and C has the largest displaced volume of fluid.

▶ Draw the free body diagram of block B in the next figure. The box sinks to the bottom. Therefore, gravity is larger than buoyancy. There needs to be a normal force upward to put the block into equilibrium.

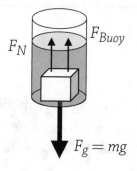

Now let's work a numerical example.

▶ A 0.50-m metal cube is suspended by a string below the surface of a container of water as seen in the following figure. The mass of the cube is 440 kg. What is the tension in the string?

▶ The cube is in static equilibrium, so the forces must cancel out.

$$\sum F = 0$$

$$F_T + F_b - F_g = 0$$

$$F_T = F_g - F_b$$

$$F_T = mg - \rho Vg$$

$$F_T = (440\,\text{kg})(9.8\,\text{m/s}^2) - (1{,}000\,\text{kg/m}^2)(0.5\,\text{m})^3\,(9.8\,\text{m/s}^2)$$

$$F_T = 3{,}100\,\text{N}$$

Moving Fluids

Moving fluids are actually quite complex to model. However, we can still use two conservation laws that we already know to show some interesting behavior of moving fluids. We will have to make a few assumptions:

1. The fluid is **incompressible**, which means the density remains constant. This is a really good assumption for liquids but not the greatest for gases since their volume can change.

2. The fluid flow is steady without any fluctuations or turbulence.

3. The fluid is **nonviscous**, which means it doesn't have any internal friction. So, no "thick" gooey fluids like molasses.

This type of idealized moving fluid is called **laminar** flow.

Conservation of Mass for a Moving Fluid

When a fluid moves through a pipe (**flow tube**), the mass that enters at one end must be the same mass that exits the other end of the pipe: $m_{in} = m_{out}$. That seems straightforward enough. Since we are assuming the density of the fluid is constant, we can rewrite the equation as $A_1 v_1 = A_2 v_2$, where A is the cross-sectional area of the pipe and v is the velocity of the fluid. This is called the **continuity equation**. The big consequence of the continuity equation is that the flow is slower in wider parts of the piping system and faster in narrower parts of the pipe. The following figure shows a fluid entering a flow tube at a slow speed. As the flow tube constricts, the velocity of the fluid must increase so that the same mass of fluid exits the tube in the same amount of time.

BTW

The continuity equation applies to any flow tube: water pipes, air conditioner ductwork, wind tunnels, the human circulatory system, sprinkler systems, water in a river, air flowing around building in a city, etc.

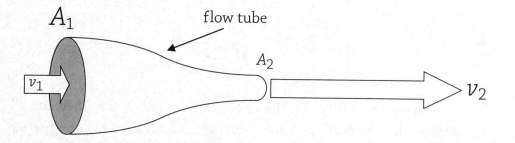

Conservation of Energy for a Moving Fluid

Applying the idea of energy to our idealized laminar fluid flow gives us the longest equation of the book, called Bernoulli's equation:

$$\left(P + \rho g y + \frac{1}{2}\rho v^2 \right)_1 = \left(P + \rho g y + \frac{1}{2}\rho v^2 \right)_2$$

P is pressure, ρ is the density of the fluid, g is the acceleration due to gravity, y is the height of the fluid, and v is the velocity of the fluid. This equation looks quite intimidating, but if you look closely you will see that it looks vaguely familiar. The last two terms on both sides look suspiciously like gravitational potential energy and kinetic energy ($U_g = mgy$ and $K = \frac{1}{2}mv^2$). In fact, that's exactly what they are. This is simply conservation of energy rewritten for moving fluids, including fluid pressure.

Let's apply Bernoulli's equation to our flow tube from earlier. (See the following figure.) Notice the average height of the left and right sides of the tube are the same. Therefore, we can subtract our potential energy term from both sides because they are identical.

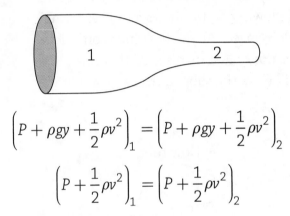

$$\left(P + \rho gy + \frac{1}{2}\rho v^2\right)_1 = \left(P + \rho gy + \frac{1}{2}\rho v^2\right)_2$$

$$\left(P + \frac{1}{2}\rho v^2\right)_1 = \left(P + \frac{1}{2}\rho v^2\right)_2$$

We know from continuity that the velocity at location 2, the small end, must be larger than at location 1. The only way for both sides of the equation to be equal is for the pressure in the faster-moving fluid to be less than the pressure in the slower-moving fluid. Bernoulli tells us that the pressure in a moving fluid decreases as the speed of the fluid increases.

▶ You can show this to be true. Take a long strip of paper and blow across the top as shown in the following figure. The low-speed/high-pressure below the paper will push the strip of paper upward into the high-speed/low-pressure air.

High speed-
low pressure air

Stationary-
high pressure air

▶ Other examples of this phenomena are:

- Wings on a plane. (See the following figure.)

▶ Due to the shape of the wing, the air flow above a wing gets compressed into a smaller flow tube, while below the wing, the flow tube stays the same size. This increases the velocity of the air above the wing, creating a low-pressure area. The pressure difference between the top and bottom of the wing produces an upward lift to keep the plane in the air.

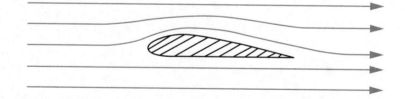

- A curveball in baseball. (See the following figure.)

▶ Friction between the ball and the air causes the ball to drag the air faster past itself on one side while hindering the air flow on the other side. This velocity difference creates a pressure difference and curves the trajectory of the ball. Try this yourself with a beach ball or ping-pong ball. Put a lot of spin on the ball when you throw it for the maximum effect.

Spinning ball pulls air past, making the air faster on top.

Air moves past slower on bottom. Ball curves up.

REVIEW QUESTIONS

Let's test our fluid knowledge by answering the following questions.

1. What properties do gases and liquids have in common?

2. How are gases and liquids different?

3. Where will you experience a greater pressure: 1 m below the surface in a hot tub or 1 m deep in an Olympic-size swimming pool?

4. How deep would you have to dive into a lake to double the pressure on your body?

5. You have just been in an accident, and the paramedics need to give you a unit of blood. (See the following figure.) Your blood pressure is 10,600 Pa above atmospheric pressure. How high must the IV bag be lifted above your body so that the blood flows into your veins? The density of blood is 1,050 kg/m^2.

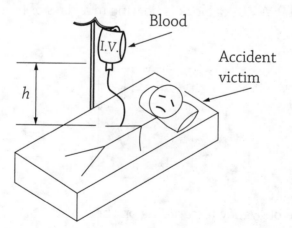

6. Why does the fluid stay up in a barometer tube instead of falling out?

7. A hydraulic system is made of two pistons, as shown in the following figure, with the smaller piston having an area of 0.01 m² and the larger one having an area of 0.20 m². What force would be required to hold up a 1,500-N football player?

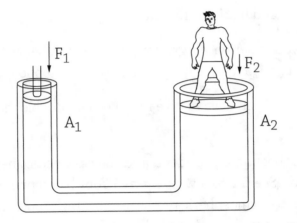

Questions 8 and 9: A 7.0-kg block is floating in a glass of liquid as shown in the next figure.

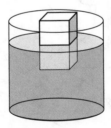

8. Calculate the buoyancy force acting on the block.

9. How much of the block's volume is under water?

10. You are floating in a swimming pool. You inhale as much air as you can into your lungs and then exhale all the air. In which case do you float higher in the water: when your lungs are full or empty?

11. Why does a ship sink when it gets a hole in the hull? Use the concept of buoyancy to make your argument.

12. Use physics to explain how you can squirt your friend with water from a garden hose across the yard.

7 Gases

 Gas molecules are in constant random motion that depends on the temperature.

 The ideal gas law describes the system behavior of a gas using only four variables: temperature, volume, pressure, and the number of molecules (measured in moles).

There are three common states of matter: solid, liquid, and gas. All of these states of matter are made of atoms and therefore have common properties of mass, inertia, and density; are affected by gravity; and can have a net charge, which we will discuss in Chapters 10 and 11. As we saw in the last chapter, fluids (liquids and gases) have additional properties that solids do not have, including the ability to exert pressure, create an upward buoyancy force, and change speed as they move and flow.

Gases have additional properties all their own. Among the three common states of matter, only gases have the unique ability to change volume. This gives gases some special properties that solids and liquids don't have. Our goal in this chapter is to start small by studying the microscopic motion of atoms in a gas and work our way up to the macroscopic properties of the system.

Gases Are a Chaotic Mess of Kinetic Energy Action

Molecules in a gas are like free agents. They are not connected to the other molecules in the gas in any way. This means that gases have kinetic energy due to the motion of the molecules but no potential energy between the molecules. They fly around free. The motion of the molecules is **random** in all directions. Visualize a room filled with pool balls flying around in all different directions at arbitrary speeds in straight lines until they collide with each other or the walls. When they collide, they bounce without any energy loss because the collisions are all **elastic**, so the motion never ever stops. That wouldn't be a room you would want to walk through, but that's exactly what a gas is like. It's a good thing the gas molecules are so very tiny.

The "fly-free" nature of gases is why gases don't have a defined volume and makes them hard to contain. Imagine our room filled with pool balls again. If you open the door, some of the particles will fly right out the door. Gases do the same thing and will expand to fill any container completely. But what about the Earth's atmosphere? It isn't constrained by any walls or lid. Why does it stay on the Earth? Well . . . gravity holds it to the Earth but some of it gets away. To understand why some of the atmosphere escapes, we need to talk more about the motion of gas molecules.

Random Motion of Molecules, Kinetic Energy, and the Temperature of the Gas

The motion of gas molecules is random, with each molecule doing its own thing. Amid all this chaos is a surprisingly predictable pattern. When we plot a speed distribution of the gas molecules, we get a curve similar to a normal distribution curve from statistics, as shown in the following graph.

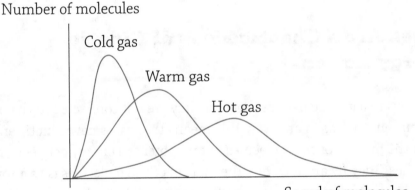

Notice that most of the molecules of the warm gas are moving at a speed in the middle. When the temperature of the gas changes, the speed distribution of the gas molecules shifts to the right or left on the graph. Molecules in "hot" gases, as a whole, move faster. Atoms in "cold" gases, on average, move slower. In fact, the temperature of the gas is a direct measure of the average kinetic energy of the atoms: $K_{average} = \dfrac{3}{2}k_B T$,

where T is the temperature of the gas in kelvin and K_B is **Boltzmann's constant:** $k_B = 1.28 \times 10^{-23}\,\text{J/K}$. This equation tells us that the temperature of the gas is a direct measure of the average kinetic energy of the gas molecules. The higher the temperature, the greater the kinetic energy of the molecules and the faster, on average, they move. Notice that the cold gas has a higher peak because all the molecules are bunched up at the low speeds. The hot gas has a lower peak because the molecules have a wider range of speeds, creating a long, mushed-out curve. The average speed of the atoms in the gas is given by the equation $v_{gas\ molecules} = \sqrt{\dfrac{3k_B T}{m_{gas\ molecule}}}$, where m is the mass of the gas molecule. Notice that $v \propto \sqrt{\dfrac{T}{m}}$. As the temperature of the gas goes up, the velocity increases, but as the mass of the gas molecules gets bigger, the speed of the gas molecules slows down.

This brings us back to the atmosphere. The gas molecules in our atmosphere don't all move the same speed. Heavier molecules, like carbon dioxide, move slower. The lightest molecules, like hydrogen and helium, move the fastest. In fact, they move so fast in the upper atmosphere that they are going faster than the escape velocity to leave Earth. This is why we don't have much hydrogen and helium in the atmosphere. They "leak away" and fly off into space.

BTW

The atoms inside solids and liquids are also in constant random motion, but the atoms are bound together and are not able to fly around free like in a gas. Therefore, solids and liquids have internal potential energy within the bonds that hold the atoms together. That means temperature is not a direct measure of the average kinetic energy inside solids and liquids like it is in gases. But the general behavior of the atoms holds true: the hotter the object is, the faster the atoms inside are moving about. We'll look at this in Chapters 8 and 9.

IRL Large planets, like Jupiter, have enough gravity to hold onto light gas molecules, which is why they are gas giants.

EXAMPLE

▶ A mole of helium, neon, and argon are all trapped in a container at room temperature.

▶ Rank the average speeds of the molecules. (The appendix has the periodic table for your reference.)

▶ Helium is the lightest and argon is the heaviest. Therefore, the speed of helium > neon > argon.

▶ Now let's sketch the speed distribution curves of the three gases. All gases have the same number of atoms. Argon's peak should be higher and to the left because all the atoms are bunched up at low speeds. Hydrogen's peak should be to the right and lower because the speeds of the atoms are more spread out.

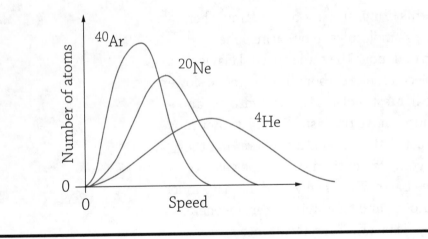

Now let's work an example with numbers.

EXAMPLE

▶ The average room temperature is 295 K. The mass of oxygen is 2.66×10^{-26} kg.

▶ What is the average speed of oxygen at room temperature?

$$v_{gas\,molecules} = \sqrt{\frac{3k_B T}{m_{gas\,molecule}}} = \sqrt{\frac{3(1.28 \times 10^{-23}\,\text{J/K})(295\text{K})}{(2.66 \times 10^{-26}\,\text{kg})}} = 652\,\text{m/s}$$

Ideal Gas Law

After looking at the microscopic behavior of a gas, let's examine the bulk behavior. Gases have four main properties: temperature measured in Kelvin (K), pressure measured in pascal (Pa), volume measured in meters cubed (m^3), and the number of molecules measured in moles (mol). The connection between these state variables is called the ideal gas law: $PV = nRT$, where $R = 8.31\,\text{J/mol} \cdot \text{K}$ is the ideal constant.

We can use the ideal gas law to calculate the number of atoms in a room. Let's say the room is about the size of a bedroom 9 m square with a ceiling that is 3 m tall. The volume of the room will be 243 m^3. Assume the room is at atmospheric pressure of 100,000 Pa and a temperature of 295 K.

BTW

The temperature of a gas must be given in Kelvin. The conversion between Celsius and Kelvin is $T_K = T_C + 273$.

$$PV = nRT$$

$$n = \frac{PV}{RT} = \frac{(100{,}000\,\text{Pa})(243\text{m}^3)}{(8.31\,\text{J/mol} \cdot \text{K})(295\text{K})} = 9{,}910\,\text{mol}$$

Using **Avogadro's number**, $N_A = 6.02 \times 10^{23}$ atoms/mol, we can find the number of gas molecules:

$$(9,910 \, \text{mol})(6.02 \times 10^{23} \, \text{gas molecules/mol}) = 5.97 \times 10^{27} \, \text{gas molecules}$$

That's approximately 6,000 trillion-trillion molecules. A truly mind-numbing number. And that is just the number of molecules in one bedroom! Let's use the ideal gas law to calculate the temperature of a gas.

> A sealed container with a volume of 0.07 m³ has 2.0 moles of gas inside at a pressure of 250,000 Pa. What is the temperature of the gas inside the container?
>
> $$PV = nRT$$
>
> $$T = \frac{PV}{nR} = \frac{(250{,}000 \, \text{Pa})(0.07 \, \text{m}^3)}{(2.0 \, \text{mol})(8.31 \, \text{J/mol} \cdot \text{K})} = 1{,}050 \, \text{K}$$

Most of the time, two of the variables in the ideal gas law equation can be held constant while the other two change. For instance, if we have a closed container, the number of molecules will be constant because none of the gas can escape. The temperature can also be held constant by keeping the container in an oven or an ice bath. When both the number of moles and temperature are held constant, our equation become: $PV = $ constant. Written another way: $P_1 V_1 = P_2 V_2$. The pressure and volume are inversely related. As one goes up, the other has to decrease in order to maintain the same constant.

Let's say you have a balloon that is at room temperature and you squeeze it to half its original volume. What happens to the pressure inside the balloon? Well, none of the gas can escape, and as long as the balloon stays at room temperature, the only variables that are changing are pressure and volume. Cutting the volume in half causes the pressure to double: $P_1V_1 = (2P_2)\left(\dfrac{V_2}{2}\right)$. This will probably pop the balloon.

When we can do this same thing for different pairs of variables, we see that:

- Pressure is inversely proportional to volume: $P \propto \dfrac{1}{V}$

This is how an air pump works. You push down on the handle to decrease the volume, and the pressure goes up high enough to push the air into a bike tire.

- Pressure is directly proportional to the temperature: $P \propto T$

Soda cans sometimes explode if you leave them in a broiling hot car in the summer. The carbonated gas inside the soda expands with temperature.

- Volume is directly proportional to the temperature: $V \propto T$

A balloon will expand if you heat it up and shrink if you put it in the freezer. Try it!

Let's look at the graphical relationships between the ideal gas law variables.

EXAMPLE

▶ Sketch the relationship between gas pressure and temperature using the following figure.

P

T

▶ Pressure is directly proportional to the temperature: $P \propto T$.

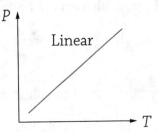

▶ Now let's sketch the relationship between gas pressure and volume in the figure.

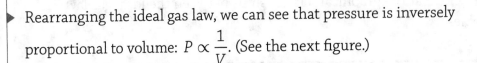

▶ Rearranging the ideal gas law, we can see that pressure is inversely proportional to volume: $P \propto \dfrac{1}{V}$. (See the next figure.)

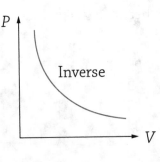

REVIEW QUESTIONS

Let's see what we've learned about gases by answering the following questions.

1. The following figure shows the speed distribution of gas molecules in a container. The container is placed into the oven, which heats up the gas. Sketch the new speed distribution of the gas.

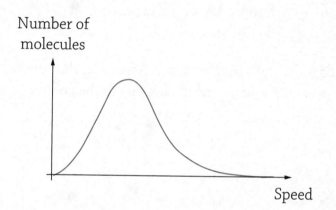

Number of molecules

Speed

2. There are oxygen and carbon dioxide gas in the room where you are reading this book. Which molecules are moving faster: oxygen or carbon dioxide? Explain why this is so.

3. In the room where you are reading this book, which molecules have the greatest kinetic energy: oxygen or carbon dioxide? Explain.

4. A 0.006-m³ container has 1.5 moles of gas inside at 400 K. What is the pressure inside the container?

5. Two identical rooms are connected by an open door. The left room is hotter than the right room. Which room has the most air molecules in it? Explain your reasoning.

6. Sketch the relationship between volume and temperature in the following figure.

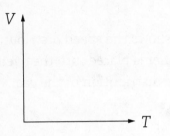

7. A sealed balloon is put in the freezer. Assuming that the pressure inside the balloon stays the same, what happens to the volume of the balloon?

PART THREE
Thermodynamics

So far, we have been talking about large-scale things. Now, we are going to apply what we have learned to the atomic scale of objects. We'll look into the internal working of objects to investigate the internal energy of an object and discover why and how thermal energy moves from hot to cold.

8 The First Law of Thermodynamics

MUST KNOW

 The internal energy of an object is the sum total of every atom's kinetic and potential energies combined.

 Temperature is the measure of the average kinetic energy of the atoms inside an object.

 The first law of thermodynamics is an extension of the law of conservation of energy that takes thermal energy into account. It describes how energy can be added to or subtracted from an object through work (W) and/or heat (Q).

PV diagrams are used to visualize the relationships between work, heat, and internal energy.

n Chapter 4, we learned about energy conservation for mechanical systems, but the study of energy is so much bigger. In this chapter, I'll expand the concept of energy to include the random motion of atoms inside of objects. The atoms inside objects are always in constant random roiling motion. Even in stationary solid objects that appear to just be sitting there doing nothing, the atoms inside hide a frenetic energy of motion. Our goal in this chapter is to understand this new form of energy, see how it connects to temperature, and show how this energy can be changed and moved. But through it all, conservation of energy still holds true.

Internal Energy

Take all the atoms inside an object. Find each atom's kinetic energy of motion. To this, add the potential energy stored in the bonds that hold the atoms to other nearby atoms in the object's structure. All of this added together is the **internal energy** of an object. The symbol for internal energy is U, and is measured in joules. There are a lot of atoms, so this can add up to a whole lot of energy. The more atoms the object is made of, the more internal energy it will have.

In all states of mater, the atoms are in constant random motion. In solids, the bonds are strong enough to hold the atoms together in a specific shape. So, the atoms vibrate around an equilibrium position in the structure. Thus, the potential energy inside solids is high. In liquids, the bonds holding the atoms together are fluid. Strong enough to hold the liquid together but not strong enough to hold a structure. Thus, the potential energy in liquids is less than that of the solid form of the material. Gases, like we saw in the last chapter, have no bonds between the atoms and the atoms fly free of each other with no internal potential energy at all.

▶ Which has more energy, a glass of water at 50°C or a sink full of 50°C water?

▶ The sink has more atoms so it will have more internal energy.

Phase Change

As the temperature goes up, the motion of the atoms increases. Get a solid object hot enough and it will eventually melt. The amount of internal energy required to convert a solid into a liquid is called the **heat of fusion**. If energy continues to be added to the liquid, the atoms will eventually break free and turn into a gas. The energy input required to convert a liquid into a gas is called the **heat of vaporization**. Every material has a unique heat of fusion and vaporization. The heat of vaporization is always larger than the heat of fusion because it is harder to completely separate atoms from each other than to simply break the ridged bonds in the solid to liquid transition. Take a gas and cool it down, and the process can be reversed. By taking enough internal energy away from a gas, it turns into a liquid; take more away, and it turns into a solid.

▶ Which has more energy, a kilogram of ice at 0°C or a kilogram of water at 0°C?

▶ Both have the same mass and number of atoms. Both have the same exact temperature, but liquid water has more internal energy equal to the heat of fusion.

IRL Steam, the gaseous state of water, is much more dangerous than boiling water because it carries much more energy than water. Steam at 100°C has to give up its heat of vaporization just to turn into 100°C boiling water before it can cool off. So, be careful around steam—just a little can burn you.

Temperature

We use temperature for so many things. We take our temperature when we are sick. We set the oven to the correct temperature when we cook, and we check the temperature on our phones before we pick out our clothes for the day. But is that temperature? What does it actually measure? Turns out, temperature measures the average kinetic energy of the atoms inside the object. We actually learned this last chapter with gases. So, for a gas, all we have to do is take the temperature, convert it to the average kinetic energy with the equation $K_{average} = \frac{3}{2}k_B T$, and finally multiply by the total number of atoms inside the gas, and we will get the total internal energy of the gas. This equation turns out to be $U_{gas} = \left(\frac{3}{2}k_B T\right)nN_A = \frac{3}{2}nRT$, where n is the number of moles and N_A is Avogadro's number.

For liquids and solids, temperature still measures the average kinetic energy of the molecules, but these two states of matter also have internal potential in the atomic bonds that hold the atoms to each other. So, temperature is an incomplete picture of the internal energy of liquids and solids. Scientists use a new term called **specific heat** to describe the internal potential energy of materials. A higher specific heat means that it takes more energy to raise the substance's temperature because there is more internal potential energy in the atomic bonds. Water has a high specific heat. It takes a lot of energy input to raise the temperature of water because of its powerful atomic bonds. Metals, on the other hand, have low specific heats and are much easier to make warmer and cooler.

EXAMPLE

▶ A metal chair by a swimming pool will get very hot on a sunny day, but it takes weeks for a swimming pool to warm up enough to swim in. Why is this?

> Two reasons: 1) The pool has a lot more atoms to heat up, so it will take longer. 2) Water has a high specific heat and will take longer to heat up.

Thermal Expansion

As the temperature goes up, the motion of the atoms increases. This causes objects to expand when they get warmer. Just in time for swimsuit season, our bodies expand a little. Some of you may have noticed this with the rings you wear. They tend to get tighter in the summer and loose in the winter because our bodies expand and contract at a faster rate than the ring does. The next time you are walking on a sidewalk, across a bridge, through a parking garage, or a large building, keep an eye out for expansion joints in the structure. These expansion points are used to accommodate the change in the shape of the structure through the seasons.

IRL When the iconic 630-ft-tall Gateway Arch was being built in St. Louis, Missouri, the north and south legs were built separately. When it came time to connect the two legs with the final top piece, it wouldn't fit because the morning sun shining on the south leg of the arch had expanded it too much. Resourceful workers used fire hoses to cool the south leg, shrinking it enough so that the two legs were close enough to connect with the last section of the arch.

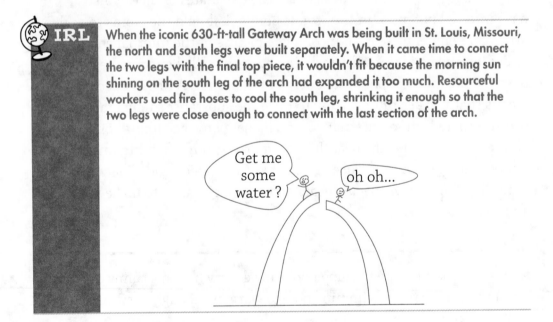

Temperature Scales and Absolute Zero

Our temperature scales of choice in physics are Celsius and Kelvin. Celsius is based on the phase changes of water. Freezing = 0°C and boiling = 100°C. Kelvin is based on the motion of the atoms themselves. The lowest temperature possible would be when the atoms stop moving all together. This is absolute zero. No atomic kinetic energy at all. This point is designated zero in Kelvin. The conversion between Celsius and Kelvin is $T_K = T_C + 273$.

BTW

While plotting the change in volume of a gas versus temperature, scientists noticed that as the gas cooled, the gas contracted. The graph of volume versus temperature approaches a curious point where the gas will have no volume at all. The temperature of this point is –273°C: absolute zero. The same experiment can be performed for gas pressure versus temperature to find absolute zero.

Heat and Thermal Equilibrium

The atoms in hot objects are moving faster than in colder objects. If we put a hot and cold object in contact, internal energy will be transferred from the hot object to the cold object. This transfer of thermal energy from one object to another is called **heat** or **heat transfer**. Heat is designated by the symbol Q and has units of joules. When two objects have the same temperature, the atoms inside the objects have the same average atomic kinetic energy and there is no heat transfer. When this occurs, the objects will be in **thermal equilibrium**, which means they won't exchange any internal energy, or heat. This doesn't mean that the two objects will have the same total energy; just the same average kinetic energy. For example, drop a hot rock in a swimming pool. The rock will cool off as it transfers heat to the pool and the pool warms up. The rock will end up at the same temperature of the pool, but the rock will not have the same total energy as the pool full of water. The pool has much more internal energy because it has so many more atoms.

BTW

Heat always flows from the object with the higher temperature to the object with the lower temperature until thermal equilibrium occurs. We'll learn more about this in Chapter 9. We'll also look at how heat transfers between objects are accomplished in the same chapter.

Conservation of Energy and the First Law of Thermodynamics

Let's review conservation of energy, from Chapter 4. The energy an object starts with plus any energy transferred into or out of the objects through work equals the final energy of the object:

$$E_1 + W = E_2$$

Now let's expand the equation to include all the possible energies we have already discussed: translational and rotational kinetic, spring, and gravitational potential.

$$(K_T + K_R + U_g + U_s)_1 + W = (K_T + K_R + U_g + U_s)_2$$

Now let's add our new form of energy, internal energy U, and heat Q to the equation:

$$(U + K_T + K_R + U_g + U_s)_1 + Q + W = (U + K_T + K_R + U_g + U_s)_2$$

This is the long form of the first law of thermodynamics. Don't be scared of the equation! It's just conservation of energy. To make it look simpler, I'm going to take out all the mechanical energy so we can concentrate on the thermal properties of the system and rearrange like so:

$$U_1 + Q + W = U_2$$
$$Q + W = U_2 - U_1 = \Delta U$$
$$\Delta U = Q + W$$

This is the short form of the first law of thermodynamics and the one we will use.

Sign Conventions and Definitions

The first law of thermodynamics tells us how the system interacts thermally with the environment. But what are the **system** and **environment**?

- A system is simply the object or collection of "stuff" we are interested in.

- The environment is everything outside of the object we are interested in. Everything other than the stuff in the system is the environment.

- The energy of the system only changes through interactions with the environment through the exchange of heat (Q) and work (W).

- Heat is positive when it enters the system because heat entering the system will increase the energy of the system. Heat flowing out of the system is negative.

- Work done on the system is positive because this increases the system energy. Work done by the system on the environment to move it around is negative, because the system has to expend energy to move the environment.

This all sounds hard, but it's not. Here are a couple of examples of how to define a system and identify how the system interacts with the environment.

EXAMPLE

System: a gecko

Environment: everything else but the gecko

- The gecko wakes up and notices that its "system" is cold. It sees a hot rock in the environment and snuggles up to it. The gecko absorbs heat from the environment/rock ($+Q$). The gecko/system has a positive internal energy change ($+\Delta U$). The heat exchange ceases when the gecko/system and the rock/environment reach the same temperature (thermal equilibrium).

$$\Delta U = Q + W$$

- Since no external force did work on the gecko: $W = 0$. Thus, $\Delta U = Q$.

- The gecko's internal energy increase is equal to the amount of heat that was added to its system.

Now let's look at an example that is a little more abstract, involving a gas trapped in a sealed container with a movable piston that allows the volume of the container to change.

System: gas confined in a cylinder with a movable piston

Environment: everything else outside of the gas, including the cylinder and the piston

▶ A blowtorch is held below the cylinder. The gas/system absorbs heat $(+Q)$ from the blowtorch/environment. The gas expands as it heats up, forcing the piston upward into the environment and thus doing negative work $(-W)$.

$$\Delta U = Q - W$$

▶ The system gains heat energy from the blowtorch. The system loses energy as it expands into the environment and does work. The net change in energy of the gas in the cylinder depends on which factor is largest $(-W$ or $+Q)$.

That wasn't so hard! Don't make thermodynamics problems any harder than they need to be. It's just energy.

Gases and the First Law of Thermodynamics

A gas turns out to be a convenient tool to turn thermal energy into useful work. Adding heat to a gas can make it expand. An expanding gas can be harnessed to move things around and do work, like making your car move down the road or generate electricity.

From here on out, in this chapter we will only be considering gases and their thermodynamic behavior. This has three important implications:

■ **Implication 1 Work (W)**

The basic equation for work is $W = \Delta E = Fd \cos \theta$. But for a gas, the equation becomes $W_{gas} = -P\Delta V$ where pressure is in pascals (Pa) and the volume is in m^3. Notice that the equation is negative because when a gas expands into the environment, the gas expends its own energy to move the environment. An expanding gas, $+\Delta V$, transfers energy from itself to the environment and work is negative. With a contracting gas, $-\Delta V$, energy transfers into the gas from the environment and work is positive.

■ **Implication 2 Internal energy change (ΔU) is directly related to temperature change (ΔT)**

Remember from the last chapter that a gas has no internal potential energy. Gases only have internal kinetic energy because the gas molecules are not connected to each other. (The average kinetic energy of a gas $K_{average} = \dfrac{3}{2}k_B T$ and the internal energy of a gas $\Delta U = \dfrac{3}{2}nR\Delta T$ were covered in Chapter 7.) This means that as the temperature of the gas goes up, the internal energy of the gas also goes up proportionally: $\Delta U \propto \Delta T$.

■ **Implication 3 The ideal gas law**

Since we will be dealing with gases, we will be using the ideal gas law to connect pressure, volume, number of moles, and temperature: $PV = nRT$.

PV Diagrams and Finding Temperature (T)

There are quite a few variables to keep track of in thermodynamics. To help us with this, we use a pressure-volume graph called a **PV diagram**.

▶ The following figure shows point D on the PV diagram that represents 4.0 moles of gas.

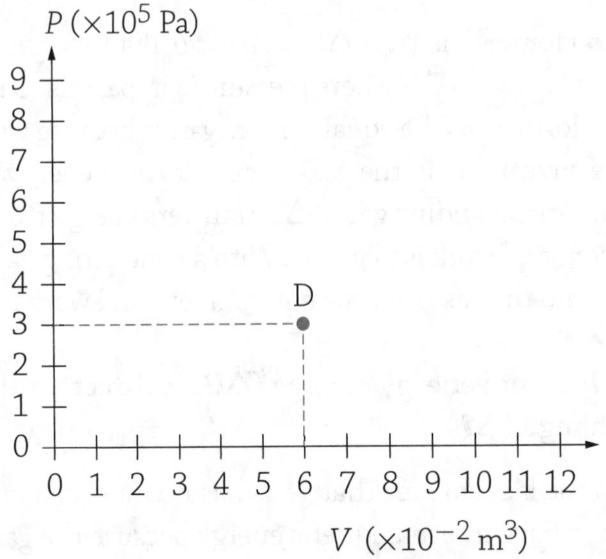

$P\,(\times 10^5\,\text{Pa})$

$V\,(\times 10^{-2}\,\text{m}^3)$

▶ From the graph we can read that the pressure is 300,000 Pa and that the volume is 0.006 m³. (Be careful to read the axis properly.) We can use the ideal gas law to calculate the temperature of the gas:

$$PV = nRT$$

$$T = \frac{PV}{nR} = \frac{(300{,}000\,\text{Pa})(0.06\,\text{m}^3)}{(4.0\,\text{mol})(8.31\,\text{J/mol} \cdot \text{K})} = 540\,\text{K}$$

▶ So, as long as we know the number of moles, we can calculate the temperature of a point on the PV diagram.

Now, let's take the same gas through five different locations on the PV diagram, listed A through E, as seen in the next figure.

EXAMPLE

▶ At which of the locations on the graph will the gas have the highest temperature?

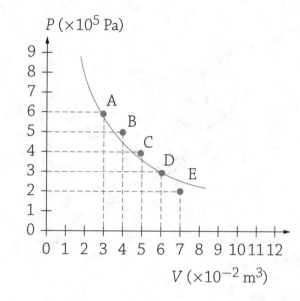

▶ We could calculate the temperature for all five states, but that's a lot of work, so let's use our physics brain. Remember that the number of moles is staying the same. Temperature only depends on the pressure and volume: $T \propto PV$. So, all we need to do is find out which location has the highest pressure times volume.

▶ Points B and C have a PV value of 20,000 and will be the same temperature. A and D have a PV value of 18,000. E has a PV value of only 14,000 and will be the coldest location. Our ranking would be $T_B = T_C > T_A = T_D > T_E$.

▶ Any location on the PV diagram where the pressure times volume value is constant will have the same temperature.

A line drawn through locations with the same temperature is called an **isothermal** line. The 540 K isothermal line is drawn through points A and

D in the graph from the previous example. There will actually be a whole family of isothermal lines to represent every possible temperature on the graph. Each isotherm will be a hyperbolic curve (also called an inverse curve):

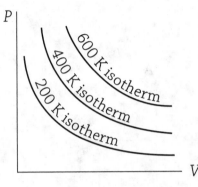

Along which pathway would you move the gas on the PV diagram to make the gas warmer or colder?

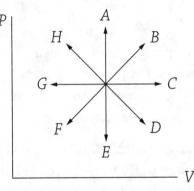

Paths A, B, and C all move up in pressure and/or volume, so the temperature would increase. These three paths will move to higher isothermal locations. Paths E, F, and G all move to lower temperature isotherms and lower PV values. It is hard to say what happens along paths D and F without knowing more information because pressure and volume are moving in opposite directions.

PV Diagrams and Calculating Work (W)

We know that the pressure inside a gas is always positive. From the work equation for a gas, $W_{gas} = -P\Delta V$, we can see that if the change in volume is also positive, the work must be negative. Therefore, movement to the right on a PV diagram will produce negative work. This makes sense because an expanding gas moves the environment, which transfers energy to the environment; this will lower the internal energy of the gas. The following figure shows a gas moving to the right along three different pathways.

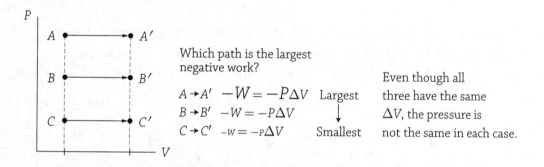

Which path is the largest negative work?

$A \to A'$ $-W = -P\Delta V$ Largest

$B \to B'$ $-W = -P\Delta V$ $\downarrow$

$C \to C'$ $-w = -P\Delta V$ Smallest

Even though all three have the same ΔV, the pressure is not the same in each case.

Movement to the left will produce positive work, and the gas is compressed by an outside force transferring energy into the gas. In the next figure, a gas is compressed along three different paths with different pressures and volume changes.

Which path is the largest positive work?

$D \to D'$ $W = -3P(-3V) = +9PV$

$E \to E'$ $W = -2P(-2V) = +4PV$

$F \to F'$ $W = -P(-V) = +PV$

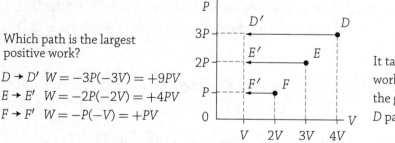

It takes 9 times more work to compress the gas along the D path than the F path.

When a gas moves up or down on a PV diagram, there is no volume change; therefore, the work must be zero. The next figure shows a gas taken along two paths where there is no volume change and work equals zero.

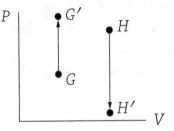

In both cases, $\Delta V = 0$
So...$W = -P\Delta V = 0$

No work is done, because the gas does not expand or contract— work *requires* motion!

Another way to think about work is to visualize the area under the curve. Paths #1 and #2 in the following figures both have the same starting and ending point, but different amounts of work.

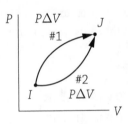

Both paths start and end at the same place. The work is negative in both cases, but the magnitudes are not the same! They don't have the same *average* pressure!

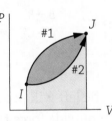

Think area under the curve— Path #1 has more area under it. That means a larger magnitude for work.

PV Diagrams and Calculating Heat (Q)

How do we find heat (Q) from a PV diagram? Well...we don't. Not directly anyway. We use the first law of thermodynamics: $\Delta U = Q + W$. Here is the procedure:

1. We either find W and ΔU from the PV diagram or they are given values in a problem.

2. We plug these values into $\Delta U = Q + W$ and calculate Q.

EXAMPLE

Here is a practice problem with no numbers. For each path in the PV diagram, determine if the value is positive, negative, or zero and fill in the table with a +, −, or 0.

Remember the rules:

- To find the temperature change, ΔT, we see how the path moves through the isothermal lines. Or we see if P times V increases or decreases.

- To find the change in internal energy, ΔU, we see what ΔT does because $\Delta T \propto \Delta U$ as seen in the equation $\Delta U = \dfrac{3}{2} nR\Delta T$.

- To find work, W, we see if the path moves right ($-W$), left ($+W$), or up and down ($W = 0$) and we use the equation $W_{gas} = -P\Delta V$.

- To find the heat, Q, we take what we know about both ΔU and W and use the first law of thermodynamics, $\Delta U = Q + W$, to find the value of heat.

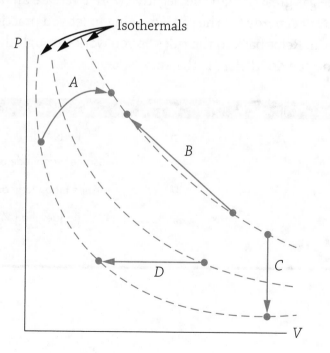

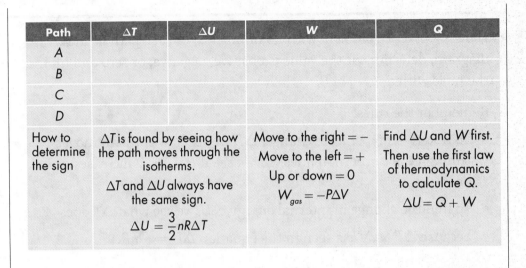

Path	ΔT	ΔU	W	Q
A				
B				
C				
D				
How to determine the sign	ΔT is found by seeing how the path moves through the isotherms. ΔT and ΔU always have the same sign. $\Delta U = \dfrac{3}{2}nR\Delta T$		Move to the right $= -$ Move to the left $= +$ Up or down $= 0$ $W_{gas} = -P\Delta V$	Find ΔU and W first. Then use the first law of thermodynamics to calculate Q. $\Delta U = Q + W$

For path A, the temperature must be rising because it is moving to a warmer isothermal line. Therefore, the internal energy must also be increasing. Since the pathway is moving to the right, the work must be negative. Using the first law of thermodynamics, we see that heat must be added to the gas greater than the negative work to create an overall increase in internal energy for the gas. I'm going to let you practice the rest on your own. Refer back to the rules as you work through the rest of the table. You can do it! Here is the answer key:

Path	ΔT	ΔU	W	Q
A	+	+	−	+ (bigger than W)
B	0	0	+	− (same magnitude as W)
C	−	−	0	− (same magnitude as ΔU)
D	−	−	+	− (bigger than W)

Four Special Processes (Paths) on a PV Diagram

One last thing before we close the book on the first law of thermodynamics. There are four special processes that we see over and over again:

- **Isobaric: Constant Pressure**   In an isobaric process, the pressure remains constant ($\Delta P = 0$). They move right and left on a PV diagram.

- **Isovolumetric/Isochoric: Constant Volume**   In these processes, there is no volume change ($\Delta V = 0$). Therefore, the work is zero ($W = 0$) and the first law of thermodynamics become: $\Delta U = Q$. These processes move up and down on a PV diagram.

- **Isothermal: Constant Temperature**   We have already discussed these processes. The temperature remains constant ($\Delta T = 0$) and therefore the internal energy of the gas also stays the same ($\Delta U = 0$). Thus, the first law becomes $Q = -W$. These processes move along the hyperbolic constant PV lines.

- **Adiabatic: No Heat (Q) Transfer Between the System and the Environment**   This process usually occurs so fast that there is no time for heat to move between the gas and environment ($Q = 0$). This is a curved path similar to an isothermal but steeper.

▶ What do all four processes look like on the PV diagram? See if you can pick them out in the following figure.

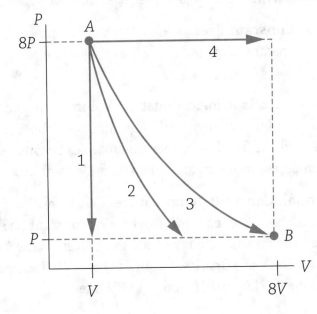

1: Isovolumetric/isochoric ($\Delta V = 0$)

2: Adiabatic (curved and steeper than isothermal)

3: Isothermal (a hyperbolic curve that starts and ends at the same $8PV$ value on the graph)

4: Isobaric ($\Delta P = 0$)

Note that each of the processes shown in the figure could move in the opposite direction. I just happened to drawn them moving to the right and downward.

PV diagrams can be challenging, so let's make sure we get it. Here is another practice problem with no numbers.

EXAMPLE

▶ Fill in the table with a +, −, or 0 for each of the four special processes, as shown in the previous figure. I know you can figure this out, but don't be afraid to go back to the rules in the previous example if you need a reminder of how it all fits together.

Path	ΔT	ΔU	W	Q
1				
2				
3				
4				
How to determine the sign	ΔT is found by seeing how the path moves through the isotherms. ΔT and ΔU always have the same sign. $\Delta U = \dfrac{3}{2}nR\Delta T$		Move to the right = − Move to the left = + Up or down = 0 $W_{gas} = -P\Delta V$	Find ΔU and W first. Then use the first law of thermodynamics to calculate Q. $\Delta U = Q + W$

▶ For path 1, the temperature must be decreasing because it is moving to a lower PV value. Therefore, the internal energy of the gas must also be decreasing. Since the pathway is moving up and down with no change in volume, the work must be zero. Using the first law of thermodynamics, we see that heat must be leaving the gas equal to the decrease in internal energy for the gas. Keep going. I'll let you do the rest on your own. You're doing

BTW

PV diagrams are used to model heat engines (gasoline-powered engines and nuclear power plants) as well as refrigeration cycles (air conditioners and refrigerators).

great! Refer back to the rules as you work through the rest of the table. Here is the answer key:

Path	ΔT	ΔU	W	Q
1	–	–	0	– (same magnitude as ΔU)
2	–	–	–	0
3	0	0	–	+ (same magnitude as W)
4	+	+	–	+ (bigger than W)

REVIEW QUESTIONS

Let's exercise our thermodynamics skills by answering the following questions.

1. An ice cube placed in a drink will melt. Did the internal energy of the ice cube go up or down? Explain.

2. Which is the only state of matter that does not have any internal potential energy?

3. Which has more energy: a pot of water at 100°C or the same mass of steam at 100°C?

4. When I was young, I couldn't unscrew the lid off a jar. My grandmother told me to run hot water over the lid. Why did this help me remove the lid from the jar?

5. The normal internal temperature of a human is 37°C. Convert this to Kelvin.

6. When my children were little and seemed sick, I would place my forehead against theirs to see if they were running a fever. Why does this work?

Questions 8 to 9: The following figure shows a PV diagram with a point marking the condition of 5.0 moles of gas.

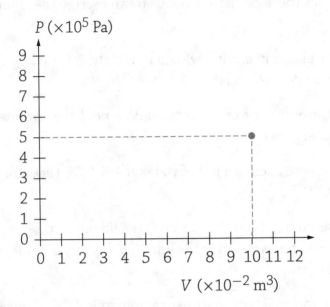

7. Calculate the temperature of the gas.

8. Sketch a path on the PV diagram that will increase the temperature of the gas.

9. Sketch a path on the PV diagram that will decrease the internal energy of the gas.

10. Rank the temperatures of the points on the PV diagram shown next from highest to lowest temperature. Explain how you arrived at your ranking.

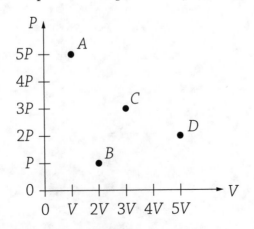

11. A gas is taken through the four processes shown in the following figure. For each path, identify the name of the process and determine if the values of ΔT, ΔU, W, and Q are positive, negative, or zero, and fill in the table with a $+$, $-$, or 0. The table is started for you. Process 1 is adiabatic.

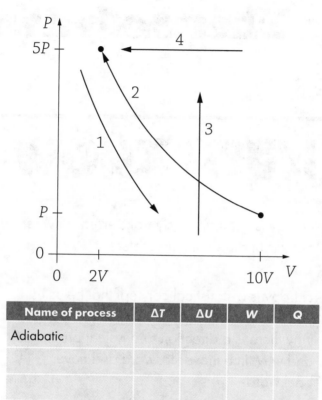

Name of process	ΔT	ΔU	W	Q
Adiabatic				

The Second Law of Thermodynamics

MUST KNOW

⚡ Energy is spontaneously transferred between objects of different temperatures.

⚡ Entropy is a statistical concept that helps us understand why heat flows in only one direction, why there is a direction to physical processes, and why time moves forward and never backwards.

⚡ Heat transfer occurs by three methods: conduction, convection, and radiation.

kay, so we know that atoms are in constant, random, frenzied motion. But how does this random motion lead us to the very predictable behavior of nature? Why does heat always flow from hot to cold and not the other way around? How does heat even move from hot to cold? And the biggest question of all, why does time always move forward and never backward?

In this chapter, our goal is to understand why the random behavior of atoms leads to the inevitable conclusion that energy is moving in one direction and so are we.

Random Motion of Atoms Leads to Spontaneous Transmission of Internal Energy

Molecules are in constant random motion that is spread out over a distribution, as seen in the following figure. On average, "hot" objects have faster-moving molecules than "cold" objects. As you can see in the figure, it is possible for some of the "cold" molecules to be moving faster than some of the "hot" molecules. However, on average, the "hot" molecules are moving faster. When put in contact, molecules in a hot object tend to collide and transfer more energy to the molecules in a cold object because they are moving faster. (Remember conservation of momentum from Chapter 3.)

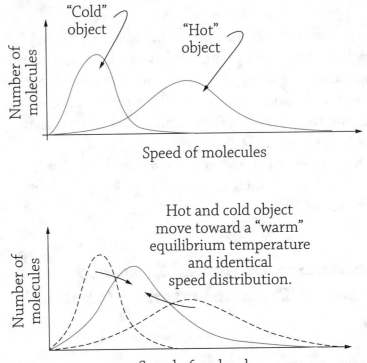

Is it possible for a "cold" molecule to collide and transfer energy to the "hot" molecule? Sure! But on average, it is much more likely for energy to transfer from "hot" to "cold." Just like it is much more likely for a speeding car to transfer energy to a slow-moving car in a collision instead of the other way around. For *net* heat to transfer from a cold object to a hot object, more of the "cold" molecules would have to transfer energy to the "hot" molecules. While this might be theoretically possible in the magic world of physics, it is statistically and practically impossible. This is simply a statistical matter of what is more likely to happen given a very large group of atoms. And remember that we have moles and moles of atoms inside objects. The net real-world result is that heat always flows from hot objects to cold objects because atoms are so small and there are so many of them that it is simply more likely to happen that way.

When does the heat transfer between objects stop? In reality the heat transfer between objects never really stops. Hot objects transfer lots of heat to cold objects. But remember that "cold" objects have a few fast-moving molecules that can transfer heat to the "hot" object. Overall, the *net* heat transfer is from hot to cold. (See the next figure.) Once the two objects reach the same temperature, the average molecular motion is the same for both. So, they transfer heat back and forth between each other at equal rates. When two objects have the same temperature, they are in thermal equilibrium and the *net* heat transfer between them is zero.

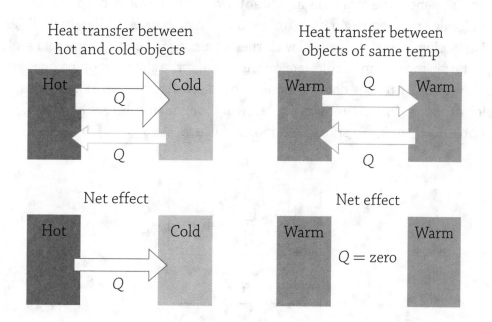

Entropy

The laws of physics, in general, don't have a preferred direction. What I mean is that a ball can roll up a hill and its kinetic energy can turn into potential energy just as easily as the ball starting at the top of the hill can repeat this transformation in the other direction. A video of a cue ball colliding with an 8-ball could be run forward or backward and in both cases the collision would look okay. The physics of the collision would work in either direction.

But...heat always flows from hot to cold, not the other way around. Entropy expresses the idea that the natural world is not reversible but is on a one-way street. All physical processes in nature actually have a preferred or more likely order of events.

Entropy (S) is often described as a measure of disorder. The following figure shows two identical rooms connected by a doorway. Originally, the left room has hot air inside and the right room has cold air inside. When the door is opened, the air mixes and heat flows until the air in both rooms is randomly arranged in a mostly uniform warm distribution. The original gas distribution is more "ordered," with the hot air separated from the cold air. The faster-moving molecules are on the left and the slower-moving ones are on the right. After the door is opened, the air is all mixed up in a random blob of warm. The fast-moving and slow-moving molecules are all in a jumble of "disorder": a spread-out mush of warm.

Before

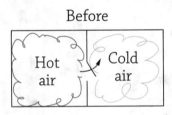

After

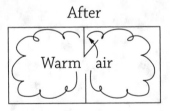

We can even calculate the change in entropy of the system with the equation $\Delta S = \dfrac{Q}{T}$, where S is the entropy of the system in units of joules/kelvins, Q is the heat exchanged between the objects in units of joules, and T is the temperature of the object.

Let's calculate the increase in entropy for this situation where a hot gas mixes with a cold gas and both reach equilibrium as a warm temperature in between.

EXAMPLE

▶ Let's say the average temperature of our hot gas is 400 K and the cold gas has an average temp of 200 K. When mixed, they exchange 1,200 J of heat and end up at a final equilibrium temperature of 300 K. The change in entropy of the hot gas is negative because it loses heat:

$$\Delta S = \frac{Q}{T} = \frac{-1,200 \text{ J}}{400 \text{ K}} = -3.0 \text{ J/K}$$

▶ The entropy change for the cold gas will be positive because it gained heat:

$$\Delta S = \frac{Q}{T} = \frac{+1,200 \text{ J}}{200 \text{ K}} = +6.0 \text{ J/K}$$

▶ Notice that the net change in entropy of the system that includes both gases is positive: +4.0 J/K. The entropy of the system rose when the two gases mixed.

Objects that are colder have less entropy because they have less random motion in their molecules. Hot objects have higher entropy. When a hot "high-entropy" object exchanges heat with a cold "low-entropy" object, the hot object loses entropy and the cold object gains entropy, but the overall entropy of the system of interacting objects always goes up. The **second law of thermodynamics** is a statement of this fact: In a closed system, entropy always goes up. But why? We need a bigger picture of entropy to really understand entropy.

Entropy has to do with statistics and the probability that a particular arrangement of atoms is likely to happen. Let's take a look at a single atom. We want this atom to move to the right and, to make things as simple as possible, let's say the atom can only move right or left. So, we have a 50/50 chance that the atom will do what we want it to do. Now consider a two-atom system. There are four possible motion arrangements, as seen in the following figure: one where both atoms move to the right, two where the atoms move in opposite directions, and one where they both move to the left. However, we want all the atoms to move to the right. We have a one in four chance of getting what we want. You see where this is going? With three atoms there are eight possible velocity arrangements and only one of them has all the atoms moving to the right. With three atoms, we have a one-eighth chance of all the atoms moving to the right. With four atoms, our odds get worse, with a one-sixteenth chance.

EXAMPLE

▶ Remember that real objects are made of many moles of atoms. What are the chances that all the atoms in a real object, like a baseball, will all move to the right at the same time?

▶ Never! It is a statistical improbability that *all* the atoms of an object like a baseball will randomly move to the right *all* at the same time. It takes outside work to make all the baseball atoms move to the right at the same time. You have to pick up the ball and throw it to make all the atoms do what we want them to do.

Number of atoms	Possible atom movements	Numerical value of possible atom movement combinations	Chance of all atoms moving to the right at the same time
1 atom		$2^1 = 2$	$\frac{1}{2}$
2 atoms		$2^2 = 4$	$\frac{1}{4}$
3 atoms		$2^3 = 8$	$\frac{1}{8}$
4 atoms		$2^4 = 16$	$\frac{1}{16}$
Approximately 10 moles of "baseball" atoms 10(6.02 × 10²³) = 6.02 × 10²⁴ atoms		$2^{6.02 \times 10^{24}} =$ insanely unimaginable number	Never ever ever!

> **IRL** Shuffle a deck of cards and look at them. You may find a few cards that are grouped together in order or by suits, number, or color but, overall, the cards will be in an arrangement of high disorder (high entropy) because high disorder is statistically more likely than an orderly (low-entropy) arrangement.

Entropy is simply a statement that it is much more likely that the atoms in an object will be in a state of randomness than order, because it's much more likely to happen that way. To put atoms in highly unlikely arrangements of low entropy takes energy or work from the outside, because atoms are more likely to be arranged in high-entropy states. For example: There are a billion ways for your house to be in a state of messy disarray (high entropy) but only a few states where the house is clean and orderly (low entropy). It takes work on your part to reduce the entropy of your house because the universe wants your house to move toward higher entropy—a messy and dirty state.

Entropy is always on the rise. All other forms of energy will completely and easily convert to random thermal energy, but not the other way around because it's just more likely that the energy will move toward the disorder of internal energy than stay organized. This is why a ball dropped from a table will bounce lower and lower as potential and kinetic energy degrade into thermal energy until all the original organized energy is random thermal energy. We never see the opposite. A ball never spontaneously cools off and flies into the air. Is it theoretically possible? Sure. But will it ever happen? No, because too many atoms have to randomly and spontaneously work together at the same time to make it happen.

Entropy and the Flow of Time

Take a series of photos, mix them up, and then try to place them in order. It's actually not that hard to do because we have a built-in sense of time. Certain things happen after others. A hot cup of coffee cools off. Not the other way around. The empty cup on the table falls off and breaks to pieces.

Not the other way around. Entropy is always on the rise. Scientists think that entropy gives us the direction of time. We can move forward in time but not backward because moving backward in time means moving backward in entropy—the second law of thermodynamics doesn't seem to allow that.

Entropy and Life

A living organism is a highly ordered system and, to the universe, a highly unlikely arrangement of atoms. In order for an organism to maintain homeostasis and live, the organism must constantly interact with its environment by taking in energy in order to keep its own entropy state low. As a consequence, this raises the entropy of the environment around the living organism.

 **IRL** In order for you to stay alive, you must constantly take in usable energy from the environment by eating food, drinking water, maintaining your internal body temperature with air conditioning and heating and clothes, etc. This keeps our entropy at a stable level but raises the entropy of the environment.

Entropy and the Fate of the Universe

Constantly rising entropy has some big consequences for the Universe. Once heat flows from hot to cold and thermal equilibrium has been established, the system is in a state called **heat death**. This means that the energy is all spread out disorderly and all the usable energy of the system has been lost to randomness. In order for a system to lower its entropy, it must gain new energy from the environment.

But as far as we can tell, the Universe is a closed system (even though it might be infinite!). If that's true, all the stars in the Universe will eventually burn out and all the energy in the Universe will be spread out in a random mess of heat death, and the Universe will die. Wow! That turned dark.... When scientists discovered this, it was quite depressing because they had just predicted the end of the Universe. But not to worry. The Universe is so big and there is so much energy in it that the latest estimate for when heat death occurs is 10^{100} years from now, an incomprehensibly long period of time. So don't worry—we have plenty of time to enjoy the Universe!

Heat Transfer

Now that we know why heat always moves from hot objects to cold objects and we have waded through the deep waters of entropy, it's time to learn how exactly heat flows. There are only three methods: conduction, convention, and radiation. See the following figure.

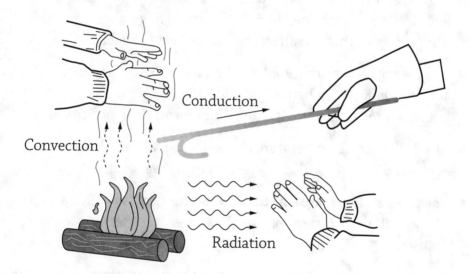

Conduction

Have you ever noticed that some materials feel colder than others even though they have been sitting in the same room and are at the same temperature? For example, you could have a block of wood and a block of metal sitting on a table in your home. Both are the same temperature after sitting in the same room. If you picked up the wood, it wouldn't feel particularly warm or cold, but the metal feels cool to the touch. Well, the reason the metal feels colder than the wood is that it is a better conductor of heat than the wood. It draws heat away from your hand at a faster rate than the wood, making it feel colder.

 IRL Humans are not really that good at telling the temperature of something by touching it. What we really sense when we touch something is the heat flow (Q) between ourselves and the other object.

Conduction is the physical transfer of the atomic vibration energy from atom to atom through an object from the hot end to the colder end. Whenever two objects with different temperatures touch each other, heat will transfer from the hotter object to the colder one through atomic collisions until they reach the same temperature, or equilibrium. There are a couple of properties that can affect the rate of heat transfer:

- The thermal conductivity, k, of the material. For example, metals tend to conduct heat better and will have a high thermal conductivity compared to nonmetals. The units for thermal conductivity are J/m·K·s.

- The difference in temperatures between the cold and hot objects, ΔT. A bigger difference will give a larger rate of heat transfer.

- The cross-sectional area of the material the heat is being transferred along, A. The larger the cross-sectional area, the greater the rate of heat transfer.

- The length, l, of the material the heat is being transferred through. The longer the length, the lower the rate of heat transfer.

This gives us the thermal conduction equation:

$$\frac{\Delta Q}{t} = \frac{kA\Delta T}{l}$$

$\Delta Q/t$ is the rate of heat transfer in joules per second, or watts.

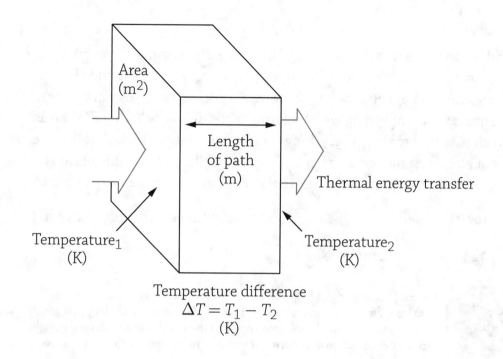

Area
(m^2)

Length
of path
(m)

Thermal energy transfer

Temperature$_1$
(K)

Temperature$_2$
(K)

Temperature difference
$$\Delta T = T_1 - T_2$$
(K)

 IRL To keep you warm in the winter, you wear a coat with a low thermal conductivity. The thicker the coat, the better job it does at keeping you warm.

Convection

Convection is the transfer of thermal energy from one place to another through fluid flow. Remember that hotter objects grow in size because the atoms are moving faster. This affects density; thus, when a fluid is hot, it is less dense and will naturally rise due to the buoyancy force.

 IRL Convection is why you can put your hand to the side of a flame but not over the flame. Above the flame, the added heat from convection will burn you.

Radiation

Radiation is the transfer of energy through electromagnetic waves. It turns out that the vibration of charged particles, like protons and electrons, creates electromagnetic waves (EM waves), which carry energy away from the object. Thus, anything with vibrating molecules is radiating EM waves, which means absolutely everything is radiating energy. (We will talk more about electromagnetic radiation in Chapter 16.) The rate at which heat is transferred through radiation is proportional to the temperature raised to

the fourth power: $\dfrac{Q}{\Delta t} \propto T^4$. This means that an object that is twice as hot

will radiate 16 times more heat.

 IRL Infrared vision goggles help you to see in the dark. At night there isn't any light from the sun, but everything is radiating EM waves. Including you! Most of these waves are in the infrared region of the EM spectrum. Infrared goggles pick up this radiation and convert it into a false light image that we can see.

 IRL Microwave ovens use radiation to cook food. Convection ovens force the hot air that rises and is trapped at the top of the oven to circulate and heat food faster.

Let's take a look at how conduction, convection, and radiation affect a car when it sits in the sun.

EXAMPLES

▶ **Example 1**   On a cold but sunny day, the inside of a car is nice and toasty warm. Why is this?

▶ The car is warmed up by radiation from the sun that enters through the windows.

▶ **Example 2** On a hot, sunny day, the inside of a car is smoking hot. Why is it that the metal seat belt latch will burn you but the cloth seat belt won't?

▶ Both the metal latch and the cloth seatbelt are the same temperature. They both will transfer heat to your hand. Metal is a better conductor and transfers heat to your hand at a faster rate, which burns you.

▶ **Example 3** Why does cracking your windows by rolling them down a little help to keep the car from getting so hot inside on a sunny day?

▶ Cracking the windows allows convection currents to rise through and out the cracked window and transfer some of the heat away from the inside of the car.

IRL Let's consider how to design a house to be super-thermally efficient. How would we design the house?

To reduce heat conduction to/from the environment, the house needs to have thick insulation and a small surface area with the environment. So, the house should be round with thick, well-insulated walls.

To reduce convection heat losses, the house should be tightly sealed. Perhaps we should even have double-entry doors to form an airlock so that air does not get in or out.

To reduce radiation heat transfer, the house should have a highly reflective surface. This will reflect unwanted outside heat away and reflect wanted inside heat back into the house so that it does not escape.

Guess what? You just designed what energy-efficient habitation modules will probably look like on the Moon and Mars!

REVIEW QUESTIONS

Let's show our skills with the second law of thermodynamics by answering the following questions.

The following figure shows three blocks at different temperatures. Refer to it to answer questions 1 to 5.

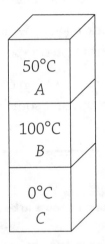

1. Between which blocks does heat flow and in which direction does the heat flow?

2. Between which blocks will the heat flow the fastest?

3. When will the heat transfer stop?

4. Why does the heat transfer stop?

5. Can the entropy of a system ever go down?

6. Why is the second story of a house always warmer than the first floor?

7. If you want to boil water faster, should you keep the lid on the pot or leave it uncovered?

8. Many houses are now built with a silvery wrapping before the brick or siding is added to the house. What is the purpose of this reflective wrapping?

9. You just finished cooking a pizza in the oven. Why can you touch the crust without any problem but when you bite into the pizza it burns your mouth?

10. Why does blowing on a hot cup of coffee cool it off?

PART FOUR

Electricity and Magnetism

Now, we're going to apply what we've learned about physical laws to positively and negatively charged objects. We'll investigate how charges affect each other, the energy of charged objects, and how to make charges move where we want them to go. Finally, we will see how magnetism connects to electricity and learn how to generate electricity.

Electric Force and Electric Fields

MUST ⚡ KNOW

⚡ There are two types of charges, positive and negative, and their arrangement in an object can be manipulated.

⚡ Charges create electric fields in the space around them.

⚡ Charges exert forces on each other.

verything is made of atoms, which in turn are made of protons and electrons. These particles have the property of electric charge. In this chapter, our goal is to focus on charge and its properties, as well as how charge can move around and how normally neutral objects can gain a net charge. We'll also learn how charges exert forces on each other.

A Quick Review of the Atom and Charges

Atoms are made of protons and neutrons that are locked in the nucleus. The electrons zip around the outside and take up most of the atom's space in an electron cloud.

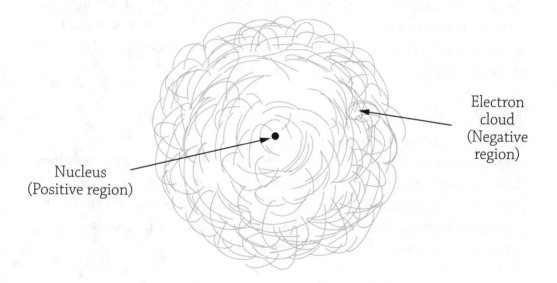

Electron cloud (Negative region)

Nucleus (Positive region)

Protons have a positive charge, while electrons have a negative charge of the same magnitude. Charge is quantized, meaning that all net charge comes in multiples of this smallest magnitude of charge, called the "magnitude of

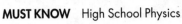

the electron charge," which is $e = 1.6 \times 10^{-19}$ C. The unit of charge is called a **coulomb** (C). Charge is carried by real particles, so it won't just appear or disappear. Charge can move around from object to object and be rearranged on the same object, but the net charge must be conserved at all times. Charges create electric fields that produce forces on other charges. "Like" charges repel, and "opposite" charges attract each other. The force between two charges is called the electric force.

Charge

Most of the objects that we encounter in our daily lives are electrically neutral, meaning they have the same number of protons as electrons. Things like a house, flower, car, squirrel, and you are all mostly neutral. If an object has more protons than electrons, it has a net positive charge. If an object has more electrons than protons, it has a net negative charge. But how exactly do mostly neutral objects become charged? There are three methods to charge an object.

Friction

Some materials have more of an affinity for electrons than others do. Remember your chemistry? Atoms on the left side of the periodic table lose their outer electrons easily, while those on the right side of the periodic table grab electrons to fill in their almost complete outer electron shells. The same thing happens in everyday materials like clothes, hair, and plastic. The following figure shows a triboelectric series.

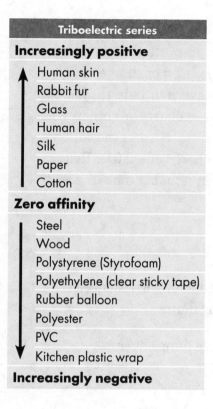

Objects nearer the top of the chart tend to lose electrons and become positively charged. Those at the bottom tend to grab electrons and become negatively charged. When put in contact, electrons from materials nearer the top of the list jump ship to the electron grabbers closer the bottom of the list. Rubbing the materials together with friction increases the contact area between the two materials and facilitates the transfer. For example, rubbing a balloon on your hair will cause your hair to lose electrons to the balloon. Your hair becomes

BTW

A quick reminder about the difference between **conductors** *and* **insulators**. *Conductors allow charge to move easily through themselves and across their surface. Insulators don't allow the movement of charge across or through themselves. Metals, the Earth, and wet objects are good conductors. Plastic, wood, ceramics, and cotton make good insulators.*

positively charged, while the balloon becomes negatively charged. It is important to note that in accordance with conservation of charge, if 2 coulombs of charge leave your hair, the balloon will pick up 2 coulombs of charge. Your hair will end up with a charge of +2.0 C, and the balloon will have the opposite charge of −2.0 C. The net charge of the system remains the same as it started: neutral.

Conduction/Contact

Let's say we have a conductor with extra electrons, as seen in the following figure.

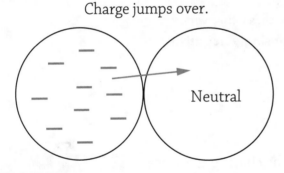

Charge jumps over.

Neutral

Remember, electrons repel each other. When we touch a charged object to a neutral object, the excess electrons repel each other and some move onto the neutral object. They share the excess charge and become charged with the same sign charge. If the objects are the same size, they both end up with equal amounts of the charge and share equally. If one is bigger, it will end up with more charge than the smaller object because there is more room for the charges to spread out.

The Earth is a good conductor. When we touch or connect a charged object to the Earth, it will share its charge with the Earth. Since the Earth is so huge compared to the charged object, all of the excess charge flows to the Earth and neutralizes the object. This is called **grounding**.

What if the charged object is positively charged? Positives can't jump to the neutral objects because they are trapped in the nucleus. In this case, negatives are attracted to the positive object, causing it to become less positive and the neutral object to become positive. It looks exactly like positive charges moved to the right, but in reality, negative charges moved to the left. So, sometimes we will say things like "positive charge moved from the left object to the right object." (See the following figure.) Actually, the negative charge moved in the reverse direction—from the right object to the left object.

Negative charge jumps over.

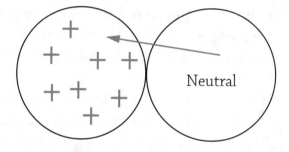

Neutral

One last comment on conduction: remember that insulators do not let charge move. So, if you touch a charged insulator, you will only share the charge right where you touched it because the rest of the charges on the object are locked in place.

Induction

When a charged object is brought close to a neutral object, it repels the like charge and attracts the opposite sign charge. The following figure shows a positively charged object brought close to a conductor.

BTW

You experience charging by friction when you rub your shoes on carpet when you walk. You pick up electrons from the carpet, making you negatively charged and the carpet positively charged. When you touch something metal, you feel a shock as you and the metal object share the charge by conduction.

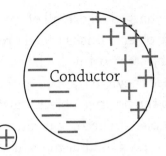

Notice how the protons in the conductor are repelled and the electrons are attracted. The conductor is still neutrally charged. The charge has just been separated because a group of electrons have moved from the right side to the left side, which leaves a net charge on either side of the object. This is called **charge polarization**, or we just say that the object has been **polarized**. This is a temporary condition. When the positive charge is removed, the electrons on the conductor revert to their normal positions, and the conductor will be neutral all over again.

If a charge is brought close to an insulator, positives and negatives will again try to move about, but they can't because the insulator won't allow the charge to flow. Instead, the atoms inside the insulator become polarized with the electron cloud shifting due to the electric forces. The next figure shows an insulator being polarized on an atomic scale.

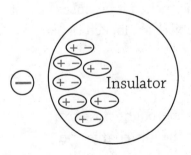

An **induced** electric charge can be created when an electrically neutral object is first polarized. The following figure shows how you can create an induced charge in an object.

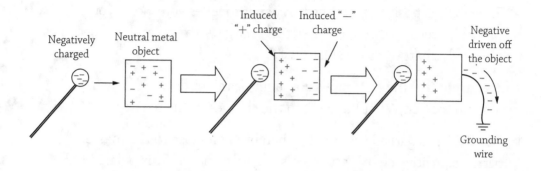

Once the object is polarized, we must supply an escape route for the unwanted charge to be driven off the object, as seen in the figure. Usually we connect an escape wire to the Earth or other large conducting object. Then we must disconnect the escape route, and like magic, we just gave the object a permanent positive charge by the process called **induction**. Note that the negatively charged sphere was brought close to, but did not touch, the neutral metal object.

> **IRL**   During a storm, the bottom side of a cloud can build up a negative charge. This polarizes the ground underneath the cloud, causing an opposite sign (positive) induced charge on the ground. When this charge difference between the cloud and the ground gets large enough, it ionizes the air, creating a conductive pathway between the two. Lightning!

EXAMPLES

▶ **Example 1**   A rabbit rubs up against a PVC pipe. What happens to the charge of each?

▶ Looking back at our triboelectric chart, we see that electrons transfer from the rabbit to the PVC pipe. The rabbit becomes positively charged, and the PVC becomes negatively charged. Conservation of charge dictates that the rabbit and the pipe have equal magnitudes of charge.

▶ **Example 2**   The rabbit comes over to you and touches its nose to yours. What happens?

▶ The rabbit is positively charged. When it touches your nose, you probably feel a spark as it shares this charge with you and you become positively charged by conduction.

▶ **Example 3**   The PCV pipe is brought near, but does not touch, a metal toaster. What happens to the charges in the toaster?

▶ The PVC is negatively charged. When brought close to the toaster, it attracts positive charge and repels negative charge. This polarizes the toaster.

Charge Distributions on Conductors and Insulators

Conductors, like metals, allow charge to move. So, when a metal object becomes charged, the net charge moves to the very outside surface of the conductor. The net charge inside the object is zero. (See the following figure.) On the other hand, insulators don't allow charge to move. So, when an insulator becomes charged, the charge is stuck wherever it was placed. For this reason, we put electrically sensitive items like electronics in protective metal cases so that any stray charge will stay on the outside and keep the inside protected from static shocks.

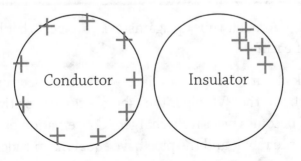

Electroscopes

A metal leaf **electroscope** is used in physics labs to test the behavior of electric charges. One electroscope design consists of a metal knob attached to a conducting rod. At the bottom of the rod are thin metal aluminum foil leaves. All of this is insulated from its surrounding by a rubber stopper and a glass flask.

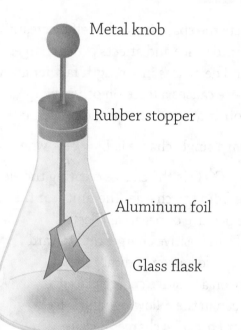

Metal knob

Rubber stopper

Aluminum foil

Glass flask

If a negatively charged balloon touches the top of the electroscope, it will share some of its negative charge with the metal knob by conduction. This charge spreads over the entire metal of the electroscope. The two aluminum foil leaves now share the same charge and repel each other, and they will move apart indicating that the electroscope is now charged.

When the charged balloon is brought close to but does not touch the metal knob, the metal leaves will also move apart. But why? The negatively charged balloon polarizes the metal of the electroscope, pulling positive charge upward and forcing negative charge to the bottom. Both leaves now have the same negative electric charge. The leaves repel each other and move apart. When the balloon is removed, the leaves fall back down because the electroscope is no longer polarized.

Electric Fields

Every charge affects the space around itself by creating an **electric field**. This field extends out into space and affects other charges. The field is strongest near the charge and decreases in strength farther away from the charge. An electric field in space causes a force on other charges. This is how one charge can exert a force on another charge without even touching it. The electric field strength from a single charge is $E = k\dfrac{Q}{r^2}$, where k is a constant: $k = 9.0 \times 10^9 \, \text{Nm}^2/\text{C}^2$, Q is the charge creating the electric field, and r is the radius or distance from the charge. The units for the electric field are N/C.

Charges can be positive or negative. Electric fields point away from positive charges and toward negative charges. E-field vectors indicate the direction of a force on a positive charge placed at that location, as seen in the following figure. Field vectors point toward negative charges because they pull positive charges toward themselves. Field vectors point away from positive charges because they will push positive charges away.

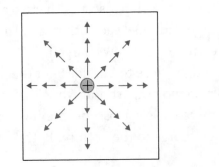
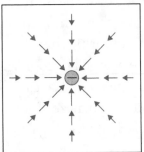

The electric field around multiple charges will be the **superposition** (vector combination) of the individual E-fields. Look at the electric vector field in the next figure. What is going on at points X and Y?

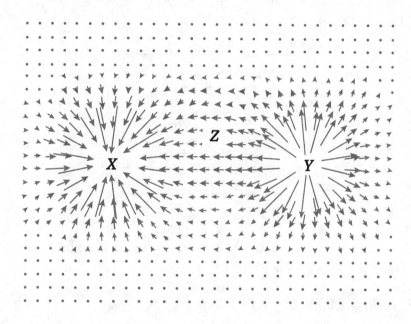

Electric field arrows are pointing inward toward X and are getting bigger as they get closer. X must be a negative charge location. Y must be a positive charge location because the E-field vectors are pointing away from it. What direction is the electric field at Z? That's easy—it's to the left. What's the direction of the force on a charge placed at Z? Careful, it's a trick question! Is the charge positive or negative? It makes a difference. If it is negative, it gets pushed to the right in the opposite direction of the electric field. If it's positive, it gets forced to the left in the same direction as the E-field.

The force that a charge feels in an electric field is $F_E = Eq$.

▶ An electron, proton, and a neutron are each placed in a uniform electric field with a magnitude of 8,000 N/C, directed to the right, as shown in the following figure. What are the magnitude and direction of the force exerted on each particle?

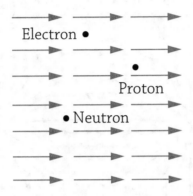

▶ First of all, the neutron will not receive any force at all because it is neutral. The force on the proton is:

$$F_E = Eq = (8,000\,\text{N/C})(1.6 \times 10^{-19}\,\text{C}) = 1.3 \times 10^{-15}\,\text{N to the right}$$

▶ Electrons have the same magnitude charge as protons so, the force on the electron will be the exact same magnitude but in the opposite direction of the field because the charge is negative.

▶ Remember that on a conductor, all the excess charge moves to the outside surface of the conductor with none left inside. The previous figure shows a metal sphere with positive charge on the surface. The electric fields from each individual charge point away from themselves.

BTW

Electromagnetic shielding is used to protect sensitive objects from outside electric fields. It can also be used to keep electromagnetic fields trapped inside. Take a look at your microwave. The inside is a metal box, and the door window has a metal screen embedded inside to block the escape of the microwaves.

This means that all of the charges on the surface produce an overlapping, inward-pointing E-field that completely cancels itself out. This happens on any conducting surface. The E-field inside the object will always be zero. This is called electromagnetic shielding.

Let's look at an example where we have a single charge producing an electric field.

A charge with a magnitude of 7.5×10^{-6} C is located at the center of the electric field in the following figure. Points N and P are a distance of 2.0 m from the central charge. Point O is closer to the central charge than N and P.

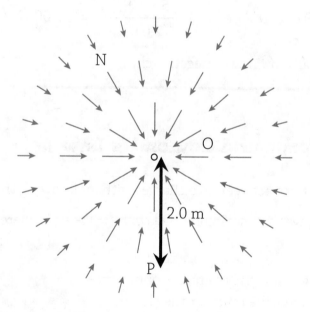

What is the sign of the charge at the center of the electric field? Explain how you know this.

▶ The charge must be negative because the electric field vectors are pointing inward toward the charge.

▶ Rank the electric field strength at the points N, O, and P.

▶ $E_O > E_N = E_P$. The electric field strength around a single charge is $E = k\dfrac{Q}{r^2}$. The E-field is stronger closer to the charge.

▶ What is the direction of the electric force on an electron placed at point N?

▶ It is in the opposite direction of the E-field because the electron is negative. Remember that the electric field is set up for positive charges.

▶ Calculate the electric field strength at point P.

$$E = k\frac{Q}{r^2} = (9.0 \times 10^9 \text{ Nm}^2/\text{C}^2)\left(\frac{7.5 \times 10^{-6} \text{ C}}{(2.0 \text{ m})^2}\right) = 17{,}000 \text{ N/C}$$

pointing upward toward the negative charge.

Electric Force and Coulomb's Law

To calculate the force between two charges without knowing the E-field strength, we can use Coulomb's law: $F_E = \left| k\dfrac{q_1 q_2}{r^2} \right|$, where

$k = 9.0 \times 10^9 \text{ Nm}^2/\text{C}^2$, q_1 and q_2 are the two interacting charges, and r is the radius or distance between the charges. The units for the electric force are N. Notice there is an absolute value sign around the right side of the equation indicating that it only gives you the magnitude of the electric

BTW

Remember Newton's Third Law. The two interacting charges will always exert an equal but opposite directional force on each other no matter the relative size of the two charges.

force, not the direction. You will need to decide what the direction of the force is based on whether the two charges are attracting or repelling.

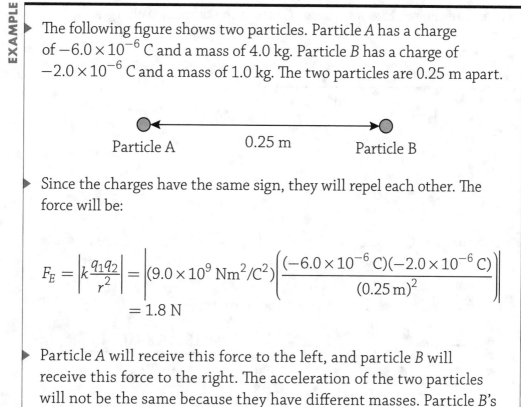

The following figure shows two particles. Particle A has a charge of -6.0×10^{-6} C and a mass of 4.0 kg. Particle B has a charge of -2.0×10^{-6} C and a mass of 1.0 kg. The two particles are 0.25 m apart.

Particle A   0.25 m   Particle B

▶ Since the charges have the same sign, they will repel each other. The force will be:

$$F_E = \left| k \frac{q_1 q_2}{r^2} \right| = \left| (9.0 \times 10^9 \text{ Nm}^2/\text{C}^2) \left(\frac{(-6.0 \times 10^{-6} \text{ C})(-2.0 \times 10^{-6} \text{ C})}{(0.25 \text{ m})^2} \right) \right|$$
$$= 1.8 \text{ N}$$

▶ Particle A will receive this force to the left, and particle B will receive this force to the right. The acceleration of the two particles will not be the same because they have different masses. Particle B's acceleration will be four times greater because it has four times less mass.

What if we have more than two charges? How do we figure out the electric force then?

The following figure shows three electric charges. All three charges exert forces on each other. In this example, we are just going to calculate the net force on charge Q_3 from the other two charges.

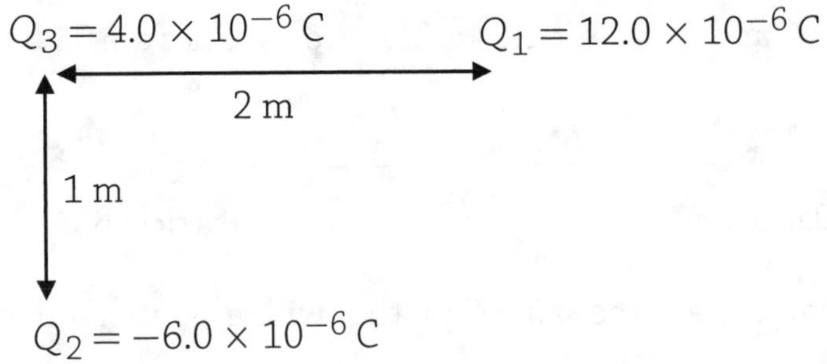

$Q_3 = 4.0 \times 10^{-6}$ C $\qquad\qquad$ $Q_1 = 12.0 \times 10^{-6}$ C

2 m

1 m

$Q_2 = -6.0 \times 10^{-6}$ C

Two charges, $Q_1 = 12.0 \times 10^{-6}$ C and $Q_2 = 16.0 \times 10^{-6}$ C, are placed to the right and below a third charge, $Q_3 = 4.0 \times 10^{-6}$ C. First let's calculate the forces that Q_1 and Q_2 exert on Q_3. Then we will add these two force vectors to determine the net force on Q_3:

$$F_{13} = \left| k \frac{q_1 q_2}{r^2} \right| = \left| (9.0 \times 10^9 \text{ Nm}^2/\text{C}^2) \left(\frac{(12.0 \times 10^{-6} \text{ C})(4.0 \times 10^{-6} \text{ C})}{(2\text{ m})^2} \right) \right|$$

$$= 0.108 \text{ N to the left}$$

$$F_{23} = \left| k \frac{q_1 q_2}{r^2} \right| = \left| (9.0 \times 10^9 \text{ Nm}^2/\text{C}^2) \left(\frac{(-6.0 \times 10^{-6} \text{ C})(4.0 \times 10^{-6} \text{ C})}{(1\text{ m})^2} \right) \right|$$

$$= 0.216 \text{ N downward}$$

Adding these two forces as vectors we get:

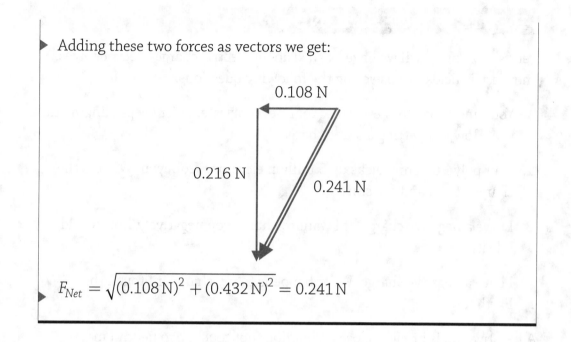

$$F_{Net} = \sqrt{(0.108\,\text{N})^2 + (0.432\,\text{N})^2} = 0.241\,\text{N}$$

REVIEW QUESTIONS

Let's show how much we have learned about electric charge, electric forces, and electric fields by answering the following questions.

1. You time-travel to the 1,970s. Your new polyester shirt keeps clinging to you. How is it getting a static charge?

2. Your polyester shirt picks up 5 million electrons from you. What is the charge of you and the shirt?

3. I have a negative charge. If I want to charge you negatively, how would I do it?

4. I have a negative charge. What happens to my charge if I touch the Earth?

A negative rod is brought close to, but does not touch, two neutral metal spheres on insulating stands, as shown in the following figure. The two spheres are touching. Use this information to answer questions 5 to 7.

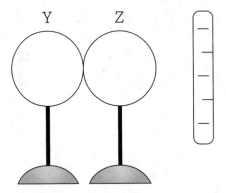

5. What does this do to the two spheres?

6. What happens to the charge of the spheres if they stay in contact and the rod is removed?

7. What happens to the charges if the spheres are moved apart from each other first and then the rod is removed?

The next figure shows an electric field about three charges labeled 1, 2, and 3. Two locations in the electric field, A and B, are also indicated. Refer to the figure to answer questions 8 to 11.

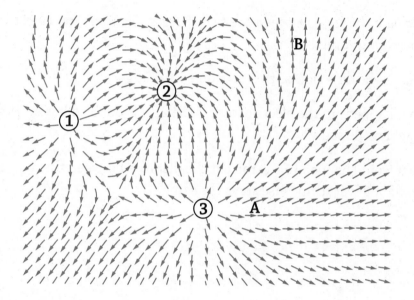

8. What are the signs of the three charges? What indicates the sign on the charge?

9. A proton is placed at point A. In which direction will it receive an electric force?

10. In which direction will an electron placed at location B receive a force?

11. If the electric field strength at B is 50,000 N/C, what is the electric force on the electron?

Two charges sit on a number line as shown in the following figure. Use this information to answer questions 12 to 15.

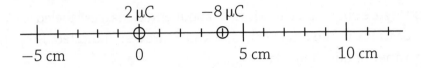

12. Calculate the force between the two charges. Caution! You need to convert your units!

13. What is the direction of the force on each charge?

14. What happens to the electric force if the $2\,\mu C$ is doubled but everything else stays the same?

15. What would happen to the force if the distance between the force is cut in half but everything else stays the same?

11 Electric Potential Energy and Electric Potential

MUST ⚡ KNOW

⚡ The locations of electric charges relative to each other in space create electric potential energy.

⚡ Electric potential is the energy per charge in units of joules/coulombs, which equals volts. Electric potential can be increased through the input of work. Electric potential will naturally turn into kinetic energy, which causes the movement of charges.

⚡ Equipotential lines can be used to visualize the electric potential field.

n the last chapter, we applied the ideas of forces to electric charges. In this chapter, our goal is to apply the concepts of energy from Chapter 4 to electric charges. We'll see how electric potential energy can be converted into kinetic energy to make charges move around. A new and very useful concept, called electric potential, will also be introduced. So, let's get started.

Electric Potential Energy

Imagine a large positive charge that is held stationary with two small charges near it, as seen in the following figure. Where do the smaller charges want to naturally go?

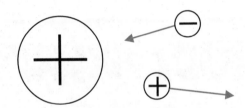

The tiny negative charge is attracted to the large positive charge, and the tiny positive charge is repelled. Now put on your energy glasses. What kind of energy do you see? The system has electric potential energy that converts into kinetic energy when the small charges are free to move.

Now look at the large negative charge in the next figure. Where would you move the small positive and negative charges to give them more electric potential energy? Which way would the charges naturally move if released?

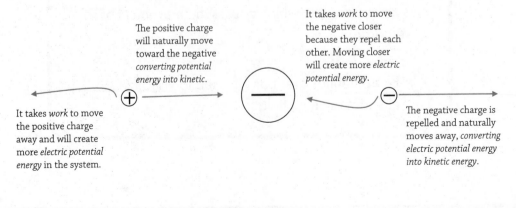

The positive charge will naturally move toward the negative converting potential energy into kinetic.

It takes *work* to move the negative closer because they repel each other. Moving closer will create more *electric potential energy.*

It takes *work* to move the positive charge away and will create more *electric potential energy* in the system.

The negative charge is repelled and naturally moves away, *converting electric potential energy into kinetic energy.*

Notice that electric potential energy naturally converts into the kinetic energy of motion: $K = -\Delta U_E$. But it takes work to move the charges where they don't "want" to go. Thus, the work equals the increase in potential energy of the system: $W = +\Delta U_E$.

Electric Potential

Now let's make this idea of electric potential energy a little more abstract but also more useful. An electric potential scalar field tells us the effect that a charge has on the space around itself. It shows where the energy has the potential to be greater or smaller. Look at the following figure.

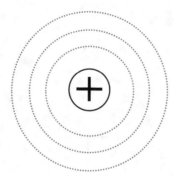

Concentric rings are drawn around a charge. A charge placed anywhere along one of the concentric rings will have the exact same energy as anywhere else on the same ring. If we place a larger charge on the same ring as before, it will have more electric potential energy. If we place a smaller charge on the ring, it would have less potential energy. But any charge placed on the ring will have the same *potential for energy* or the same **electric potential**.

Electric potential is defined as $V = \dfrac{U_E}{q}$, where U_E is the electric potential energy of the system in joules and q is the charge placed at that location in coulombs. The units for electric potential are volts (V). The electric potential around a single charge is $V = k\dfrac{Q}{r}$, where k is the constant: $k = 9.0 \times 10^9 \ \text{Nm}^2/\text{C}^2$, Q is the charge creating the potential field, and r is the radius from the charge. The electric potential is inversely proportional to the radius $V \propto \dfrac{1}{r}$, which means that the potential decreases with distance.

Equipotential Lines

This may all seem overly confusing, so let me try to demystify it. In the next figure, a positive charge has a potential field around it.

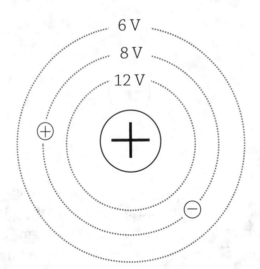

Three locations of equal potential are shown. These are called **equipotential lines**. A +2 C and a −3 C charge are placed on the 8 V equipotential. Look at the positive charge. Being positive, it naturally wants to move away to a lower potential location. Doing so will convert electric potential energy into kinetic energy. *The decrease in potential energy equals the gain in kinetic energy.*

EXAMPLE

▶ If the positive charge in the previous figure moves from its current location on the 8 V equipotential line to the 6 V equipotential, it will gain 4 J of kinetic energy:

$$K = -\Delta U_E = -\Delta Vq = -(6\,V - 8\,V)(2\,C) = 4\,J$$

▶ What about the negative charge? It is attracted to the central charge. What is its kinetic energy when it reaches the 12 V equipotential line?

$$K = -\Delta U_E = -\Delta Vq = -(12\,V - 8\,V)(-3\,C) = 12\,J$$

▶ That's actually quite easy, except for all the negative signs.

Now, let's look at how much work it takes to move the charges in the opposite direction they naturally want to move.

EXAMPLE

▶ Let's take the positive charge and move it from the 8 V equipotential to the 12 V line. This will bring it closer to the central positive charge, and it will not naturally move to that location. Moving the little positive charge to the 12 V equipotential requires work. How much work did that require? Work will equal the gain in electric potential energy:

$$W = \Delta U_E = \Delta Vq = (12\,V - 8\,V)(2\,C) = 8\,J$$

▶ Now, move the negative charge from the 8 V to the 6 V equipotential:

$$W = \Delta U_E = \Delta Vq = (6\,V - 8\,V)(-3\,C) = 6\,J$$

▶ Easy! Just remember that $\Delta V = V_f - V_i$ and you shouldn't have any problems.

Here is the general rule of thumb for charges moving through potential fields:

- Positive charges naturally move to lower potential locations, but it takes work to move them to a higher potential location.

- Negative charges naturally move to higher potential locations, but it takes work to move them to a lower potential location.

Let's mix things up and practice with an opposite-sign charge in the middle.

EXAMPLE

▶ A −1 C negative charge and a 3 C positive charge are located in a potential field produced by a large negative charge, as seen in the following figure.

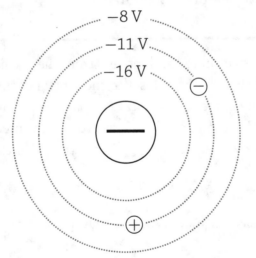

▶ Calculate the work required to move the positive charge to the −8 V equipotential line.

$$W = \Delta U_E = \Delta Vq = (-8\,\text{V} - (-11\,\text{V}))(3\,\text{C}) = 9\,\text{J}$$

Be very careful of all the negative signs!

▶ Calculate the kinetic energy of negative charge released from rest when it reaches the –8 V line.

$$K = -\Delta U_E = -\Delta Vq = -(-8\,\text{V} - (-11\,\text{V}))(-1\,\text{C}) = 3\,\text{J}$$

Multiple Charges

When there are multiple charges, it is helpful to think of the charges creating a topographic map with peaks for positive charges and negatives represented by valleys. The following figure shows an arrangement of three charges—two positive and one negative.

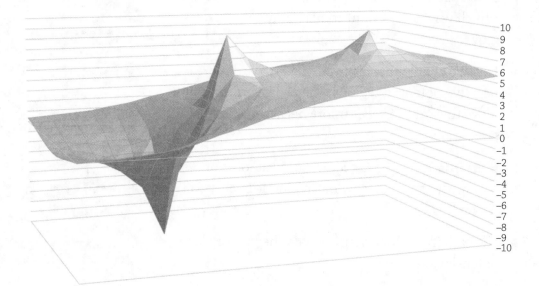

If you took a knife and sliced horizontally through the map, you would get a series of lines that would represent lines of equipotential. This is shown in the next figure that represents the exact same information.

BTW

All the negative signs can be quite irritating. I tell my students to just follow what the equation says to do but always check your answer at the end to see if it makes sense. The math is not that hard...it's just annoying!

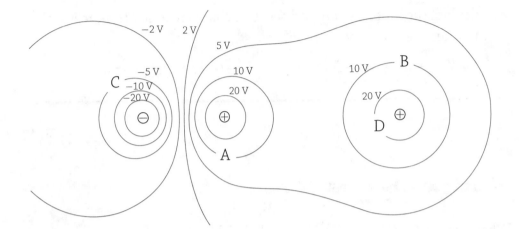

EXAMPLES

▶ Use the previous figure to answer the following questions:

▶ **Example 1**   A positive charge is placed at location A. Which way does it want to naturally move?

▶ Positive charges are always forced toward more negative potential regions. The positive charge will move toward the lower (more negative) electric potentials.

▶ **Example 2**   A +5 C charge placed at location A moves to point C. How much kinetic energy will it gain?

▶ The gain in kinetic energy equals the loss of electric potential energy:

$$K = -\Delta U_E = -\Delta Vq = -(-5\text{ V} - (10\text{ V}))(5\text{ C}) = 75\text{ J}$$

▶ **Example 3**   Does it take work to move a negative charge from position A to position B? Explain your answer.

▶ No work at all. There is no change in electric potential.

BTW

It helps many of my students to look at a topographic map and visualize positive charges wanting to roll downhill to more negative locations. It would take work to lift a positive charge to higher, more positive, volt locations. Negative charges will do just the opposite. Negative charges want to roll uphill to higher volt locations, and it takes work to push them back downward to negative volt locations.

▶ **Example 4**   Which would take the most work, moving a positive charge from point C to position B, or moving the same positive charge from C to position D?

▶ Moving from point C to point D requires more work because there is a larger gain in electric potential.

Electric Potential and Electric Field on the Same Diagram

This next figure shows two charges with the electric field and electric potential field overlaid together.

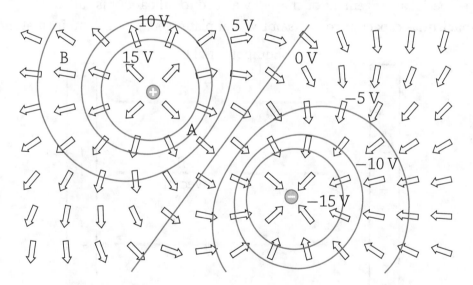

Notice how the electric field is perpendicular to the equipotential lines. This will always be true. Electric field and electric potential are connected by this equation: $E = \left| \dfrac{\Delta V}{d} \right|$, where ΔV is the electric potential difference in volts and d is the distance in meters. This equation tells us that the more the volts change with distance, the larger the electric field will be. Look at the figure again. See how the equipotential lines are packed closer together at *A* than at *B*? This means the electric field will be stronger at *A* and weaker at *B*.

If a positive charge is placed at location *A*, notice how the electric field vectors show that the charge will be pushed away from the positive region and toward the negative region by an electric force. Also notice that the electric equipotential lines indicate that the positive charge would "fall" toward the decreasing volt region near the negative charge. The two fields show complementary ideas in different ways. If a negative charge is placed at *A*, it would receive a force in the opposite direction of the electric field and would "fall" toward the positive voltage region.

Capacitors

One special arrangement of charges we need to talk about is called a **capacitor**. A capacitor consists of two flat plates separated by an insulator.

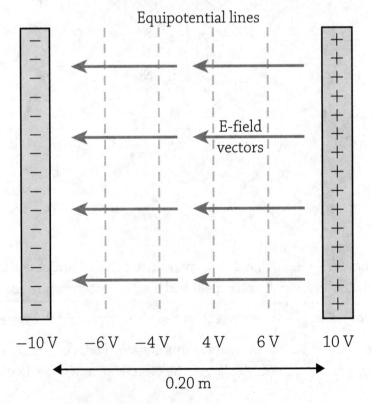

Opposite charges can be stored on the two plates. The opposite charges will attract each other and create a uniform electric field between the plates, as shown. Notice how the electric field is perpendicular to the equipotential lines and that they are evenly spaced. Between the plates, the electric field and potential difference of the plates are related by the equation $E = \left|\dfrac{\Delta V}{d}\right|$, where ΔV is the electric potential difference between the plates in volts and d is the distance between the places in meters. The electric field strength between the capacitor plates in the figure is:

$$E = \left|\frac{\Delta V}{d}\right| = \frac{20\,\text{V}}{0.2\,\text{m}} = 100\,\text{V/m} = 100\,\text{N/C}$$

The units of volts per meter are the same as newton/coulomb from the last chapter.

Capacitors are rated by a unit called farads (F). Farads measure the number of coulombs the capacitor can hold per volt of potential difference between the plates: $C = \dfrac{Q}{\Delta V}$. For a parallel plate capacitor, like the one in the following figure, the **capacitance** depends on how it is built: $C_{\parallel} = \dfrac{\kappa \varepsilon_0 A}{d}$, where k is the **dielectric constant** of the insulator between the places, ε_0 is the **vacuum permittivity constant**: $\varepsilon_0 = 8.85 \times 10^{-12}\,\text{C}^2/\text{N} \cdot \text{m}^2$, A is the area of one plate in meters squared, and d is the distance between the plates in meters. (The dielectric constant for air is 1.0; for other insulating materials, like plastic, the constant will need to be given or looked up in a table.)

Capacitors store electric potential energy as well as charge, which can be calculated using this equation: $U_C = \dfrac{1}{2}Q\Delta V = \dfrac{1}{2}C(\Delta V)^2$, where U_C is the energy in joules stored inside the capacitor, Q is the charge in coulombs, ΔV is the potential difference between the plates in volts, and C is the capacitance in farads.

EXAMPLE

▶ Let's calculate some numbers for a capacitor: A capacitor in the following figure has air between the plates and is charged to a potential difference of 80 V. The area of the plate is 0.04 m² and the plates are separated by a distance of 0.02 m.

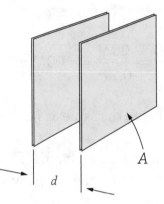

▶ What is the electric field strength between the plates?

$$E = \left|\frac{\Delta V}{d}\right| = \frac{80\,\text{V}}{0.02\,\text{m}} = 4,000\,\text{V/m} = 4,000\,\text{N/C}$$

▶ What is the capacitance of the capacitor?

$$C_{\parallel} = \frac{\kappa\varepsilon_0 A}{d} = \frac{(1.0)(\varepsilon_0 = 8.85 \times 10^{-12}\,\text{C}^2/\text{N} \cdot \text{m}^2)(0.04\,\text{m}^2)}{(0.02\,\text{m})}$$
$$= 1.8 \times 10^{-11}\,\text{F} = 18\,\text{pF}$$

▶ How much charge is being held on the plates?

$$C = \frac{Q}{\Delta V}$$

$$Q = C\Delta V = (1.8 \times 10^{-11}\,\text{F} = 18\,\text{pF})(80\,\text{V}) = 1.4 \times 10^{-9}\,\text{C} = 1.4\,\text{nC}$$

▶ Calculate the energy stored in the capacitor.

$$U_C = \frac{1}{2}Q\Delta V = \frac{1}{2}(1.4 \times 10^{-9}\,C)(80\,V) = 1.1 \times 10^{-7}\,J = 110\,nJ$$

Charge Flow and Equilibrium

In the last chapter, we talked about objects sharing charge by conduction. In the following figure, sphere A has an excess of negative charge. When it is touched to sphere B, it will share this charge, but how much of the overall charge ends up on each sphere?

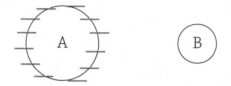

Sphere B is only half the size of sphere A. It turns out that the charge will flow from A to B until the two spheres have the same electric potential. The equilibrium condition is when they have the same volts. Using our equation for electric potential, $V = k\dfrac{Q}{r}$, we can find out exactly how much each sphere will end up with:

$$V_A = V_B$$

$$\left(k\frac{Q}{r}\right)_A = \left(k\frac{Q}{r}\right)_B$$

$$\frac{Q_A}{r_A} = \frac{Q_B}{r_B}$$

However, sphere A is twice the size of sphere B: $r_A = 2r_B$

$$\frac{Q_A}{2r_B} = \frac{Q_B}{r_B}$$

Therefore, after the charge has been shared between the spheres and a new static equilibrium has been established, sphere A will end up with a final charge that is twice the charge of sphere B: $Q_A = 2Q_B$.

Remember that charge must be conserved. The final charge must be equal to the original charge. This means that the original charge that resided on sphere A is now spread between both spheres. Thus, the original charge of sphere A equals the final charges of sphere A and B combined: $(Q_A)_{initial} = (Q_A + Q_B)_{final}$.

EXAMPLE

▶ Metal sphere A has a negative charge. Metal sphere C is neutral and is three times as large as sphere A. The two spheres are connected by a wire until electrostatic equilibrium is reached (see the following figure).

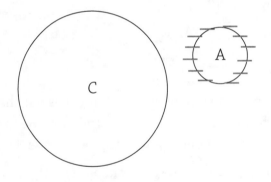

▶ Which way does charge flow through the wire?

▶ The negative charge will be shared between the two spheres. Thus, negative charge flows from A to C.

▶ When will the charge stop flowing?

▶ Charge will stop flowing when equilibrium is established. This is when both objects have the same electric potential: $V_C = V_A$.

▶ How much charge ends up on each of the spheres?

$$V_A = V_C$$

$$\left(k\frac{Q}{r}\right)_A = \left(k\frac{Q}{r}\right)_C$$

$$\frac{Q_A}{r_A} = \frac{Q_C}{r_C}$$

▶ Sphere C is three times the radius of sphere A: $r_C = 3r_A$

$$\frac{Q_A}{r_A} = \frac{Q_C}{3r_A}$$

▶ Therefore, sphere A will end up with one-third as much charge as sphere C: $Q_A = \frac{1}{3}Q_C$.

▶ Or, by rearranging the equation, we could also say that the final charge of sphere C will be three times as much charge as that of sphere A: $Q_C = 3Q_A$.

▶ How does the original charge of sphere A compare to the final charges of spheres A and C?

▶ Remember that charge must be conserved. The original charge of sphere A is now spread out between both spheres A and C. Therefore, the final charge of each sphere is less than the original charge of sphere A. And the final charges of spheres A and C must equal the original charge of sphere A.

BTW

This idea—that charge will flow until the electric potential is the same everywhere—will be a key concept in Chapter 12.

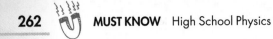
REVIEW QUESTIONS

Let's practice what we have learned about electric potential by answering the following questions.

The following figure shows three unknown charges and the electric potential field that surrounds them. Use this information to answer questions 1 to 3.

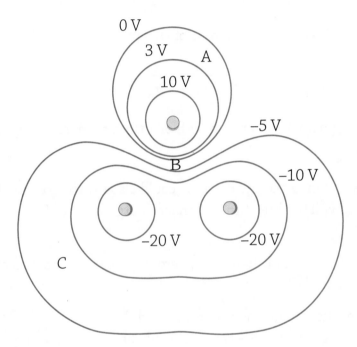

1. What are the signs on the three unknown charges? What is it about the figure that indicates the sign on the charges?

2. Rank the electric field strength at the three points *A*, *B*, and *C*.

3. In which direction does the electric field point at locations *A*, *B*, and *C*?

Questions 4 to 8: This figure shows three charges and the electric potential field that surrounds the charges.

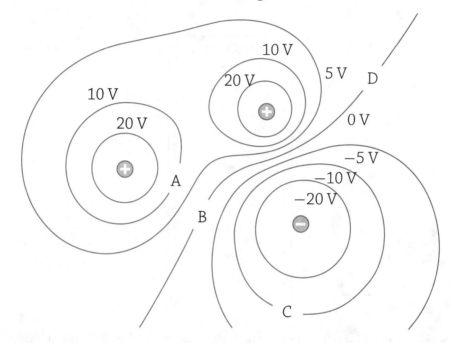

4. Rank the electric potentials at locations A, B, C, and D from most positive to most negative.

5. A positive charge is located at point B. Moving the charge to which of the other points would require the most work?

6. A proton is released at point A. Calculate its kinetic energy when it gets to point B.

7. How much kinetic energy would an electron gain moving from point C to point A?

8. Does it take any work to move an electron from point D to point B? Explain.

The following figure shows an air-filled parallel plate capacitor. The capacitor has a plate area of 0.3 m². Refer to the figure to answer questions 9 to 13.

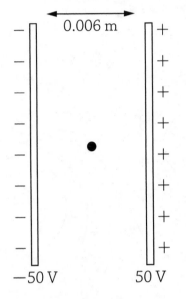

9. Calculate the electric field strength between the plates. In what direction is the electric field?

10. Calculate the capacitance of the capacitor.

11. How much charge is being held on the plates?

12. Calculate the energy stored in the capacitor.

13. The dot between the plates is an electron. Which way will it receive a force?

Questions 14 to 17: Metal sphere A has a negative charge. Metal sphere B is neutral and has a radius three times smaller than sphere A. The two spheres are connected by a wire:

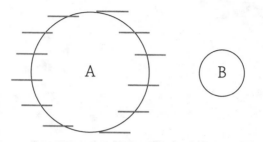

14. Which way does charge flow through the wire?

15. When will the charge stop flowing?

16. How much charge ends up on each of the spheres?

17. How does the original charge of sphere A compare to the final charges of spheres A and B?

Flashcard App

  Circuits

MUST ⚡ KNOW

⚡ A circuit consists of a power source, a conducting pathway, and at least one device (load). Circuits are wired in two primary ways: series and parallel.

⚡ Ohm's law relates electric current in a circuit to the potential difference and resistance of the circuit.

⚡ The laws of conservation of charge and conservation of energy hold true in every circuit and are expressed in Kirchhoff's junction and loop rules. Kirchhoff's rules help us understand how series and parallel circuits behave.

We know that charges exert forces on each other and will naturally move from high potential to low potential. Now it's time to make these charges do our bidding to power our life and all the great things that we know and love. Electricity rules our lives. Just about everything we own or use requires electricity. Don't believe me? Have you ever been through a power outage, or an Internet outage, or both? Imagine your life without electricity, the Internet, or battery power. It would be like the Civil War era. Yikes! So, take a few philosophical moments to be thankful for the amazing little electron.

Our goal in this chapter is to understand how we make individual charges move in unison in a current, how to wire a pathway to carry this current, and how to connect devices to make them run and the basics of electrical circuits. So, let's get started.

The Circuit

To make charge go where we want, we need a difference in electric potential, ΔV, commonly referred to as **voltage**. In the following figure, there is a charged sphere and a neutral sphere with a wire connecting them.

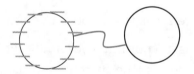

Charge will flow to the neutral sphere through the conducting pathway until the electric potential of the spheres is the same and equilibrium is established. If we put a lightbulb in the wire, it will light up, but only while the charge is flowing.

What we need is a permanent supply of potential difference or voltage, referred to as an **electromotive force**, also called an **EMF source** (ε) or a **power source**. You are already familiar with these devices: batteries, capacitors, **generators**, and **photovoltaic cells**. Batteries convert chemical energy into electrical energy. Capacitors use stored charge, as seen in the last chapter. Generators convert mechanical energy into electrical energy and will be discussed in Chapter 14. Photovoltaic cells, or **solar cells**, convert light energy directly into electricity and will be explained in Chapter 19. We will be using batteries in this chapter, but any power source could be used. What all of these devices have in common is that they keep one side of the device at a higher electric potential than the other so we have a continuous flow of charge. The following figure shows a battery with a high potential (+) end and a low potential end (–) and a wire connecting the two. Now charge flows continuously around in a circuit. (At least until the battery dies.)

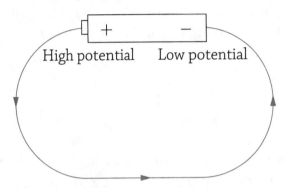

This flow of charge is called **current**. Current (I) is defined as $I = \dfrac{\Delta q}{\Delta t}$, where Δq is the amount of charge in coulombs that flows past a point in the circuit per time, Δt, measured in seconds. The current is measured in units of amperes, or amps (A) for short. Notice in the figure that the current is moving from the positive end of the battery to the negative end. From high voltage to low. This is called **conventional current** because it is what scientists' thought was happening—positive charges moving—before they discovered the atom. We have discovered that it's actually electrons moving from the negative end of the battery to the positive in an **electron current**. But it really doesn't matter until you start your electrical engineering classes in college, so I'm going to be drawing conventional current for convenience in this chapter.

Wires carry the current from the power source to whatever it is we want to power. Wires can have **resistance**, which is a restriction to the flow of current. The equation of resistance (R) in a wire is $R = \dfrac{\rho L}{A}$, where ρ is **resistivity** measured in ohm $\cdot$ meters ($\Omega \cdot$ m), L is the length of the wire in meters, and A is the cross-sectional area of the wire in meters squared (m^2). Resistivity is a property of the material that can be looked up in a table. Metals have a low resistivity, and insulators have a very high resistivity. The resistivity of copper is $1.7 \times 10^{-8}\ \Omega \cdot$ m; for insulators like plastic, it's between 10^{12} and $10^{20}\ \Omega \cdot$ m so there is a huge difference between the two types of materials.

IRL Semiconductors used in electronics have a resistivity that is not too high or too low. This "Goldilocks" resistance allows engineers to precisely control the movement of charge in integrated circuits and computers. This has allowed for the miniaturization of electronic devices to tiny scales that would have taken an entire building full of equipment in the past.

Ohm's Law and Power

A potential difference causes the current to move, but resistance hinders its motion. This battle is described by Ohm's law: $I = \dfrac{\Delta V}{R}$, where I is the current in amps (A), ΔV is the voltage measured in volts (V), and R is the resistance measured in ohms (Ω). Ohm's law is also written as $\Delta V = IR$.

The battery pushes current through the circuit and is supplying power. The devices in the circuit that receive the current consume the power. The power supplied by the battery is equal to the power consumed by the circuit. This power can be found with one of the equations: $P = I\Delta V = I^2 R = \dfrac{\Delta V^2}{R}$, where I is the current in amps (A), ΔV is the voltage measured in volts (V), and R is the resistance measured in ohms (Ω). Remember that power is measured in watts (W), which is a joule per second.

EXAMPLE

▶ Let's use Ohm's law and the power equation to analyze an electric clock. A 9 V battery supplies 0.04 A of current to a clock. What is the resistance of the clock?

▶ Rearrange Ohm's law so that it solves what we are looking for:

$$\Delta V = IR$$

$$R = \frac{\Delta V}{I} = \frac{9\,\text{V}}{0.04\,\text{A}} = 225\,\Omega$$

▶ How much power does the clock use?

$$P = I\Delta V = (0.04\,\text{A})(9\,\text{V}) = 0.36\,\text{W}$$

Here is one more practice example using Ohm's law and the power equation.

EXAMPLE

▶ You turn on a lamp that has a 40 W lightbulb. The lamp is plugged into the wall socket for power. The wall plug at your house supplies 120 V. What is the resistance of the bulb?

▶ This time we know the power and the voltage. Let's use the power equation to solve for the resistance. Rearrange the power equation so that it solves for what we are looking for:

$$P = \frac{\Delta V^2}{R}$$

$$R = \frac{\Delta V^2}{P} = \frac{(120\,\text{V})^2}{40\,\text{W}} = 360\,\Omega$$

▶ What is the current to the bulb?

▶ Let's use Ohm's law to find out:

$$I = \frac{\Delta V}{R} = \frac{120\,\text{V}}{360\,\Omega} = 0.33\,A$$

Circuit Schematic Drawings

To draw pictures of a circuit so they are simple and easy to understand, we use **schematic symbols**:

Wire (conductive path)	────────	**Capacitor**	─┤├─
Switch	──╱──	**Light**	(light symbol)
Battery (EMF source)	─┤├─ ─┤│├─	**Voltmeter** (measures voltage, ΔV)	Ⓥ
Resistors (electrical device)	─/\/\/─	**Ammeter** (measures current, I)	Ⓐ
Ground (connected to the Earth)	┴		

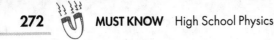

A circuit will always involve three things:

- A closed conducting pathway

- An EMF source like a battery

- Some device in the pathway that you wish to power

The device in the pathway can be anything that runs on electricity: a TV, phone, toaster, microwave, electric car, light, computer, etc. We will draw all these devices as resistors because they all consume energy.

Two examples of complete circuits are shown in the following figure. The top schematic is a **series** circuit because it has only one pathway for the current to flow through, and the current will move through each resistor. The bottom schematic is a **parallel** circuit. Notice how this circuit has several different pathways for the current to move through. Circuits can be a **combination** of these two basic arrangements.

Series circuit

Parallel circuit

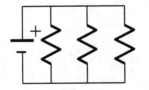

Conservation of Charge

The current that moves around the circuit is made of real electrons, and they don't just disappear when they move through the circuit elements. Your phone does not consume electrons—it consumes energy. Therefore, all the charge that moves into the battery—resistor, light, or phone—comes out the other side. This idea was stated by Gustav Kirchhoff in what is now called Kirchhoff's junction rule, which says that all of the current that enters a junction or device must be equal to the current that leaves the junction or device: $\sum I_{junction} = 0$. Current that enters the junction is considered positive and those that leave are negative. Look at the following figure of a series circuit with two resistors.

EXAMPLE

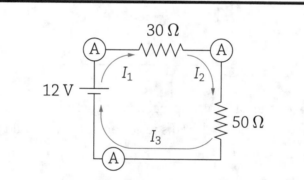

▶ I have labeled the current that leaves the battery and enters the 30 Ω resistor as I_1. The current leaving the 30 Ω resistor is I_2, and the current leaving the 50 Ω resistor and returns to the battery is I_3.

▶ Kirchhoff's junction rule tells us these all must be the same. This is the hallmark of devices connected in series: the current is the same through every component. Notice I have drawn three **ammeters** in the circuit. Ammeters measure the current that passes through

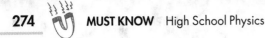

themselves; therefore, they must be placed in series with the components you want to measure the current for.

▶ All the ammeters in this circuit will read the same value because they are in series with each other.

Now take a look at a parallel circuit in the next figure.

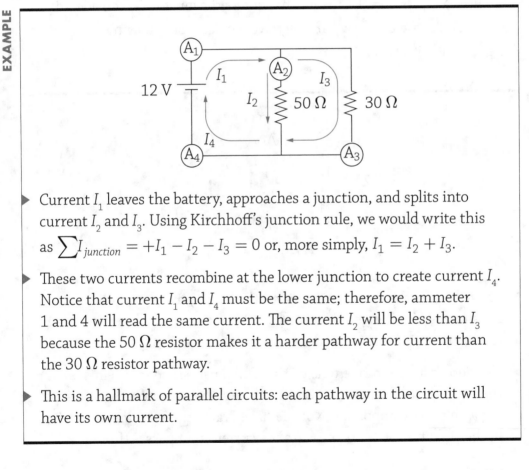

▶ Current I_1 leaves the battery, approaches a junction, and splits into current I_2 and I_3. Using Kirchhoff's junction rule, we would write this as $\sum I_{junction} = +I_1 - I_2 - I_3 = 0$ or, more simply, $I_1 = I_2 + I_3$.

▶ These two currents recombine at the lower junction to create current I_4. Notice that current I_1 and I_4 must be the same; therefore, ammeter 1 and 4 will read the same current. The current I_2 will be less than I_3 because the 50 Ω resistor makes it a harder pathway for current than the 30 Ω resistor pathway.

▶ This is a hallmark of parallel circuits: each pathway in the circuit will have its own current.

Finally, look at a circuit that is a combination of both series and parallel.

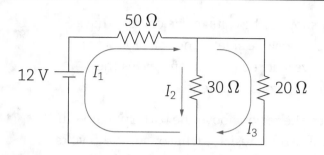

EXAMPLE

▶ Notice that the battery and the 50 Ω resistor must have the same current because they share the same pathway in series and $I_1 = I_2 + I_3$.

Conservation of Energy

The **Electromagnetic Field** (EMF) source supplies the energy to the circuit, and the components in the circuit consume that energy as described by Kirchhoff's loop rule: $\sum \Delta V_{closed\ loop} = 0$. This rule simply states that if you follow a loop around the circuit from a starting point back to where you began, add up the voltage supplied, and subtract the voltage consumed, it must sum to zero. Look at the complicated-looking figure of our series circuit shown next.

EXAMPLE

▶ This time I have connected voltmeters and have a dashed line to show our Kirchhoff loop.

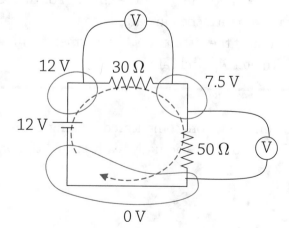

▶ The battery supplies 12 volts to the circuit. That's 12 volts at the positive side of the battery and zero on the negative side. As the current leaves the battery, it has 12 volts of potential.

▶ I have circled the region above the battery that will be 12 volts. The 30 Ω resistor will consume 4.5 volts of potential, leaving 7.5 volts left over, as shown in the next circled area in the top right of the circuit. The 50 Ω resistor will use up the final 7.5 volts, leaving no volts of potential for the final region of the circuit, which is labeled 0 V at the bottom of the circuit.

▶ Let's walk around the circuit in the direction of the dashed arrow. Notice I started the loop in the zero potential region and we end up back at zero. Using the loop rule, we can write:

$$\sum \Delta V_{closed\ loop} = +\Delta V_{battery} - \Delta V_{30\ \Omega} - \Delta V_{50\ \Omega}$$
$$= +12\,V - 4.5\,V - 7.5\,V = 0$$

▶ I have included two voltmeters in the drawing. Voltmeters measure the difference in electric potential between two regions of the circuit. For this reason, voltmeters must be attached in parallel so that they can connect two separate regions of the circuit. The top voltmeter will read 4.5 V, and the right voltmeter will indicate 7.5 V.

BTW

The energy supplied in the circuit by the battery is what has to be consumed by the devices in the circuit. A lightbulb will consume exactly the number of volts that are supplied to it. So, if too few are supplied, the bulb will be dim. If too many are supplied, it will shine too brightly and burn out. This is also true of any electronic device. I'm typing on a computer that uses a 11.4 V battery pack. If I try to use a 9 V battery, it won't work up to specifications or maybe at all. If I plug it directly into the 120 V wall plug, my computer will fry. I have to use the cord with an adapter that converts the volts to what the computer needs.

Now, look at an even more complicated-looking figure of a parallel circuit, shown next. Don't worry! It's actually really straightforward.

EXAMPLE

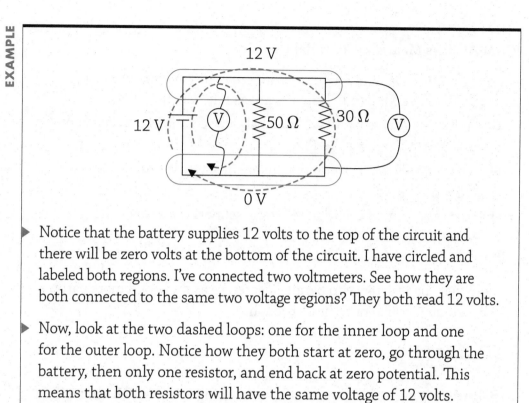

▶ Notice that the battery supplies 12 volts to the top of the circuit and there will be zero volts at the bottom of the circuit. I have circled and labeled both regions. I've connected two voltmeters. See how they are both connected to the same two voltage regions? They both read 12 volts.

▶ Now, look at the two dashed loops: one for the inner loop and one for the outer loop. Notice how they both start at zero, go through the battery, then only one resistor, and end back at zero potential. This means that both resistors will have the same voltage of 12 volts.

▶ This is a hallmark of parallel circuits: all of the components connected in parallel will have the same potential difference.

What All This Means for Series Circuits

Series circuits have only one pathway for all the components. The rules for series circuits are as follows:

- The current is the same in each component: $I_{series} = I_{R_1} = I_{R_2} = I_{R_3} \cdots$

- The voltages of the components add up to equal the total voltage of the circuit: $V_{total} = V_{R_1} + V_{R_2} + V_{R_3} \cdots$

- The total resistance is the sum of the individual resistors: $R_{series} = R_1 + R_2 + R_3 \cdots$

What All This Means for Parallel Circuits

Parallel circuits have many pathways but only one component in each pathway. The rules for parallel circuits are as follows:

- The sum of the currents in each pathway equal the total current of the circuit: $I_{total} = I_{R_1} + I_{R_2} + I_{R_3} \cdots$
- The voltages for each component are always the same: $V_{parallel} = V_{R_1} = V_{R_2} = V_{R_3} \cdots$
- The reciprocal of the total resistance equals the sum of the reciprocals: $\dfrac{1}{R_{parallel}} = \dfrac{1}{R_1} + \dfrac{1}{R_2} + \dfrac{1}{R_3} \cdots$

Note that the resistance in parallel decreases as more and more pathways are opened up for current to flow through.

VIR Chart

To help us keep everything organized, we're going to use a **VIR chart**. This chart puts the voltage (V), current (I) and resistance (R) for each circuit component in columns to help us keep all the circuit information organized. The easiest way to learn how to use it is to practice.

EXAMPLE

▶ Look at the following figure of a circuit. The resistors are already filled into the chart. The total voltage of the circuit is equal to the battery's potential difference. Knowing the values in any two columns lets you use Ohm's law to find the missing quantity: $\Delta V = IR$.

	ΔV	I	R
R_1			$1\,\Omega$
R_2			$2\,\Omega$
R_3			$3\,\Omega$
R_4			$4\,\Omega$
Total	30 V		

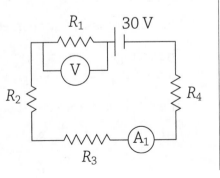

▶ This is a series circuit with only one pathway. So, look back at the series rules. We can add up the resistors to find the total: $R_{series} = R_1 + R_2 + R_3 \ldots = 1\,\Omega + 2\,\Omega + 3\,\Omega + 4\,\Omega = 10\,\Omega$ (fill in the table as you go). Now that we know the total voltage and resistance, we can find the total current:

$$I = \frac{\Delta V}{R} = \frac{30\text{ V}}{10\,\Omega} = 3\text{ A}$$

▶ Fill in the VIR chart with the values we just calculated.

	ΔV	I	R
R_1			$1\,\Omega$
R_2			$2\,\Omega$
R_3			$3\,\Omega$
R_4			$4\,\Omega$
Total	30 V	**3 A**	**10 Ω**

▶ Remember that the current in a series circuit is the same through all the components. This means all the resistors will receive 3 A of current. So, we can fill in the entire current column in our VIR chart with 3 A as shown.

	ΔV	I	R
R_1		**3 A**	$1\,\Omega$
R_2		**3 A**	$2\,\Omega$
R_3		**3 A**	$3\,\Omega$
R_4		**3 A**	$4\,\Omega$
Total	30 V	3 A	10 Ω

▶ Finally, we can calculate the potential difference for each resistor using Ohm's law: $\Delta V = IR$.

▶ Here is a sample calculation for resistor R_4:

$$\Delta V = IR = (3\,\text{A})(4\,\Omega) = 12\,\text{V}$$

▶ Our final VIR chart will look like this:

	ΔV	I	R
R_1	**3 A**	3 A	1 Ω
R_2	**6 A**	3 A	2 Ω
R_3	**9 A**	3 A	3 Ω
R_4	**12 A**	3 A	4 Ω
Total	30 V	3 A	10 Ω

▶ Notice how individual voltages add up to equal the total, just like they are supposed to in series circuits. Look back at our original circuit. What will the two meters read? The voltmeter will indicate 3 V, and the ammeter will read 3 A.

Next let's practice on a parallel circuit so you can see the difference between the two.

▶ We want to figure out what all the meters read. The beginning VIR chart is given next.

	ΔV	I	R
R_1			10 Ω
R_2			30 Ω
Total	9 V		

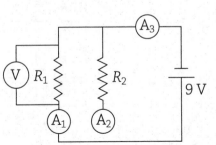

▶ This is a parallel circuit. Therefore, all of the components receive the same voltage, which means we can fill in the electric potential column with 9 V. (See the chart.)

	ΔV	I	R
R_1	**9 V**		10 Ω
R_2	**9 V**		30 Ω
Total	9 V		

▶ Using Ohm's law, we can calculate the current going through each component. A sample calculation for the 30 Ω resistor is shown next:

$$I = \frac{\Delta V}{R} = \frac{9\,V}{30\,\Omega} = 0.3\,A$$

	ΔV	I	R
R_1	9 V	**0.9 A**	10 Ω
R_2	9 V	**0.3 A**	30 Ω
Total	9 V		

▶ We also know that the sum of the currents passing through the two resistors must be equal to the total current going through the battery:

$$I_{battery} = I_{R_1} + I_{R_2} = 0.3\,A + 0.9\,A = 1.2\,A$$

▶ And finally, we can calculate the total resistance using Ohm's law:

$$R = \frac{\Delta V}{I} = \frac{9\,V}{1.2\,A} = 7.5\,\Omega$$

	ΔV	I	R
R_1	9 V	0.9 A	10 Ω
R_2	9 V	0.3 A	30 Ω
Total	9 V	**1.2 A**	**7.5 Ω**

▶ This seems a bit odd, but remember that the total resistance always decreases in parallel. In fact, the total parallel resistance is always less than the smallest resistor in parallel. We can also confirm the resistance using the parallel equation:

$$\frac{1}{R_{parallel}} = \frac{1}{R_1} + \frac{1}{R_2}$$

$$\frac{1}{R_{parallel}} = \frac{1}{10\,\Omega} + \frac{1}{30\,\Omega} = \frac{3}{30\,\Omega} + \frac{1}{30\,\Omega} = \frac{4}{30\,\Omega}$$

$$R_{parallel} = \frac{30\,\Omega}{4} = 7.5\,\Omega$$

▶ Total resistance is confirmed!

▶ Finally, let's go back and figure out what the meters located in the circuit will read. The voltmeter will read 9 V. The ammeters will read $A_1 = 0.9$ A, $A_2 = 0.3$ A, and $A_3 = 1.2$ A.

Our last example will be a combination circuit, which contains both parallel and series components.

▶ Take a look at the circuit in the following figure. Notice how it is a combination of series components and parallel components. Can you identify which is which? I have prefilled the VIR chart for the circuit shown in the following figure.

	ΔV	I	R
R_1			4 Ω on the right
R_2			6 Ω
R_3			8 Ω
R_4			4 Ω on the left
Total	12 V		

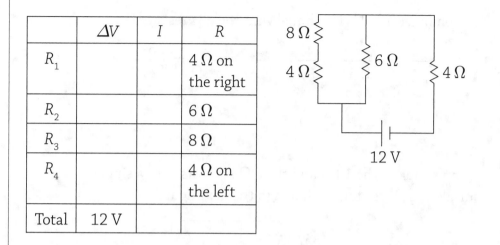

First, let's simplify and find the total resistance. Notice that there is an 8 Ω and a 4 Ω on the left side of the circuit that are in series. We can add them up to equal 12 Ω. We can swap out the 8 Ω and the 4 Ω resistors with the 12 Ω resistor they are equal to, as shown in the following figure.

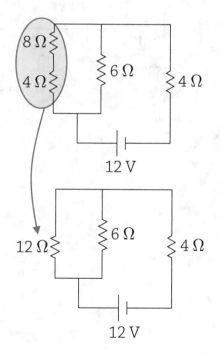

▶ This combined 12 Ω is in parallel with the 6 Ω resistor.

$$\frac{1}{R_{parallel}} = \frac{1}{12\,\Omega} + \frac{1}{6\,\Omega} = \frac{1}{12\,\Omega} + \frac{2}{12\,\Omega} = \frac{3}{12\,\Omega}$$

$$R_{parallel} = \frac{12\,\Omega}{3} = 4\,\Omega$$

▶ We can swap out the 12 Ω and the 6 Ω resistors with a single equivalent 4 Ω resistor, as shown in the next figure.

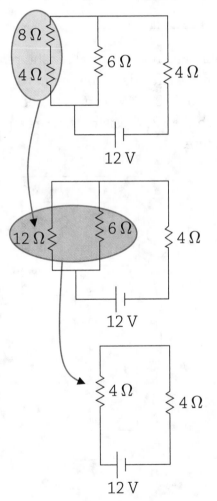

▶ This final figure is obviously in series, so we can simply add the two remaining 4 Ω resistors to calculate that the total resistance of the entire circuit is 8 Ω.

▶ Now we can begin filling in our VIR chart. Using Ohm's law, we can calculate the total current:

$$I = \frac{\Delta V}{R} = \frac{12\,V}{8\,\Omega} = 1.5\,A$$

	ΔV	I	R
R_1			4 Ω on the right
R_2			6 Ω
R_3			8 Ω
R_4			4 Ω on the left
Total	12 V	**1.5 A**	**8 Ω**

▶ The resistors in our simplified circuit are in series and will receive the same 1.5 A of current. Unfortunately, the 4 Ω on the left is a simplified resistor, so we can only put 1.5 A in our chart for the 4 Ω resistor on the right. The voltage for the 4 Ω resistor on the right will be:

$$\Delta V = IR = (1.5\,A)(4\,\Omega) = 6\,V$$

	ΔV	I	R
R_1	6 V	1.5 A	4 Ω on the right
R_2			6 Ω
R_3			8 Ω
R_4			4 Ω on the left
Total	12 V	1.5 A	8 Ω

▶ Now we can start working our way from our simplified circuits back to the original circuit.

▶ Using Kirchhoff's loop rule, we know that only 6 V remains for the rest of the circuit. (See the following figure.) This means that the 6 Ω and 12 Ω resistors in our second simplified circuit will both receive 6 V because they are in parallel. We can put this voltage in for the 6 Ω resistor, but not the 12 Ω, because it's still a simplified resistor. We can calculate the current in the 6 Ω resistor to be 1 A.

	ΔV	I	R
R_1	6 V	1.5 V	4 Ω on the right
R_2	6 V	1.0 A	6 Ω
R_3			8 Ω
R_4			4 Ω on the left
Total	12 V	1.5 A	8 Ω

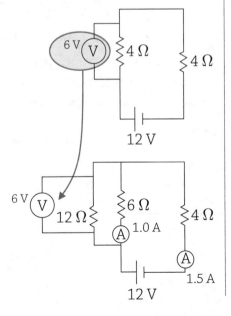

▶ Using Kirchhoff's junction rule, we see that there 1.5 A of current in the main wire going through the battery, and 1.0 A of this splits off to go through the 6 Ω resistor. This leaves 0.5 A of current left for the 8 Ω and the 4 Ω resistors on the left that are in series, which means they receive the same current.

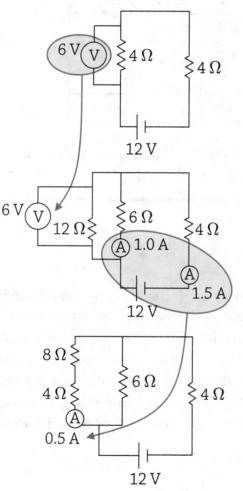

▶ Using Ohm's law, we can calculate the voltages across the 8 Ω and the 4 Ω resistors and finish out our VIR chart. Bravo!

	ΔV	I	R
R_1	6 V	1.5 V	4 Ω on the right
R_2	6 V	1.0 V	6 Ω
R_3	**4 V**	**0.5 A**	8 Ω
R_4	**2 V**	**0.5 A**	4 Ω on the left
Total	12 V	1.5 A	8 Ω

▶ Good job! This may seem complicated, but like any skill, it just takes some practice to get good at it.

Lightbulbs

Lightbulbs give off light by converting electrical energy into electromagnetic wave energy. The faster they do this, the more powerful they are and the brighter they will be. This means that bulbs of the same type with more wattage are brighter than a bulb with a lower power rating. From our power equation, $P = I\Delta V$, we can see that for a brighter bulb, we want more voltage and/or more current to the bulb. The following figure shows four identical bulbs in a circuit. Can you rank the bulbs from brightest to dimmest?

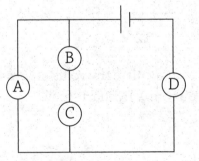

Take a look at the currents in the circuit. Bulb D is in the main line with the total current. The current splits before getting to the rest of the bulbs. Current I_2 must be larger than I_1 because there is less resistance in that pathway. Bulbs B and C must have the same brightness because they are in series and will receive the same current. The brightness ranking is $D > A > B = C$.

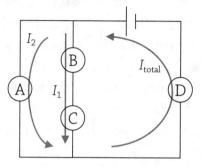

Look at the next figure with bulbs. This time multiple batteries are powering a single bulb. Using Kirchhoff's loop rule, we can look at the voltages supplied to each bulb. Notice how the second battery in B does not add any extra potential to the bulb—it simply adds another parallel pathway. Both Kirchhoff loops only have a single battery each. So, B will be the same brightness as A because they supply the same voltage to the bulb. Circuit C has two batteries in the loop to double the voltage supplied to the bulb. The ranking from brightest to dimmest will be $C > A = B$.

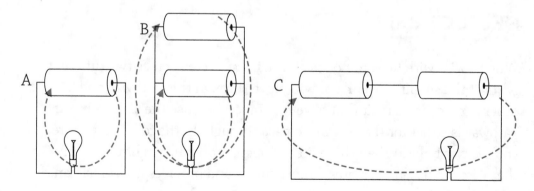

Switches

A **switch** is a disconnect in the circuit. When the switch is closed, current will flow through the line. When the switch is open, all current in that wire stops. Look at the series circuit in the following figure. When the switch is open, the current ceases in all three resistors at once because there is nowhere for the current to go. The switch could have been placed anywhere in the circuit and it would work exactly the same way because there is only one pathway. The parallel circuit has multiple pathways; however, switches 2, 3, and 4 only control the current in their specific line and only turn a single resistor on and off. Switch 1 is in the main current line that supplies all three resistors. It acts as a master switch to turn all the resistors on and off at the same time.

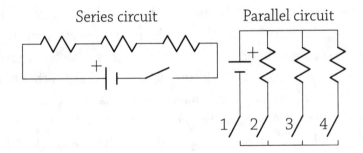

Short Circuit

When a wire with little to no resistance is connected across two different parts of the circuit, the current will move through this pathway and bypass other parts of the circuit that have resistance, because the **short circuit** pathway is easier for the current to pass through. In the following figure you see two bulbs. When the switch is open, both bulbs light up. When the switch is closed, it creates a short circuit and provides a low-resistance pathway for the current to bypass bulb B.

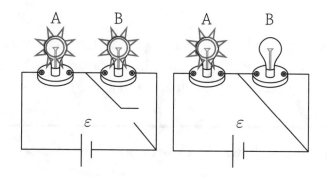

 **IRL** A short circuit will bypass resistance in the circuit. This causes excess current to flow in the circuit because the overall resistance of the circuit has been lowered. This can be dangerous, melting wires and starting fires if too much current is present. Buildings have safety devices called *circuit breakers* that shut off the circuit when they sense too much current in the line. Most residences have these built right into the wall plugs in the kitchen and bathrooms because of the higher short-circuit risk with electricity near water.

Internal Resistance of Batteries

In the real world, batteries are not perfect. The chemical reaction inside the battery is not 100 percent efficient, and they generate heat thanks to entropy. Because of this, real batteries have **internal resistance**. The ideal voltage of the battery may be designed to be 12 volts, but what comes out will always be less due to the internal resistance losses. In fact, the more power output the battery has to supply, the lower the external voltage becomes and the hotter the battery gets.

Capacitors

In a circuit, capacitors are used to store energy and charge. Remember from the last chapter:

- Charge stored in a capacitor: $Q = C\Delta V$

- Energy stored in a charged capacitor: $U_C = \dfrac{1}{2}Q\Delta V = \dfrac{1}{2}C(\Delta V)^2$

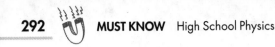

where C is capacitance in farads (F), ΔV is the voltage across the capacitor plates in volts (V), Q is the charge stored in the capacitor in coulombs (C), and U_c is the energy stored in the capacitor in joules (J).

EXAMPLE

▶ The following figure shows two capacitors connected in series.

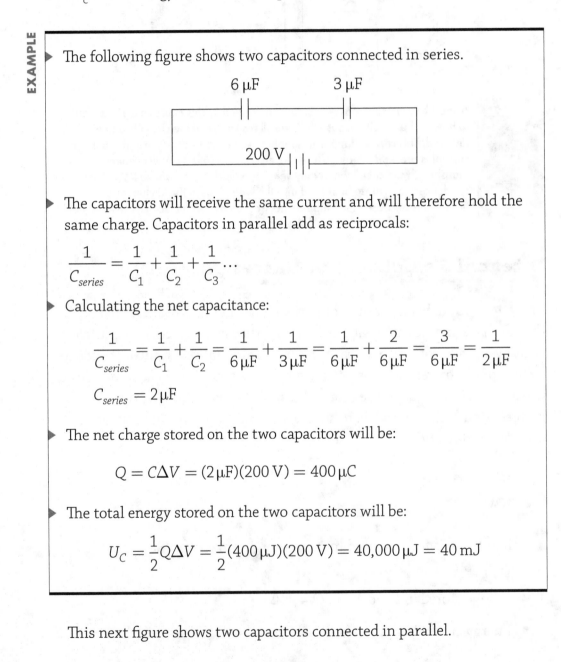

▶ The capacitors will receive the same current and will therefore hold the same charge. Capacitors in parallel add as reciprocals:

$$\frac{1}{C_{series}} = \frac{1}{C_1} + \frac{1}{C_2} + \frac{1}{C_3}\cdots$$

▶ Calculating the net capacitance:

$$\frac{1}{C_{series}} = \frac{1}{C_1} + \frac{1}{C_2} = \frac{1}{6\,\mu F} + \frac{1}{3\,\mu F} = \frac{1}{6\,\mu F} + \frac{2}{6\,\mu F} = \frac{3}{6\,\mu F} = \frac{1}{2\,\mu F}$$

$$C_{series} = 2\,\mu F$$

▶ The net charge stored on the two capacitors will be:

$$Q = C\Delta V = (2\,\mu F)(200\text{ V}) = 400\,\mu C$$

▶ The total energy stored on the two capacitors will be:

$$U_C = \frac{1}{2}Q\Delta V = \frac{1}{2}(400\,\mu J)(200\text{ V}) = 40{,}000\,\mu J = 40\text{ mJ}$$

This next figure shows two capacitors connected in parallel.

EXAMPLE

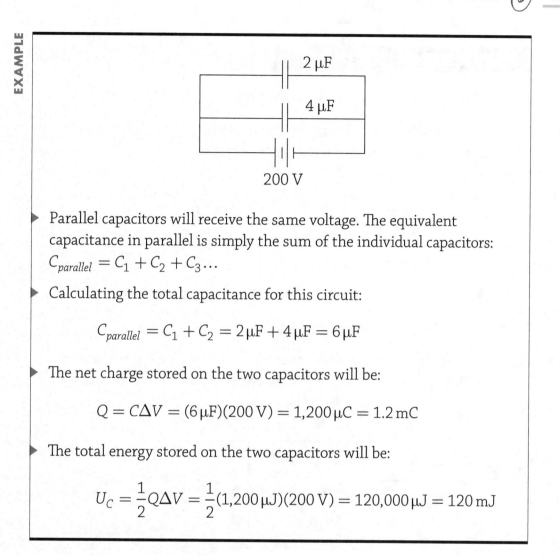

2 μF

4 μF

200 V

▶ Parallel capacitors will receive the same voltage. The equivalent capacitance in parallel is simply the sum of the individual capacitors: $C_{parallel} = C_1 + C_2 + C_3 \ldots$

▶ Calculating the total capacitance for this circuit:

$$C_{parallel} = C_1 + C_2 = 2\,\mu F + 4\,\mu F = 6\,\mu F$$

▶ The net charge stored on the two capacitors will be:

$$Q = C\Delta V = (6\,\mu F)(200\,V) = 1,200\,\mu C = 1.2\,mC$$

▶ The total energy stored on the two capacitors will be:

$$U_C = \frac{1}{2}Q\Delta V = \frac{1}{2}(1,200\,\mu J)(200\,V) = 120,000\,\mu J = 120\,mJ$$

REVIEW QUESTIONS

Let's test our skills with circuits by answering the following questions.

1. The two resistors in the figure shown are made of the same material with identical resistivity. Which has a higher resistance?

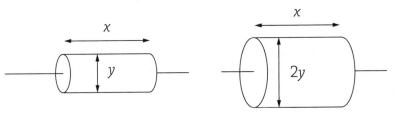

2. Seventeen coulombs of charge pass through a lightbulb over 3.4 seconds. Calculate the current through the bulb.

Use the following figure to answer questions 3 and 4. The figure shows a circuit with two switches. Both switches begin in the closed position.

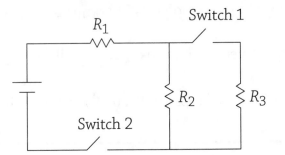

3. When switch 1 is opened and closed, which resistors does it control the current to? Explain your answer.

4. When switch 2 is opened and closed, which resistors does it control the current to? Explain your answer.

In questions 5 and 6, we have a 27 W electric pencil sharpener connected to a 12 V electrical source.

5. Calculate the resistance of the pencil sharpener.

6. Calculate the current that passes through the pencil sharpener.

7. A charger for a computer outputs 3.34 A of current at 19.5 V. Calculate the power output of the charger.

Questions 8 and 9 refer to the following figure, which shows an overly complicated combination circuit.

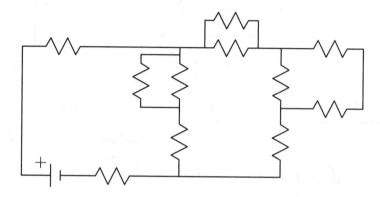

8. Box the regions of the circuit where the resistors are connected in parallel. Remember that parallel components have the same voltage but separate current pathways, with only one resistor in each pathway.

9. Circle the regions of the circuit where the resistors are connected in series. Remember that series components have the same current because there is only one pathway.

10. Why do batteries get hot when they output a lot of power?

In questions 11 and 12, we have three resistors: 40 Ω, 60 Ω, and 120 Ω.

11. What is the equivalent resistance of these resistors when connected in series?

12. What is the equivalent resistance when the resistors are connected in parallel?

Questions 13–15 refer to the following chart and circuit. The circuit has a 100 V battery and four identical resistors of 12 Ω each.

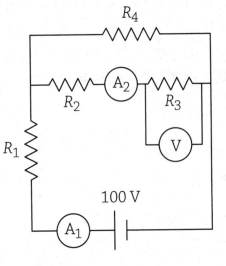

	ΔV	I	R
R_1			1 Ω
R_2			2 Ω
R_3			3 Ω
R_4			4 Ω
Total	30 V		

13. Fill in the VIR chart for the circuit.

14. What is the current through ammeters 1 and 2?

15. What voltage does the voltmeter in the circuit read?

Refer to the following figure for questions 16–18. The circuit contains four identical bulbs, and the switch in the circuit is open.

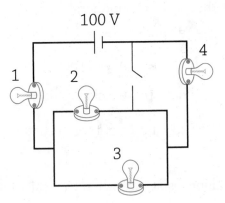

16. With the switch open, rank the bulbs from brightest to dimmest. Explain your reasoning.

17. The switch is now closed. Which bulbs turn off due to the short circuit?

18. With the switch closed, rank the brightness of the bulbs that are still lit from brightest to dimmest. Explain your reasoning.

In questions 19 and 20, we have two capacitors: 12 μF and 18 μF.

19. What is the equivalent capacitance of these capacitors when connected in series?

20. What is the equivalent capacitance when the capacitors are connected in parallel?

13 Magnetism

umans have known about magnetism for a long time. The Greeks wrote about lodestones, which are a naturally occurring magnetic iron oxide ore. It's thought that the word *magnet* comes from the lodestones found near the ancient city Magnesia in modern-day Turkey. Magnetism has many similarities to electricity, but it wasn't until the 1,800s that scientists connected the two. Our goal in this chapter is to discover how to produce a magnetic field with a current, understand how magnetic fields create forces on moving charges in strange directions, and uncover the connection between magnetism and electricity.

Electricity and Magnetism

There are "shocking" similarities between electricity and magnetism. There are two types of charge: positive and negative. There are two poles to a magnet: north and south. Like charges repel and opposites attract. The same is true for magnetic poles. North and south poles of magnets attract. But two norths or two south poles repel. The closer you get two charges together, the stronger the electric force between them. The same is true for magnets. Both charges and magnets produce force fields. There are so many similarities, but most scientists of the early 1,800s believed that electricity and magnetism were two separate phenomena. That is, until April 21, 1820, when Hans Christian Oersted, a professor of physics at the University of Copenhagen, discovered that an electric current in an electricity demonstration caused a nearby magnet to deflect. It's believed that this was probably an accidental discovery. Whatever the case, Oersted immediately understood that the moving compass was a significant event and went on to analyze the behavior and publish his results, which transformed our understanding of magnetism.

Currents Produce Magnetic Fields

Oersted discovered that a current in a wire will produce a **magnetic field**. The magnetic field forms circles in the plane perpendicular to a wire that is carrying current. That sounds confusing, but picture a wire going through the center of a bunch of circles as seen in the following figure. The direction that the magnetic field is circulating around the wire is the same direction your fingers curl in if you hold the wire with your right hand with your thumb pointing in the direction of the current. This is called a right-hand rule. Call this the right-hand "curly fingers" rule.

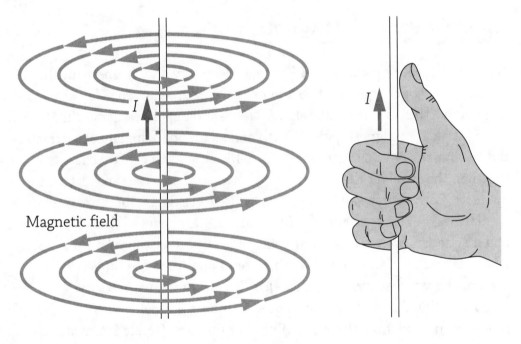

Magnetic field

You can practice the right-hand rule with your pen or pencil. Take your pencil and hold it in the direction the wire is oriented, with the tip pointing in the direction of the current. Next, grasp the pencil with your right hand and your thumb pointing in the direction of the current—the same direction the tip of your pencil is pointing. Your fingers will curl around the pencil in the same direction the magnetic field curls about the current.

This is a three-dimensional picture, so let's rotate it to get a better look at it. With the fingers of your right hand still holding the pencil, point the tip of the pencil directly at your face. See how your thumb points toward your eyes and your fingers circle around the pencil in a counterclockwise direction? When the current is directed upward, out of the page and toward your face, the magnetic field will form counterclockwise circles around the wire, as shown in the next figure.

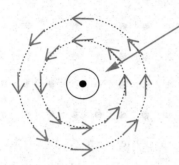

Wire with current flowing out of the page

The strength of the magnetic field around a wire depends on the current and decreases with distance away from the wire. The equation for the magnetic field strength is $B = \left(\dfrac{\mu_0}{2\pi}\right)\dfrac{I}{r}$, where B is the magnetic field strength in units of tesla (T), $\mu_0 = 4\pi \times 10^{-7}\,\text{T} \cdot \text{m/A}$ and is called the vacuum permeability, I is the current in amperes, and r is the radius from the wire in meters. Notice that $B \propto \dfrac{I}{r}$. This means that the magnetic field gets bigger when we are closer to the wire and when the current is larger.

BTW

In case you didn't catch it, a magnetic field is designated by the capital letter B. B seems like an odd letter for magnetic field, but apparently when deriving his magnificent electromagnetic theory, James Clerk Maxwell named all the variables he used somewhat randomly, A through H with B for magnetic field. Perhaps he was thinking of the Spanish word for compass, brújula, when he chose the letter B.

Magnetism is an inherently three-dimensional phenomena, but a piece of paper is two-dimensional, so it is hard for us to draw. Physicists came up with some nice ways to draw magnetic fields that are going in or out of the page. For a magnetic field going into the page, a bunch of Xs will be drawn. For a magnetic field coming out of the page, a bunch of dots will be drawn:

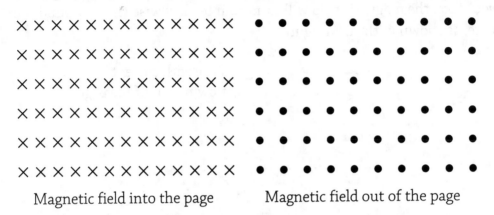

Magnetic field into the page   Magnetic field out of the page

Here's an example of how it works.

▶ What does the magnetic field look like around a wire in the plane of the page with current directed upward toward the top of the page?

▶ Using the right-hand rule, you can see that the magnetic field points out of the page on the left-hand side of the current and into the page on the right-hand side of the wire.

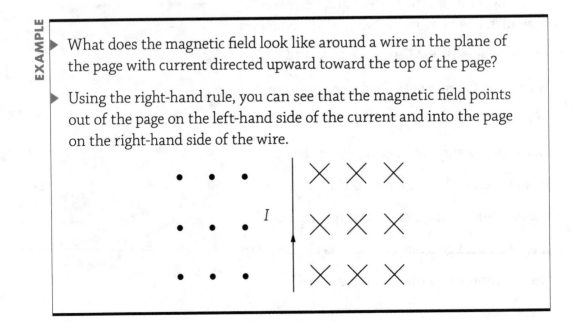

Here is another example of using Xs and dots to represent magnetic field vectors.

▶ What would the magnetic field look like in the space around a loop of wire with a counterclockwise current in it? Remember to use your right-hand rule.

▶ Grab the wire with your right hand, making sure to have your thumb pointing in the direction of the current. Notice how your finger points out of the paper on the inside of the loop no matter where you grab the loop. The magnetic field will point downward into the paper on the outside of the loop. (See the following figure.)

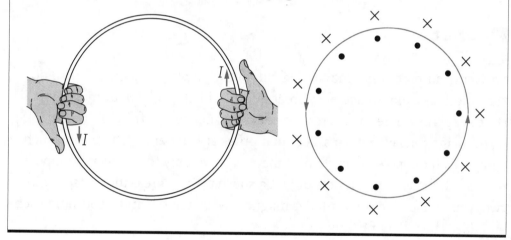

Now if we loop the wire around many times, we will create a coil called a **solenoid**. Hook the solenoid up to a battery to produce current, and you get a strong magnetic field that passes through the center and looks like the magnetic field around a **bar magnet**.

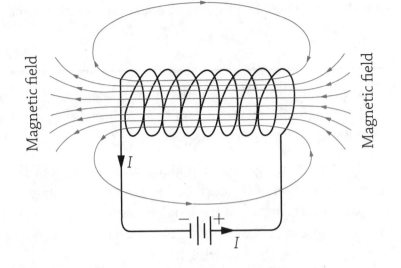

Magnets

So, a current produces a magnetic field. But what about magnets? How do they produce a magnetic field? Great question! Let's talk about magnets. I'll try to explain where the current inside a magnet comes from.

All magnets have **north** and **south poles** at their ends. There is no such thing as a monopole or single-pole magnet. If you try to create a monopole magnet with just a north pole or just a south pole by breaking a magnet in two, you will just get two smaller magnets with a north and a south on each end. (See the following figure.)

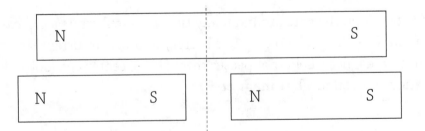

A magnet creates a magnetic field, as shown in the next figure. Unlike electric field lines, which either emanate from a positive charge, dead end on a negative charge or extend infinitely into space, magnetic field lines always form loops. The loops point away from the north end of a magnet and inward toward the south end. Near the magnet, the lines point nearly straight into or out of the poles.

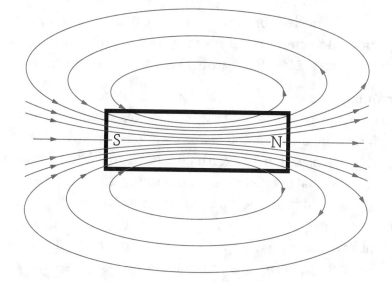

Iron filings sprinkled onto a bar magnet gather on the magnetic field lines, making them visible as we can see here.

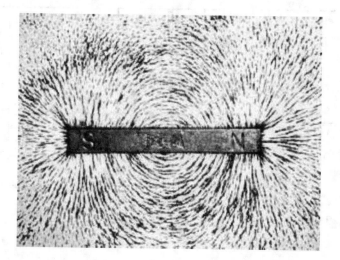

The space around a magnet contains a magnetic field. All magnetic fields are created by moving charges or current. Remember that currents are moving charges in a wire, so they create magnetic fields, as we have talked about earlier. In a permanent magnet, like the ones you stick on your refrigerator, the electrons themselves behave as tiny magnets, and their motion around the nucleus of the atom is the current. You may have noticed not every material seems to react to a magnet. In reality, all materials have a reaction to magnetic fields. There are actually three major types:

■ **Ferromagnetism**   Materials like iron, nickel, and cobalt have multiple unpaired electrons that tend to create localized regions of magnetic fields inside the material called domains. When these materials are placed in an external magnetic field, the **domains** align with the external magnetic field, amplifying it many times. In a strong enough external field, the domains can grow and merge, creating a permanent magnet, as shown in the following figure. This is why a current-carrying wire wrapped around an iron nail will produce such a strong magnet. Ferromagnetic materials are strongly attracted by magnets.

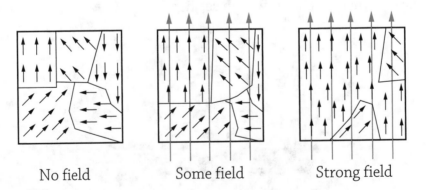

No field   Some field   Strong field

 IRL   The Earth has a liquid core of iron. This makes the Earth a very strong ferromagnet.

- **Paramagnetism** The magnetic properties of the material tend to align with an external magnetic field, but the result is weak and does not enhance the magnetic field very much. These materials will not produce a permanent magnet. Paramagnet materials are very weakly attracted to magnets. Aluminum, oxygen, titanium, and iron oxide are all paramagnetic.

- **Diamagnetism** When placed in a magnetic field, the internal magnetic properties of the material align opposite to the external field, cancelling out part of the field. Water and graphite behave this way. Diamagnetic materials are very weakly repelled by magnets.

Compass in a Magnetic Field

I hope you played with a compass as a kid. A compass is a tiny magnet mounted on a rotational pivot point so that it can turn in the Earth's magnetic field. It works like this: The north end of the magnet is forced in the direction of the magnetic field, and the south end is forced in the opposite direction. This produces a torque that causes the compass to align with the external magnetic field it is placed in. This is why the Earth's magnetic field affects a compass, and is why Oersted's compass deflected when he had a current carrying wire near it.

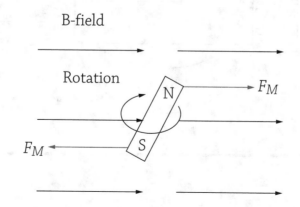

EXAMPLE

▶ Two compasses are arranged around a magnet as shown in the following figure. Indicate the direction the compasses will turn due to the magnetic field of the compass.

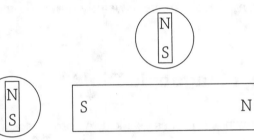

▶ The compasses will turn to align with the B-field produced by the magnet.

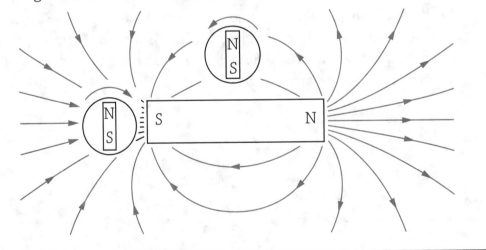

Magnetic Forces on Moving Charges

Whenever a charged particle passes through a magnetic field, if a component of its velocity is perpendicular to the magnetic field, it will experience a force. The equation for the magnetic force is $F_M = qvB\sin\theta$, where F_M is the magnetic force on the moving charge in newtons (N), q is the magnitude of the charge in coulombs (C), v is the velocity of the charge in meters per seconds (m/s), B is the magnetic field strength in tesla (T), and θ is the angle between the velocity of the moving charge and B-field directions. Notice that when the velocity and B-field are parallel, the sine will be zero and there will be no force. When the velocity of the particle is perpendicular to the B-field, the sine of 90 degrees is 1.0, and the magnetic force is at a maximum.

The really strange thing about the magnetic force is that it acts at a right angle to both the magnetic field and the velocity. This can be hard to visualize so, of course, there is a right-hand rule to help you out. (There are actually several ways to learn this right-hand rule. So, if the right-hand rule that follows is not the one that you see elsewhere or already know, don't worry! The point is for you to find one that works for you and stick with it.)

Here's the right-hand rule for a moving positive charge in a magnetic field. (You might want to call this the right-hand "flat finger" rule.) Hold your hand flat with your thumb at a right angle to your fingers, as seen in the following figure.

- Your fingers should point in the direction of the magnetic field. There are usually lots of magnetic field lines and you have lots of fingers.

- Your thumb will point in the direction of the velocity of the particle.

- Perpendicular to the palm of your hand is the direction of the force on the particle. A nice way to think of this is that the direction of the force on the particle is the same direction you would push on something with the palm of your hand.

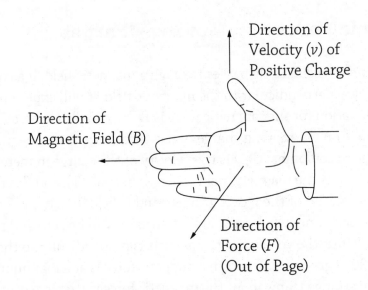

Direction of
Velocity (*v*) of
Positive Charge

Direction of
Magnetic Field (*B*)

Direction of
Force (*F*)
(Out of Page)

Notice that everything is at right angles. If the velocity is not perpendicular to the magnetic field, you can bring your thumb a little closer or farther away (if you can) from your fingers. The force is always going to be perpendicular to the two of them and will be in the direction your palm is facing.

What if there is a negative particle moving through the field? Everything is reversed for the negative particle, so simply use your left hand and follow the same rules. So, remember: right hand for positive particles and left hand for negative particles. If you find all this confusing, now is a good time for "body art." Take a pen and write a *B* on the tip of each of your fingers. Write *v* on your thumb and put a big *F* and a positive sign "+" on your right palm. Repeat this for your left hand, except put a big *F* and a negative sign "−" on your left palm. Now you have a useful physics tattoo to help you with these problems.

EXAMPLE

▶ In the following figure there are three different magnetic fields. The *B*-field on the left is directed into the page, the middle *B*-field is coming out of the page, and the final *B*-field is directed toward the top of the page as shown.

▶ Use the "flat fingers" right-hand rule for positive charges and the left-hand rule for negative charges to determine the direction of the force on each moving charge. Draw an arrow on each figure to indicate the direction of the magnetic force on each charge. If the force is directed into or out of the page or is zero, specifically state so.

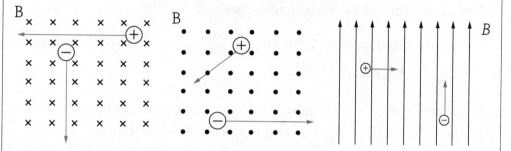

▶ These can be confusing. Use the "flat fingers" right-hand rule to determine the direction of the force on each charge. Be careful to use your right hand for positive charges and your left hand for negative charges. Take your time to line up each part of your hand correctly! The answers are in the figure shown next.

Now let's work an example with numbers.

An electron is moving at 6.0×10^5 m/s at a right angle to a 0.74 T magnetic field. Calculate the force on the electron. (The charge of an electron is $e = 1.6 \times 10^{-19}$ C. It is also in the Physics Constants table in the appendix.)

Use the magnetic force on a charge equation:

$$F_M = qvB\sin\theta = \left(1.6 \times 10^{-19} \text{ C}\right)\left(6.0 \times 10^5 \text{ m/s}\right)\left(0.74 \text{ T}\right)\sin 90°$$
$$= 7.1 \times 10^{-14} \text{ N}$$

Now calculate the acceleration of the electron. (The mass of an electron is $m_e = 9.11 \times 10^{-31}$ kg. It is also in the Physics Constants table in the appendix.)

Use Newton's Second Law:

$$a = \frac{F}{m} = \frac{7.1 \times 10^{-14} \text{ N}}{9.11 \times 10^{-31} \text{ kg}} = 7.8 \times 10^{16} \text{m/s}^2$$

One last example. Can you figure out why this is a "trick question"?

A proton enters a 2.5 T magnetic field with a velocity of 9.0×10^6 m/s parallel to the B-field. Calculate the magnetic force on the proton.

The force is zero! But why? Remember our equation has an angle in it: $F_M = qvB\sin\theta$. The velocity and B-field are parallel; therefore, theta is zero and the force is zero because $\sin 0° = 0$.

Circular Motion of Charges in Magnetic Fields

So, the magnetic forces are directed perpendicular to the velocity vector of the charge. Does that sound vaguely familiar to something you learned in Chapter 5? Circular motion! Just like the gravitational force acting perpendicular to the velocity of Earth, causing it to move in a circular orbit. Since the magnetic force is always perpendicular to the velocity of the charge, it does no work on the charge. The force gives the charge a centripetal acceleration and causes the direction of the charge's velocity to change, but not its magnitude. So, we get a nice circular path.

The following figure shows a positive charge entering a magnetic field directed into the page. The magnetic force is always perpendicular to the velocity, causing the charge to move in a circular path. The magnetic force on the charge acts as a centripetal force.

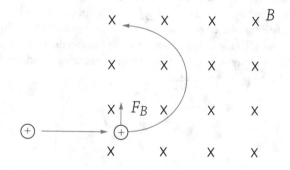

Magnetic field into the page

$$F_M = ma_c = m\frac{v^2}{r}$$
$$F_M = qvB\sin 90° = m\frac{v^2}{r}$$
$$r = \frac{mv}{qB}$$

Notice how the radius depends on the momentum (mv) of the particle. Higher momentum equals a larger radius. The radius of the path is inversely related to the charge of the particle and the magnetic field strength. A larger charge and/or larger magnetic field will result in a smaller radius.

IRL Scientists use a device called a **mass spectrometer** that uses this principle of arcing charges in a magnetic field to determine the mass-to-charge ratio, m/q, of an unknown chemical by measuring the radius of the path the molecule takes in the magnetic field. From this they can determine what the unknown chemical is. Forensic scientists can use this device to test for environmental toxins, drug use in athletes, and even to look for flammable accelerants that might have been used at the scene of a suspected arson fire.

IRL If a moving particle enters a magnetic field at an angle with part of the velocity parallel to the field, the charge will move in a helical path. (A helical path is like a stretched-out Slinky, as shown in the following figure.) Remember that velocities parallel to the field do not create magnetic forces. Only velocities perpendicular to the field create a force. Charged particles from the Sun hit the Earth's magnetic field at an angle and helix along the field lines until they harmlessly enter the atmosphere near the North and South Poles, where it ionizes gas molecules. This is what causes Earth's northern and southern lights. The Earth's magnetic field shields us from excessive radiation that would be harmful to the planet and us.

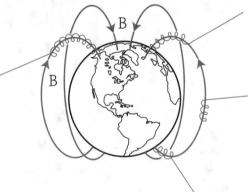

Forces on Current in a Wire

Before we get into this, let's first think about two bar magnets side by side. (See the following figure.) If the two magnets have opposite poles near each other, they're going to attract each other, and if they have like poles near each other, they're going to repel each other. But let's think about how the magnetic fields interact.

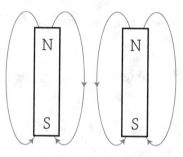

Magnetic fields in the same direction repel each other.

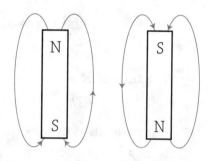

Magnetic fields in the opposite direction attract each other.

Now look at the figure below—there's a wire with current coming out of the page sitting in a magnetic field produced by two magnets going toward the right side of the page. The magnetic field around the wire is going counterclockwise. Above the wire, the magnetic field from the wire is in the opposite direction of the magnetic field from the outside magnets, so those magnetic fields cause attraction, pulling the wire up and pushing the magnets down. Below the wire, the magnetic fields are in the same direction, so they repel each other, pushing the wire up and the magnets down. So, both of these interactions are trying to push the wire up toward the top of the page.

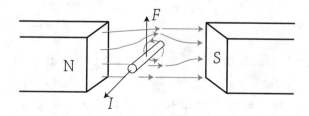

What's nice about this is that we can use the same flat fingers right-hand rule from before. Remember, we consider current to be a positive charge flow, so for a current-carrying wire, we will always be using our right hand—never the left hand.

To use the right-hand rule for a current-carrying wire in a magnetic field, once again, you are going to hold your hand flat with your thumb perpendicular to your fingers:

- Your fingers point in the direction of the external magnetic field (to the right).

- Your thumb points in the direction of the current flowing in the wire (out of the page).

- Your palm aims in the direction of the force (upward toward the top of the page).

Practice this with the figure shown next and with the previous figure and prove that it works.

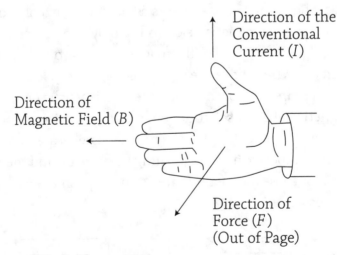

Direction of the Conventional Current (I)

Direction of Magnetic Field (B)

Direction of Force (F) (Out of Page)

The equation for force on a current-carrying wire in a magnetic field is $F_M = IlB\sin\theta$, where F_M is the magnetic force on the current in the wire in newtons (N), I is the current is amperes (A), l is the length of the wire in meters (m), B is the magnetic field strength in tesla (T), and θ is the angle between the current direction in the wire and B-field directions. Notice that when the current and B-field are parallel, the sine will be zero and there will be no force. When the current is perpendicular to the B-field, the sine of 90 degrees is 1.0 and the magnetic force is at a maximum.

IRL When you buy your electric vehicle, remember that electric motors work on the principle of magnetic fields producing force on the current that passes through the coils of wire in the motor.

EXAMPLE

▶ The following figures indicate two different magnetic fields. The B-field on the left is directed out of the page, and the B-field on the right is directed into the page as shown.

▶ Use the flat fingers right-hand rule for current in a magnetic field to determine the direction of the force on each wire with current. Draw an arrow on each figure to indicate the direction of the magnetic force on each wire. If the force is directed into the page, out of the page, or is zero, specifically state so.

▶ Remember to use the flat fingers right-hand rule to determine the direction of the force on each charge. Be sure to take your time to line up each part of your hand correctly! The answers are in the figure shown next.

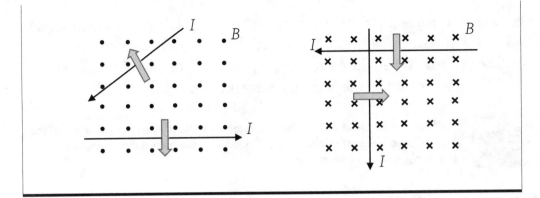

Now let's work through an example that has numbers.

▶ A 10-m-long wire with a current of 2.4 A is placed perpendicular to the Earth's 60 μT magnetic field. Calculate the force on the wire.

▶ Use the magnetic force on a current-carrying wire equation:

$$F_M = IlB\sin\theta = (2.4\,\text{A})(10\,\text{m})(60 \times 10^{-6}\,\text{T})\sin 90° = 0.0014\,\text{N}$$

▶ The same wire described earlier is now moved so that it is parallel to the Earth's magnetic field. What is the new force on the wire? Explain.

▶ Don't be fooled! The current and B-field are parallel; therefore, the force is zero because $\sin 0° = 0$.

REVIEW QUESTIONS

Some of these magnetism ideas can be a little tricky. Especially the three-dimensional nature of the current, velocity, B-field, and magnetic force. Let's exercise our magnetism muscles by answering the following questions.

1. Current is moving downward, as seen in the following figure. Sketch the direction of the magnetic field around the wire.

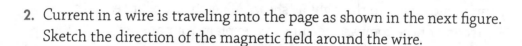

I

2. Current in a wire is traveling into the page as shown in the next figure. Sketch the direction of the magnetic field around the wire.

I

3. Current is moving around in a circular wire, as shown in the following figure. Sketch the direction of the magnetic field around the wire.

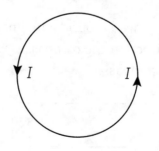

I I

4. Sketch the magnetic field around the magnet in the next figure.

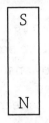

5. You want to make an electromagnet. Which material should you wrap your current-carrying wire around to produce the strongest magnet: aluminum, iron, or graphite? Explain.

6. A compass is positioned so the needle points north. It is placed on top of the wire so the wire aligns with the needle as seen in the following figure. How will the compass be affected when a strong current passes through the wire to the right as shown? Sketch any changes in the direction of the compass heading and explain your answer.

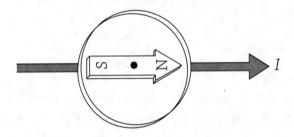

7. Two compasses are arranged around a magnet, as shown in the following figure. Indicate the direction the compasses will turn due to the magnetic field of the compass.

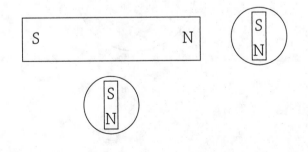

8. Below are three current-carrying wires in three separate magnetic fields. The arrows indicate the direction of the current in each case. Draw the direction of the magnetic force of each wire.

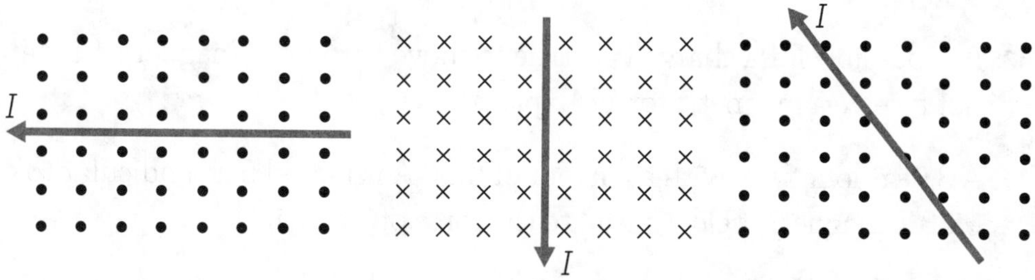

In questions 9 to 11, a particle is moving toward a magnetic field directed into the page, as shown in the following figure. Sketch and label the path taken by each of the following particles. Draw all of the pathways in proportion to the paths taken by all the other particles. All particles enter the magnetic field with the same initial velocity and direction.

9. Neutron

10. Proton

11. Electron

12. A proton is moving at 2.0×10^6 m/s at a right angle to a 0.45 T magnetic field. Calculate the force on the proton. (The charge of a proton is $e = 1.6 \times 10^{-19}$ C. It is also in the Physics Constants table in the appendix.)

13. Is it possible for a charged particle to move through a magnetic field and not receive a magnetic force? Explain.

14. A 3.0-m-long wire with a current of 0.50 A is placed perpendicular to a 0.88 T magnetic field. Calculate the force on the wire.

Flashcard
App

14 Induction

MUST KNOW

⚡ Magnetic flux is the amount of magnetic field that passes through an electrically conductive loop.

⚡ Changing the magnetic flux creates an electric field in the conductive loop that induces an electromotive force (virtual battery) and produces a current in the loop.

⚡ The direction of the induced electromotive force and current are constrained by conservation of energy.

frenzy of scientific activity was launched by Oersted's 1820 discovery that an electric current produced a magnetic field. Electricity and magnetism were all the rage in science. The big question in physics was "If a current creates a magnetic field, can a magnetic field be used to create a current?" It's a reasonable question. Newton's Third Law seems to point us in that direction: "For every action, there is an equal but opposite reaction." It was an intriguing question that many scientists set off in pursuit to answer. It took 11 years for the answer to emerge. In 1831, British Michael Faraday discovered that indeed B-fields could be used to produce a current. This same discovery was made independently by American Joseph Henry and Baltic German Heinrich Lenz. This process was dubbed **induction**.

Our goal in this chapter is to explain how exactly induction works. I will introduce a new idea called **magnetic flux**, and we will wrap up with Lenz's law. This tells us which way the induced current must flow so that we don't violate conservation of energy. So, let's get started.

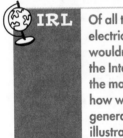

IRL Of all the technological innovations that have been implemented, generating electricity may be the most useful for our modern-day society. Without it we wouldn't have the energy to run electric lights, refrigerators, air conditioners, the Internet, or our computers. Yet most people don't really have any idea how the majority of our electricity is actually generated. This chapter will show us how windmills, nuclear power plants, coal plants, and natural gas plants use generators to create current flow and send it to where you live. Chapter 19 will illustrate how solar cells generate electricity.

Electromagnetic Induction Part 1: How Magnetic Force Can Produce a Current in a Wire

To understand how electromagnetic induction works, picture a proton moving toward the right in a magnetic field that is directed into the page. (See the following figure on the left.) Using the flat fingers right-hand rule, we can see that the force acting on the proton will be toward the top of the page. Now imagine a wire being moved to the right, as shown in the figure on the right. The wire is made up of atoms with protons and electrons. Conventional current is considered to be the flow of positive charge, so if you picture a positive charge in that wire, it will receive a force toward the top of the page as the wire is moved to the right. In fact, every positive charge in the wire will get that exact same force, which will produce a current moving toward the top of the wire. This is called an **electromotive force**, or EMF (ε), and is measured in volts.

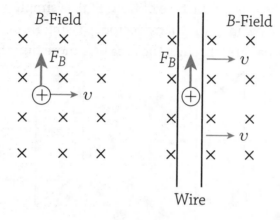

In order to use this EMF to produce a sustained current, we need to have a complete circuit loop of wire. So, dragging a wire through a magnetic field produces an EMF just like a battery does, and connecting this wire to a complete circuit loop will allow current to flow continuously. This is the essence of induction.

To find the EMF of a wire moving through a magnetic field, use the equation $\varepsilon = Blv$, where ε is the EMF, measured in volts (V); B is the magnetic field (sometimes called a B-field), measured in tesla (T); l is the length of the wire in the B-field, measured in meters (m); and v is the velocity of the wire, in meters per second (m/s). This EMF can be used just like a battery to produce current in a circuit.

BTW

An EMF (ε) is a voltage difference that can be used to produce a current. That means dragging a wire through a magnetic field produces a voltage just like a battery does. This is the basics of an electric generator: forcing wires to move through a magnetic field to produce an EMF that, in turn, creates current.

Now that we know how to produce a current in a wire with a magnetic field, we need to expand our study of what is really happening. To fully understand induction, we need to introduce a new idea called **magnetic flux**.

Magnetic Flux

Visualize an arrow being shot through a large sheet of paper, as seen in the following figure. This is our picture of **flux**. Flux is simply a vector passing through an area. Now imagine many vectors passing through the same area. The more vectors passing through the area, the more flux there will be.

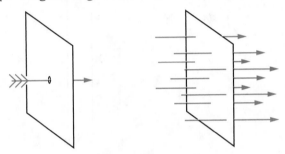

Look at the following figures. On the left I have drawn all of the arrows passing perpendicular through the area. When the vectors pass perpendicular through the area, we have maximum flux. What happens if the piece of paper is turned 90° so that the vectors pass either in front or behind the area? No vectors pass through the area. Therefore, the flux is zero.

Magnetic flux is a measure of the total magnetic field passing directly through an area. The equation for magnetic flux is $\Phi_M = BA\cos\theta$, where Φ_M is the magnetic flux measured in webers (Wb), B is the magnetic field in tesla (T), A is the area through which the magnetic field passes measured in meters squared (m²), and $\cos\theta$ is the term that tells us if the vectors are passing directly through the area at an angle or passing by the area: $\cos\theta = 1$ (max flux) when the magnetic field passes directly through the area, and $\cos\theta = 0$ (no flux) when the magnetic field passes by the area.

EXAMPLE

▶ The connection between magnetic field strength and magnetic flux can be thought of like this: picture a square area 2 meters on each side with a total magnetic field of 4 T passing directly through that area. I have drawn Xs inside of that area to represent the magnetic field, as shown in the following figure. Using our equation for magnetic flux:

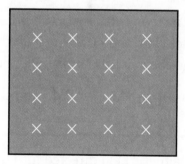

$$\Phi_M = BA\cos\theta$$

$$\Phi_M = (4\,\text{T})(4\,\text{m}^2)(1) = 16\,\text{Wb}$$

▶ Notice that $\cos\theta = 1$ because the field is passing directly through the area.

Electromagnetic Induction Part 2: How to Create an EMF by Changing the Flux

To more generally explain all forms of induction, we use the idea of flux. Or, more precisely, we must have a change in flux. Since Faraday's law states $\varepsilon = \dfrac{\Delta \Phi_M}{\Delta t}$; EMF($\varepsilon$) is produced whenever there is a change in flux with time. Looking more closely at this equation, we can substitute in our equation for magnetic flux:

$$\varepsilon = \frac{\Delta \Phi_M}{\Delta t} = \frac{\Delta(BA\cos\theta)}{\Delta t}$$

Thus, to produce an EMF, we need to change one of the variables in the numerator. We can change

- The magnetic field strength through our conducing circuit loop.

- The flux area of the loop.

- The direction that the loop is facing in the B-field. (Turning the loop so it's parallel to the field and then turning it so that it's perpendicular to the field will change the amount of magnetic flux through the loop.)

Any of these methods will produce an EMF and thus a current in a circuit.

EXAMPLE

▶ A wire loop sits in a magnetic field directed into the page:

Describe three different ways that we could change the magnetic flux to produce an EMF and electric current in the wire.

▶ There are three different methods we could use to accomplish this:

1. We could pull the loop from the field, which will change the magnetic field strength from its initial value to zero.

2. We could smash the loop flat to decrease the area inside the loop. Or if the loop is stretchy, we could expand and compress the loop to change the area either bigger or smaller.

3. We could rotate the loop about a diameter so that the magnetic flux would change from a maximum to a minimum.

IRL Remember that an EMF is like a battery. That means if you have a loop of wire in a magnetic field, all we have to do is change one of these three things:

1. The magnetic field strength
2. The size of the flux area
3. The orientation of the loop in relation to the magnetic field

The result is an EMF that will produce electric current! The electric company uses option #3. They rotate huge coils of wire and enormous magnets past each other to induce current and send it down power lines to your house—along with a monthly electric bill.

Lenz's Law

Let's take a closer look at the details of inducing an EMF and producing current. Up until now, we've just said that a changing magnetic flux creates a current, but we haven't talked about which direction that current flows. To do this, we need to understand **Lenz's law**: *The direction of the induced current opposes the increase in flux.*

When current moves through a loop, that current creates its own magnetic field. So, what Lenz said is that the current that is induced will move in such a way that the magnetic field it creates points opposite to the direction in which the already existing magnetic flux is changing.

This all sounds very confusing. It helps to see this in pictures. Let's start with a loop of wire that is next to a region of space containing a magnetic field. Initially, the magnetic flux through the loop is zero:

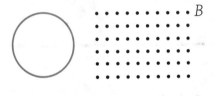

BTW

Lenz's law is an application of conservation of energy. If the direction of the induced current did not oppose the change in flux but instead enhanced the change in flux, we would have a positive feedback loop. The induced current would produce more change in flux, which would produce more induced current to produce more change in flux. This would produce more current, and on and on to infinity. Lenz realized this would be a violation of conservation of energy by creating an infinite amount of energy from nothing.

Now, let's give the loop of wire a velocity to the right and move the wire into the magnetic field, as seen in the next figure. When we move the loop toward the right, the magnetic flux will increase as more and more field lines begin to pass through the loop. The magnetic flux is increasing out of the page. At first, there was no flux out of the page, but now there is some flux out of the page passing through the loop. Lenz's law says that the induced current will create a magnetic field that opposes this increase in flux. So, the induced current will create a magnetic field into the page to oppose the increasing flux out of the page. By the right-hand curly finger rule, the current will flow clockwise:

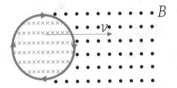

After a while, the loop will be entirely submerged in the region containing the magnetic field. Once this occurs, there will no longer be a changing flux, because no matter where it is within the region, the same number of field lines will always be passing through the loop. Without a changing flux, there will be no induced EMF, so the current will stop:

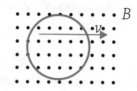

When the loop begins to exit the region of magnetic field, the magnetic flux will begin to decrease as seen in the next figure. Lenz's law says that the induced current will create a magnetic field that opposes this decrease in flux. So, the induced current will create a magnetic field out of the page to oppose the decreasing flux out of the page. By the right-hand rule, the current will flow clockwise:

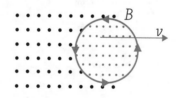

Once the loop has left the B-field, there is no longer any magnetic flux. So, the current dies. Notice that there is only an induced EMF when there is a change in flux:

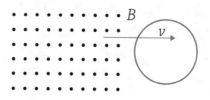

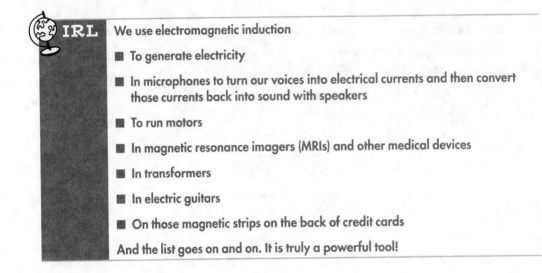

IRL

We use electromagnetic induction

- To generate electricity

- In microphones to turn our voices into electrical currents and then convert those currents back into sound with speakers

- To run motors

- In magnetic resonance imagers (MRIs) and other medical devices

- In transformers

- In electric guitars

- On those magnetic strips on the back of credit cards

And the list goes on and on. It is truly a powerful tool!

REVIEW QUESTIONS

Induction can be tricky to understand, so let's practice our skills by answering the following questions.

For questions 1 to 3, the following figures show a magnet near a loop of wire and a loop of wire moving through a magnetic field. Which of the following will create an EMF and cause current flow in the wire loop? Explain your answer in each case.

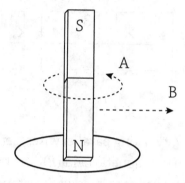

1. Rotating a magnet along its vertical axis above a loop of wire, as indicated with arrow *A*.

2. Moving a magnet from above a loop of wire to the right, as indicated with arrow *B*.

3. Moving a loop of wire to the right in a uniform magnetic field, as indicated by arrow *C*.

In questions 4 to 5, a square loop of wire is pulled through a magnetic field at a constant velocity, as shown in the next figure.

4. At which locations is current produced? Explain.

5. Where there is a current, what is the direction?

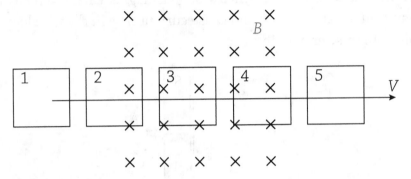

For questions 6 and 7, two long parallel wires, separated by a distance of y, pass through a region of magnetic field (B). The two wires are connected by a resistor (R) and a metal bar separated by a distance of x to produce a circuit loop, as shown in the following figure. The metal bar slides along the wires to the right at a velocity of v.

6. Explain why current is produced in the loop.

7. Which direction does the current flow through the resistor?

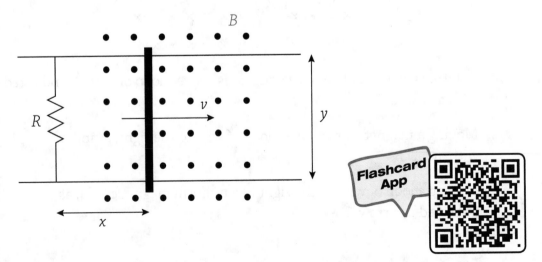

Flashcard App

PART FIVE

Waves

Waves move through a medium with a curious, coordinated behavior. They carry both energy and momentum from one place to another without transporting the medium they pass through. Mechanical waves move through physical matter, while electromagnetic waves transport through space itself. In this section, we'll learn about the many behaviors of waves, including how waves can pass through each other, bend around corners, reflect off surfaces, and change direction when they enter a new material. We'll also learn how we can use mirrors and lenses to make waves form images that we can see and project on a screen.

15 Mechanical Waves and Sound

MUST KNOW

⚡ Waves vibrate matter and have unique properties that are different from objects.

⚡ When waves meet, they pass through each other and vibrate their medium in an overlapping fashion to produce a combined effect.

⚡ When waves are trapped by a boundary, they can produce a standing wave.

S o far, we have been talking about the behavior of objects—physical things with mass that have a location. Objects that exert forces on each other when they interact. Objects that have momentum and energy. Even when we talked about thermodynamics and charge, we were talking about atoms, protons, electrons, and what they were doing. Tiny particles to be sure, but objects nonetheless. But now, we are going to study something completely different: waves. Waves don't have mass, don't have a specific location, and don't have a physical structure like a solid, liquid, or gas. Our goal in this chapter is to learn the basics of what a wave is and how it behaves and use this to explain the physical phenomena of waves you see around yourself every day. This chapter will start with behavior common to all types of waves and then apply these ideas to mechanical waves. Light and electromagnetic waves will be addressed in Chapter 16.

What Is a Wave?

Waves are all around us. Sound that we hear. Microwaves that cook our food. Water on which we surf. Musical instruments that we play. Phones that carry our texts and calls. Earthquakes that destroy our homes. Sunlight that keeps us warm and gives us a tan, or a burn. Radio waves to carry our favorite station. But what is a wave?

A **wave** is a disturbance or vibration in a **medium** that transports energy, but not the medium it travels through. For instance, a sound wave carries the energy of your voice to another person through the medium of air, but does not carry the air along with the wave. Your voice reaches the other person, but the sound wave does not create a breeze or wind.

There are two categories of waves: **mechanical waves** and **electromagnetic waves**. Mechanical waves all travel through a physical medium: sound through air, water waves through water, and earthquakes through the ground are all examples of mechanical waves. Electromagnetic waves, like light, do not require a physical medium, as they travel through an electromagnetic field. In this chapter, we will talk strictly about mechanical waves and cover electromagnetic waves in the next chapter on light.

Mechanical Waves

Mechanical waves travel through a medium by causing the atoms in the material to vibrate in a pattern about their normal equilibrium position—just like a mass on the end of a spring vibrates. Only with a wave, the atoms oscillate in a group pattern. In the following figure, you can see a row of atoms oscillating up and down on the end of springs producing a wave pattern that will be moving to the right through the atoms. This type of wave pattern is called a **transverse wave**.

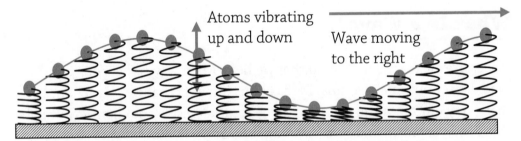

Transverse waves can be represented by a graph, as shown in the next figure. The line down the middle is the **equilibrium position** of the medium when there is no wave. The peak is called a **crest**, and the bottom is a **trough**. The distance from the equilibrium to a crest or trough is the **amplitude** (A) measured in meters. The distance between two crests or two troughs is a **wavelength** (λ) measured in meters. In transverse waves, the medium vibrates perpendicular to the direction of the wave velocity. Notice how it looks like a sine wave.

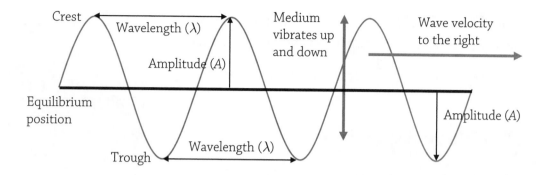

The figure provided next shows a different kind of vibration mode. Notice how the spring is scrunched together in some spots and spread apart in others. The vibration is to the right and left, while the wave travels to the right. This is a **longitudinal wave**, where the oscillation of the medium is in the same direction as the wave velocity. The compacted parts of the medium are called **compressions**, and the stretched-out parts are called **rarefactions**. A wavelength is measured between compressions or rarefactions.

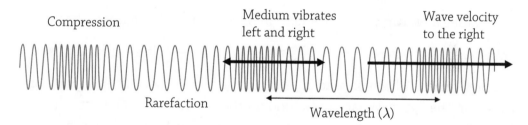

> **IRL** Sound waves, including sound that passes through water and solids, are longitudinal waves. Earthquakes produce both longitudinal and transverse waves. Examples of transverse waves include waves in strings like guitar strings, waves that are formed when you hit a drum, light, and radio waves. Waves in water are actually a combination of longitudinal and transverse waves. Molecules in a water wave oscillate in a circle, creating the classically beautiful shape of a cresting and crashing wave people use to surf.

The time necessary to complete one full wave oscillation is called the **time period** (T) measured in seconds. The number of wave oscillations that occur in 1 second is the **frequency** (f) measured in oscillations/second, which is called **hertz** (Hz). The time period and frequency of a wave are reciprocals of each other and are related with a simple equation: $f = \dfrac{1}{T}$, or if you would prefer: $T = \dfrac{1}{f}$.

The Wave Equation

Take a look at the following graph. It represents the motion of one point in the medium as it oscillates up and down when the wave moves past.

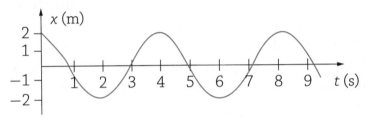

Let's see if we can write the equation of this wave. The mathematical equation of the wave is $x = A\cos(2\pi f t) = A\cos\left(\dfrac{2\pi t}{T}\right)$. It looks kind of scary, but no worries. All you have to do is fill in the blanks. Look at this graph of a wave in the figure. What can we find on the graph? Quite a bit actually. We can find the time period (T) to be approximately 4 s by reading the time difference between two crests or troughs. (Remember that knowing the time period also gives us the frequency: $f = \dfrac{1}{T}$.) We can also find the amplitude to be 2 cm by reading the distance between the equilibrium and a crest. Plug these into the wave equation:

$$x = A\cos\left(\frac{2\pi t}{T}\right) = (2\,\text{cm})\cos\frac{2\pi t}{4} = (2\,\text{cm})\cos\frac{\pi t}{2} \text{ and}$$

we are done! This equation tells us the x location of the wave at any time t.

EXAMPLE

The following figure shows the motion of a rope being oscillated back and forth in a transverse wave.

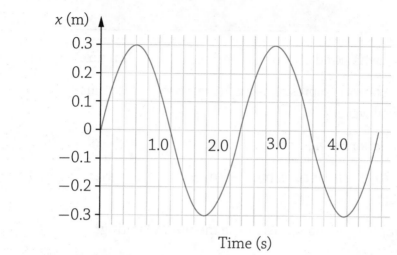

Let's determine each of the following for the wave:

Time period: Looking at the wave shape, we can see that it takes about 2.35 s to complete one wave and about 4.7 s to complete two waves. Thus, the time period is 2.35 s.

Frequency: The frequency is the reciprocal of the time period:

$$f = \frac{1}{T} = \frac{1}{2.35\,\text{s}} = 0.426\,\text{Hz}.$$

Amplitude: Remember that the amplitude is from the equilibrium position to the maximum displacement of the wave: 0.3 m.

The equation of this wave: Plug the amplitude and the time period into the wave equation:

$$x = A\sin\left(\frac{2\pi t}{T}\right) = (0.3\,\text{m})\sin\frac{2\pi t}{2.35} = (0.3\,\text{m})\sin 2.67\,t.$$

Now let's look at a different representation of a wave to see what we can figure out.

The following figure represents a wave.

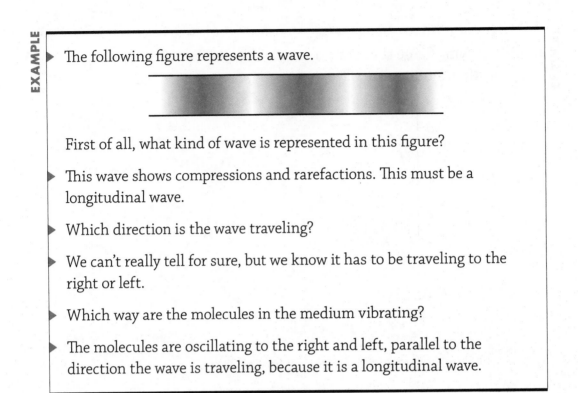

First of all, what kind of wave is represented in this figure?

This wave shows compressions and rarefactions. This must be a longitudinal wave.

Which direction is the wave traveling?

We can't really tell for sure, but we know it has to be traveling to the right or left.

Which way are the molecules in the medium vibrating?

The molecules are oscillating to the right and left, parallel to the direction the wave is traveling, because it is a longitudinal wave.

Wave Velocity

Since a wave is a disturbance that travels through a medium, the medium determines the velocity of the wave. The closer the atoms in the medium are together and the springier the bonds between the atoms, the faster the wave travels through the medium. For instance, the speed of sound in air is about 340 m/s, but in water it is nearly 1,500 m/s because water molecules are closer together than in a gas. The speed of sound in aluminum is a very brisk 5,000 m/s because aluminum has a very springy atomic structure.

To calculate the speed of a wave, we can always use our velocity equation from Chapter 1: $v = \dfrac{\Delta x}{\Delta t}$ but we know that it takes one time period T to create one wave and that single wave will have traveled one wavelength λ. Substituting this into our velocity equation we get:

$$v_{wave} = \frac{\Delta x}{\Delta t} = \frac{\lambda}{T} = f\lambda$$

You toss a rock into a lake, making a wave. The wave travels 14 meters in 10 seconds. What is the speed of the water wave?

$$v_{\text{wave}} = \frac{\Delta x}{\Delta t} = \frac{14\,\text{m}}{10\,\text{s}} = 1.4\,\text{m/s}$$

IRL The medium determines the speed of a wave. Big waves or small, high frequency or low, long wavelength or short—it does not matter. So, when two different-frequency sound waves are produced, they both travel at the same speed through the air. This is a really good thing if you like music! Think how awful it would be if the sound from the drummer, the vocalist, and the guitar player all traveled to you at a different speed. The farther from the stage you get, the more out of time the band would sound. The seats in the back of the auditorium would *really* be the cheap seats!

Wave speed remains constant in a medium. Therefore, increasing the frequency of the wave will have the effect of making the wavelength shorter. And lowering the frequency of the wave will make the wavelength longer in order to maintain the same wave velocity.

A singer belts out a 440 Hz note with a wavelength of 0.75 m. Calculate the speed of the sound wave.

$$v_{\text{wave}} = f\lambda = (440\,\text{Hz})(0.75\,\text{m}) = 330\,\text{m/s}$$

Now she sings a new note at 880 Hz. How fast will the new note travel?

Trick question! The speed of the wave stays at 330 m/s because the medium the sound is traveling through is the same, and the speed of the wave is determined by the medium. The wavelength is cut in half to maintain the same wave speed.

Sound

Sound is a vibration that travels through the medium as a longitudinal wave. We think of sound as a wave that travels through air, but sound also travels through liquids and solids. Whales can communicate with each other over vast distances up to 100 miles. Earthworms sense the sound waves of approaching animals.

In air, sound travels around 340 m/s, but it varies a little with temperature and altitude. Any vibrating object will oscillate the air around it, creating a longitudinal disturbance that is sound. Humans have a larynx, or vocal cords, in our neck that vibrate, creating our voice. Sound enters our ears as compressions and rarefactions, moving our eardrum back and forth so that our inner ear turns it into an electrical signal that our brain decodes as sound.

 IRL On average, women have shorter vocal cords than men, making their voices a higher pitch. This is also why children have high-pitched voices when young that then get deeper as their bodies grow.

Humans can hear a wide frequency range of sounds from as low as 20 Hz all the way up to 20,000 Hz, though our ears are most sensitive in the 2,000 Hz to 5,000 Hz range. Typical human speech is in the range of 80 Hz to 250 Hz.

 **IRL** As we get older, we typically lose our hearing at the high frequencies first. When I run a frequency-range hearing test with my students, my hearing drops out at around 13,500 Hz. Anything above that I can't hear, while most of my students can easily hear above 18,000 Hz. Too many loud concerts in my youth....

 IRL Elephants communicate using infrasonic sounds below the human hearing range. Dogs, cats, bats, and dolphins all hear into the ultrasonic range above what you and I can hear.

Sound Wave Energy and Intensity

Humans can hear a wide variation of **sound intensity**, or loudness. The standard unit of measure for sound intensity is the decibel (dB). Decibels are not a linear scale. An increase in 10 dB means that the sound is transporting 10 times more energy. An increase of 20 dB will be an increase in sound energy of $10 \times 10 = 100$ times more energy. The following table gives the average decibel range for various sounds.

BTW

*Two terms in sound you need to know: The **pitch** of a sound means the same thing as the frequency. Higher pitch means a higher frequency sound. The **volume**, or loudness, of a sound is related to the amplitude of the sound wave. A louder sound has a larger amplitude of oscillation.*

Sound	Decibel Level (dB)
The softest sound humans can hear	0
Breathing	10
Whisper	30
Quiet room, library	40
Light rainfall	50
Normal speech, dishwasher	60
Vacuum cleaner, TV	70
Busy street traffic, noisy restaurant, hair dryer	80
Lawnmower, power tools, shouted conversation	90
Loud music, shouting in an ear	110
Rock concert, ambulance, thunder	120
Jet taking off, fireworks, firearms	150
Rocket at launch	180

IRL   We live in an industrial world with lots of loud sounds, so protect your hearing! Sounds above 85 dB over long exposure (8 hours) can cause hearing loss. The louder the sound, the faster the damage will be. At 100 dB hearing damage happens in minutes. Most people experience pain at levels of 120 dB and higher. Earbuds set too loud can cause permanent ear damage, too!

Sound travels out from the source in all directions. As the wave energy spreads out in a bigger and bigger spherical shell, the wave energy gets spread out thinner and thinner. Therefore, the loudness decreases with distance from the source. Sound intensity (I) is inversely proportional to the distance squared: $I \propto \dfrac{1}{r^2}$. This means that when you sit twice as far from the concert stage, the sound level that gets to you will be a fourth as intense.

When Waves Meet a Boundary in the Medium

As waves travel along through one substance, they can encounter a new medium. Several things happen when this occurs:

1. The energy being transported by the wave attempts to move into the new material. If the new material has similar properties with a similar wave speed, most of the energy will transmit into the new medium. The more dissimilar the two media, the less of the wave energy transmits into the new media. For example, a sound wave that encounters drapes will mostly transmit straight through. Much less of the sound is able to transmit through a wall.

2. The wave energy that is unable to transmit into the new medium will reflect off the new medium. The more dissimilar the two media, the more of the wave energy reflects off the new medium. For example, when a sound wave traveling through air encounters a brick wall, almost all of the wave energy will reflect.

3. The part of the wave that is reflected will either reflect upright or will be inverted depending on the wave interface. When the wave hits a denser/heavier medium, the reflected wave will be inverted. This is called hitting a fixed end. When the wave hits a lighter/less dense medium, the reflected wave will reflect upright. This is called hitting a loose or free end:

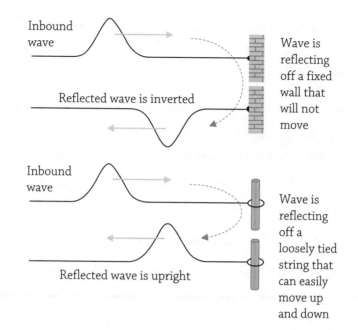

4. The part of the wave that transmits will maintain the same frequency but will move forward with a new wave speed dictated by the new medium. This means that the transmitted wave will also have a new wavelength:

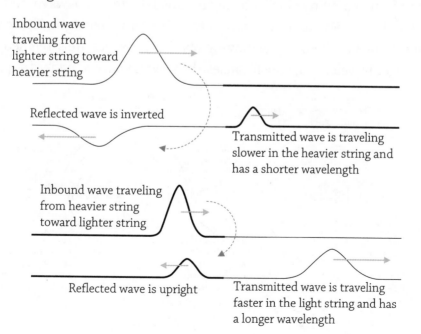

▶ A trumpet plays a loud note. The sound wave encounters a brick wall. Does more of the sound energy transmit into the wall or reflect off?

▶ Air and a brick wall are very dissimilar. Therefore, most of the wave will reflect off the wall.

▶ Describe the properties of the reflected wave.

▶ Since the wall is denser than the air, the reflected wave will be inverted.

Doppler Effect

Sound waves produced from a stationary object move out in all directions in spherical shells, as seen in the figure of a police car shown next. Waves that emanate from a moving object create an unusual effect. The waves in front of the moving object are compressed into a shorter wavelength. This creates a higher frequency. Behind the moving object, the waves are stretched out into a longer wavelength, which causes the wave to have a higher frequency.

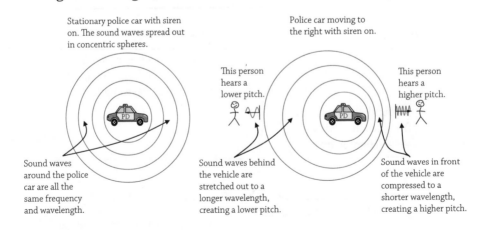

Stationary police car with siren on. The sound waves spread out in concentric spheres.

Police car moving to the right with siren on.

This person hears a lower pitch.

This person hears a higher pitch.

Sound waves around the police car are all the same frequency and wavelength.

Sound waves behind the vehicle are stretched out to a longer wavelength, creating a lower pitch.

Sound waves in front of the vehicle are compressed to a shorter wavelength, creating a higher pitch.

This is called the **Doppler effect**. We experience the Doppler effect all the time. Every time a car passes by us, we hear this change from high pitch (high frequency) to low pitch (low frequency) sound. This effect is amplified the faster the car goes. Look up a video of a car race and listen to the sound of the cars going by.

Meteorologists use Doppler radar to look for tornados. Radar is broadcast out into a storm cloud, which reflects off water in the cloud. Movement of the cloud causes the reflections to come back at different frequencies that are translated into velocity measurements of the cloud. A tightly packed portion of the cloud with high velocities in the opposite directions indicates a tornado.

 IRL   You can thank the Doppler effect for your last speeding ticket. The police shoot your car with a laser that reflects off at a shifted frequency that depends on the speed you are traveling. Bummer!

 IRL   Astronomers have discovered that the universe is expanding because light given off by distant stars (in every direction from the Earth) is shifted to longer wavelengths, indicating that they are all moving away from us at high speed.

EXAMPLE

▶ A driver in a car traveling 35 m/s is honking the horn with a frequency of 850 Hz. The speed of sound is 340 m/s. A person is standing in front of the car in the direction the car is going, and another person is standing behind the car. Calculate the wavelength of the sound wave.

▶ $$\lambda = \frac{v}{f} = \frac{340\,\text{m/s}}{850\,\text{Hz}} = 0.4\,\text{m}$$

▶ What frequency does the driver of the car hear? Less than 850 Hz, exactly 850 Hz, or greater than 850 Hz?

▶ Exactly 850 Hz because the car and driver are traveling at the same speed in the same direction. The don't have a relative speed compared to each other.

▶ What frequency does the person in front of the car hear? Less than 850 Hz, exactly 850 Hz, or greater than 850 Hz?

▶ Greater than 850 Hz because the car is approaching this person.

▶ What frequency does the person behind the car hear? Less than 850 Hz, exactly 850 Hz, or greater than 850 Hz?

▶ Less than 850 Hz because the car is moving away from this person.

▶ Which person hears the longest wavelength sound?

▶ The waves are stretched out behind the car. Therefore, the person behind the car receives the longest wavelength.

▶ At what speed does the wave travel toward the person in front of the car?

▶ Don't be fooled. This is a trick question! The medium determines the speed of the wave. The velocity of the car does not matter. The speed of the wave remains the same at 340 m/s.

Diffraction

Sometimes a wave front will meet a partial barrier. The following figure represents ocean waves traveling toward a wall. The waves on the right side of the figure will reflect off the wall, while the waves on the left side of the figure will pass by the wall. The waves on the left will bend around the corner of the wall, creating a curved wave front that travels into the space behind the wall. Solid objects do not bend around corners. This is a unique behavior of waves called **diffraction**. Diffraction occurs in all types of waves.

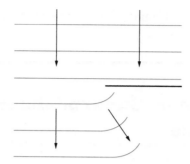

Another example of diffraction is seen in the next figure. A series of waves is traveling toward an opening in a barrier. The waves that pass through diffract, or bend, around the corner, creating a semicircular wave front. The larger the wavelength compared to the opening in the barrier, the more the waves curve around the corners of the opening.

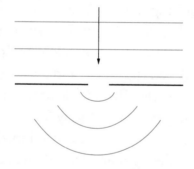

 **IRL** Diffraction occurs in all types of waves. Radio waves bend when going over the top of hills and structures. Water waves bend around sea walls and docks. Sound waves bend around the corner while passing through a doorway, making your voice more easily heard throughout the house. Musical instruments like a tuba take advantage of diffraction by having a bell-shaped end to facilitate the bending of the sound wave as it exits the open end of the instrument. This helps the sound fill up the open space around the tuba so that everyone can hear it. Even light diffracts! We will talk about that in Chapter 16.

EXAMPLE

▶ A high-pitched voice and a low-pitched voice pass through an open doorway. Which voice diffracts more?

> ▶ The low-pitched voice will have a longer wavelength, which will bend more easily around the door opening.

When Waves Meet: Superposition and Interference

When objects are traveling toward each other, they will collide and bounce off. Waves don't have mass or a physical substance. Waves are a disturbance made of energy. So, when waves meet each other, they can simply pass right through each other. The motion of the medium will be a combination of the two overlapping waves, which is called **superposition**.

Superposition happens in two ways. **Constructive interference** occurs when both waves are vibrating the medium in the same direction. This causes the motion to amplify, or build up. The net motion of the medium is the sum of the two individual waves. The following figure shows two crests approaching each other. When they meet and overlap, the net wave is the sum of the two individual waves. The middle part of the figure shows the combined wave and the original wave in gray for reference. Constructive interference always increases the size of the combined waves. The individual waves are still traveling to the right and left and will continue on as if they have never met, as seen in the bottom of the figure.

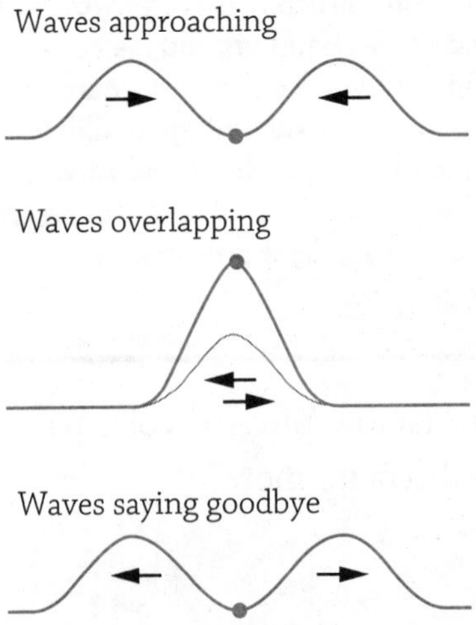

Waves approaching

Waves overlapping

Waves saying goodbye

>
> **IRL** Wave pools at waterparks use constructive interference by creating overlapping waves to produce large waves in confined pools.

Destructive interference occurs when each wave is trying to vibrate the medium in opposite directions. This causes the motions to cancel each other out. The net motion of the medium is the sum of the two individual waves. The following figure shows a crest and a trough approaching each other. When they meet and overlap, the net wave is the sum of the two individual waves. The middle part of the figure shows the combined wave and the original waves in gray for reference. Destructive interference will always decrease the size of the combined wave motion, and it looks like the wave energy has been destroyed, but it has not. The energy is still there, and the individual waves are still traveling to the right and left and will continue on past each other, as seen in the bottom part of the figure.

Waves approaching

Waves overlapping

Waves saying goodbye

> **IRL** Noise-canceling headphones use destructive interference to eliminate noise. A microphone reads the outside noise, flips the wave over so that it is an inverted match, and pumps this inverted sound into your ear. The net result is that the ambient noise from the environment and the inverted "noise" from the headphones cancel each other out. So, by making noise you can cancel noise. Genius!

EXAMPLE

▶ Two waves of the same amplitude and wavelength are traveling in the opposite directions along a string:

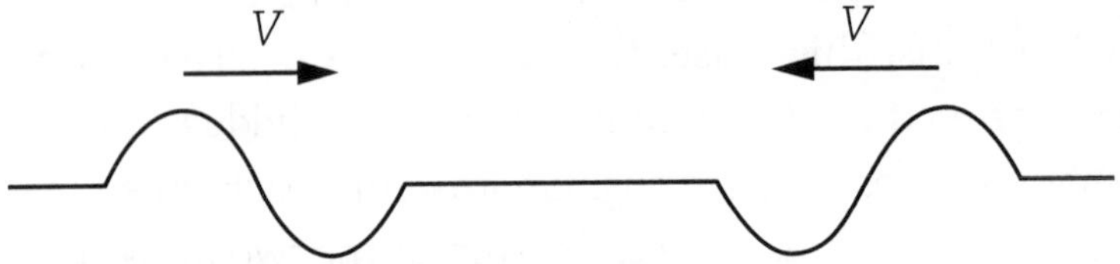

▶ What will the string look like when the two waves meet? Sketch what the resultant wave looks like when these two waves completely overlap.

▶ When the waves perfectly overlap, the crests and troughs cancel each other out. The resultant wave is a straight, flat line, as seen in the next figure.

▶ Is this constructive or destructive interference?

▶ This is destructive interference because the amplitude of the wave motion is decreased. Note that this does not destroy the two waves. The waves will pass right through each other. The wave amplitude is only diminished when the waves are right on top of each other.

BTW

Constructive and destructive interference happens in all types of waves—sound, water, and even light, which will be introduced in Chapter 16.

Beats

When two waves of slightly different frequencies overlap, there is an alternating pattern of constructive and destructive interference called **beats**. The amplitude of the combined wave alternates between high amplitude and low amplitude. In sound waves, this can be heard as a periodic variation in the loudness. The rate of the oscillation in amplitude is called the beat frequency, which is found by subtracting the two individual frequencies of the waves: $f_{beat} = |f_1 - f_2|$. The following figure shows two individual waves, one at 18 Hz and the second at 17 Hz, with the combined wave at the bottom of the figure. Notice that the two waves alternate between constructive and destructive interference, causing the amplitude to oscillate at a frequency of 1 Hz.

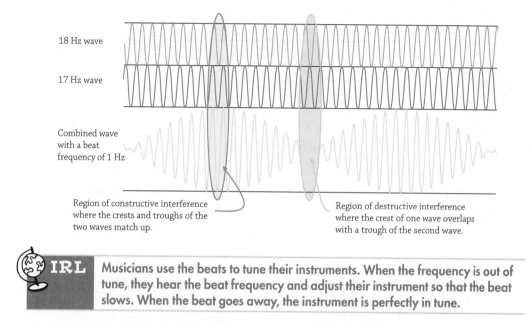

18 Hz wave

17 Hz wave

Combined wave with a beat frequency of 1 Hz

Region of constructive interference where the crests and troughs of the two waves match up.

Region of destructive interference where the crest of one wave overlaps with a trough of the second wave.

IRL Musicians use the beats to tune their instruments. When the frequency is out of tune, they hear the beat frequency and adjust their instrument so that the beat slows. When the beat goes away, the instrument is perfectly in tune.

EXAMPLE

▶ A piano tuner strikes the A key on the piano while also playing a 440 Hz note on a small speaker. The piano tuner hears a beat frequency of 2 Hz. If the A key is too high, what is its frequency?

▶ Using the beat frequency interference equation, we get:

$$f_{beat} = |f_1 - f_2|$$
$$2\,\text{Hz} = |f_1 - 440\,\text{Hz}|$$
$$f_1 = 442\,\text{Hz}$$

▶ If the A key is too low, what is its frequency?

Again, use the beat frequency interference equation:

$$f_{beat} = |f_1 - f_2|$$
$$2\,\text{Hz} = |440\,\text{Hz} - f_2|$$
$$f_2 = 438\,\text{Hz}$$

Voice

Your voice starts in your lungs with air that is forced up over your vocal cords, which are stretched horizontally across your larynx. This causes the vocal cords to vibrate in a buzzing sound. The frequency of the vibration depends on the tension in the cords. This buzzing sound resonates in your throat, mouth, and nose, which creates the unique voice that is yours. Every person's voice is a distinctively complicated sound wave. No two are exactly alike. Your voice is as singular as your fingerprint. Here are the sounds that make me unique:

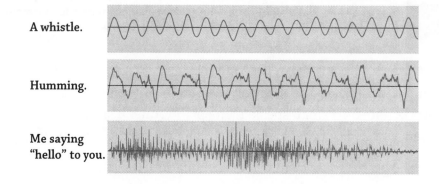

A whistle.

Humming.

Me saying "hello" to you.

Standing Waves

When waves of the correct length are trapped between two boundaries, they can produce a **standing wave**. Consider a string attached to a wall on both ends with a series of waves traveling to the right, reflecting off the boundary, and reflecting back. As the two sets of waves travel back and forth between the left and right boundary, they create constructive and destructive interference. If the wavelength matches up with the boundary conditions, the constructive and destructive interference will always be in the same physical location, making the wave appear to be standing still as seen in the following figures. The locations of total destructive interference are called **nodes**, while maximum amplitude locations are called **antinodes**.

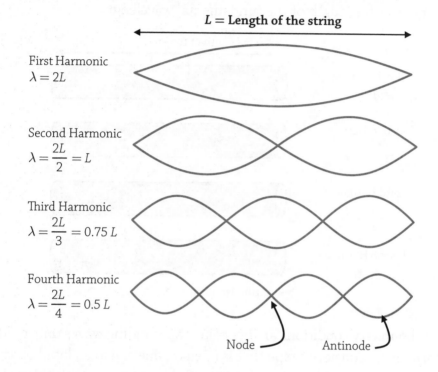

Since the string is fixed at the ends, there must be nodes on the ends of the standing wave. The longest wavelength that will exactly fit with nodes on the ends will be a wave that is twice the length of the string. This wave will produce the lowest-frequency standing wave called the **fundamental harmonic** or the **first harmonic**. Shorter waves can also fit between the boundary conditions. These are called higher harmonics, and each will be a multiple of the fundamental frequency. The **second harmonic** will be twice the frequency of the first harmonic. The **third harmonic** will be three times the first harmonic, and so on.

This pattern repeats for any wave trapped between identical boundary conditions. For example, a pipe that is open on both ends will allow air to move freely at the ends, creating an antinode boundary condition at the ends for air to vibrate between. (See the following figure.) The open-open pipe with antinodes on the ends will have the same harmonic pattern as the string with nodes on the ends. The fourth harmonic will have four times the frequency as the fundamental, and the wavelength of the standing wave will be one-quarter the length of the fundamental wavelength.

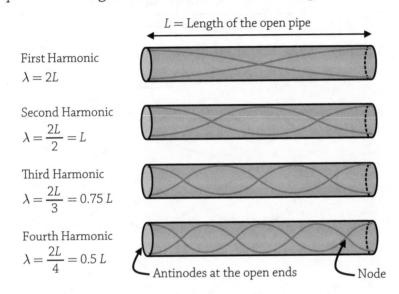

First Harmonic
$\lambda = 2L$

Second Harmonic
$\lambda = \dfrac{2L}{2} = L$

Third Harmonic
$\lambda = \dfrac{2L}{3} = 0.75\,L$

Fourth Harmonic
$\lambda = \dfrac{2L}{4} = 0.5\,L$

L = Length of the open pipe

Antinodes at the open ends   Node

If the boundary conditions are opposite, the standing waves only produce odd harmonics. Consider a pipe that is open on one end and closed on the other. The open end will produce an antinode boundary condition, but the closed end will not allow air to vibrate. Therefore, a node will be created at the fixed end of the pipe.

The longest wavelength that will exactly fit with a node on one end and an antinode on the other will be a wave that is four times the length of the pipe. This wave will produce the lowest-frequency first harmonic. Shorter waves can also fit between the boundary conditions, but the next one to exactly fit will be four-thirds the length of the pipe.

This wave is three times shorter than the fundamental frequency and will therefore be three times the frequency. Therefore, this second possible standing wave is the third harmonic. The next will be the fifth harmonic, seventh harmonic, and so on. This type of resonating pipe only produces odd harmonics:

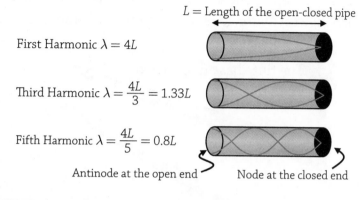

L = Length of the open-closed pipe

First Harmonic $\lambda = 4L$

Third Harmonic $\lambda = \dfrac{4L}{3} = 1.33L$

Fifth Harmonic $\lambda = \dfrac{4L}{5} = 0.8L$

Antinode at the open end ⟍   Node at the closed end

 IRL   Musical instruments are built on this idea of standing waves. Guitars and pianos have strings that oscillate as standing waves. Drums have skins that oscillate as a 2D vibrating standing wave. Wind instruments operate as open-ended or closed-ended pipes.

IRL   Find a video of the Tacoma Narrows Bridge collapse. A strong wind caused a vibration in the bridge that just happened to be the right wavelength and frequency to set up a standing wave. This twisting harmonic wave can be seen in the following figure. The standing wave became so large—over 10 feet by some accounts—that the bridge was eventually destroyed. Seriously, watch the video. It's so bizarre it looks fake!

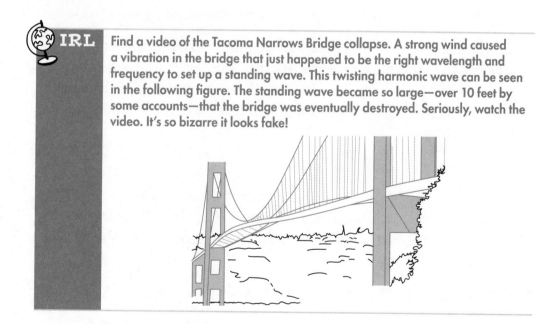

A flute and a clarinet are approximately the same length: 60 cm. The flute operates as an open-ended pipe, while the clarinet behaves as a closed-ended pipe oscillator. (See the following drawings.) Assume the speed of sound to be 340 m/s.

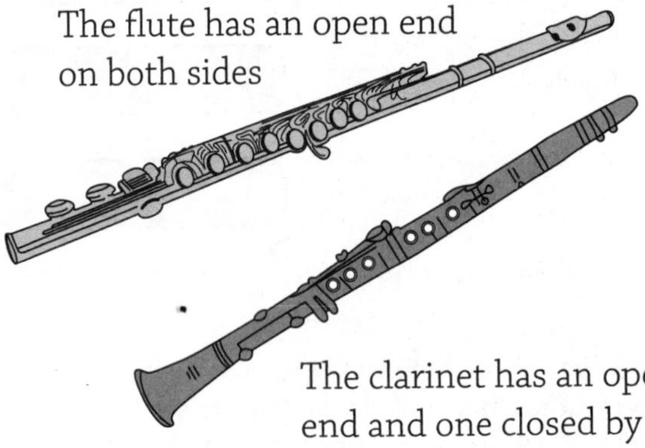

The flute has an open end
on both sides

The clarinet has an open
end and one closed by
the performer's mouth

Which instrument has the longest fundamental standing wave?

▶ The clarinet, because it is a closed-ended pipe with a fundamental wavelength that is four times the length of the pipe. This means the longest wave produced by the clarinet will be 240 cm long. The longest wave produced by the open-ended flute will be twice the length of the tube, or 120 cm.

▶ Which plays the highest fundamental frequency?

▶ The flute because it will play a fundamental frequency note that is twice the length of the open pipe, which is 120 cm long. Since it plays a note that is half the length of the clarinet, its frequency will be twice as high.

▶ Calculate the fundamental frequency played by the flute.

▶ The fundamental wavelength is twice the length of the open pipe: 120 cm = 1.2 m.

$$f = \frac{v}{\lambda} = \frac{340\,\text{m/s}}{1.2\,\text{m}} = 283\,\text{Hz}$$

▶ What are the second and third harmonics of the flute?

▶ The second harmonic will be twice the first harmonic: 566 Hz. The third harmonic will be three times the first harmonic: 849 Hz.

▶ Calculate the first harmonic of the clarinet.

▶ The clarinet is a closed-ended pipe with a wavelength four times the pipe itself: 240 cm.

$$f = \frac{v}{\lambda} = \frac{340 \text{ m/s}}{2.4 \text{ m}} = 142 \text{ Hz}$$

▶ What are the next two higher harmonics of the clarinet?

▶ Be careful! This is a closed-ended pipe that only plays odd harmonics. Therefore, we will only hear the third and fifth harmonics: 426 Hz and 710 Hz.

REVIEW QUESTIONS

All right, we learned a lot about waves. Let's show what we know by answering the following questions.

1. What determines the speed of a wave?

Questions 2 to 4: a boy sings a low-pitched note while a girl sings a high-pitched note.

2. Which waves travels faster: the girl's sound wave or the boy's sound wave?

3. Which has a higher-frequency sound wave?

4. Which has the longer wavelength sound wave?

5. What is the difference between mechanical and electromagnetic waves?

Questions 6 to 9: a wave is shown in the following figure.

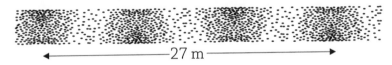

6. What kind of wave is it?

7. What is its wavelength?

8. Which direction is the wave traveling?

9. Which direction is the medium vibrating?

Questions 10 to 15: a child takes a coil spring and shakes it back and forth 12 times in 4 seconds to create the wave shown in the next figure.

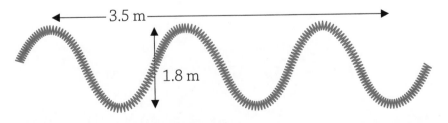

10. What kind of wave is this?

11. Determine the amplitude of the wave.

12. What is the wavelength of the wave?

13. Calculate the velocity of the wave.

14. What direction is the wave traveling?

15. Which way is the spring oscillating?

Questions 16 to 17: a sound wave has a frequency of 512 Hz. The speed of sound in air is 330 m/s.

16. Determine the time period of the wave.

17. Calculate the wavelength of the wave.

Questions 18 to 22: the following figure shows the motion of a string being oscillated back and forth in a transverse wave. The wave has a wavelength of 3.7 m. Determine each of the following for the wave:

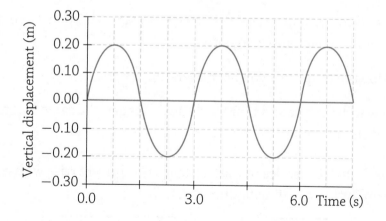

18. Time period

19. Frequency

20. Amplitude

21. The velocity of the wave

22. The equation of the wave

23. A bat makes a 22,000 Hz sound. Can you hear the sound?

24. At what sound intensity level will humans begin to have hearing loss?

25. A teacher is talking at the front of the room. Student A sits 3 m away. Student B sits 9 m away. How much more sound intensity does student A get compared to student B?

Questions 26 to 27: a wave in a string is traveling to the right toward a rope, as shown in the following figure.

26. Describe the transmitted portion of the wave.

27. Describe the reflected portion of the wave.

Questions 28 to 29: a fire truck is driving toward you with its siren blaring while traveling toward an emergency.

28. Describe the frequency you hear from the fire truck.

29. Is the wavelength sound that reaches you longer, the same length, or shorter than normal?

Questions 30 to 31: a series of waves travels toward a barrier with a hole in it, as shown in the following figure.

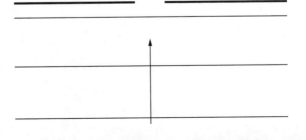

30. Sketch the waves on the other side of the barrier.

31. What property does this demonstrate, and do all waves have this same behavior?

Questions 32 to 33: two waves are traveling in a string toward each other, as shown in the next figure.

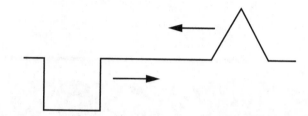

32. Sketch what the combined wave will look like when the two are aligned with each other.

33. Explain what kind of interference this is.

34. Two musicians play notes of 881 Hz and 878 Hz. What beat frequency is heard?

Questions 35 to 37: a musician plucks a guitar string that is 648 mm long. The note that is played is the first harmonic at 196 Hz.

35. Calculate the wavelength of this note.

36. What is the speed of the wave in the guitar string?

37. What are the second and third harmonics of this guitar string?

Flashcard App

 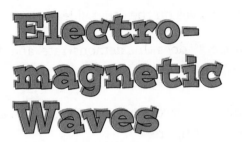

Electro-magnetic Waves

MUST KNOW

- All moving charges radiate electromagnetic waves that travel through space at the speed of light and have the same properties.

- All electromagnetic waves, including visible light, exhibit wave properties just like mechanical waves.

hen radio waves are beamed through space, or when X-rays are used to look at your bones, or when microwaves cook your food, or when visible light travels from a lightbulb to your eye, electromagnetic waves are at work. All of these types of radiation fall into what we call the **electromagnetic spectrum**. (See the following figure.)

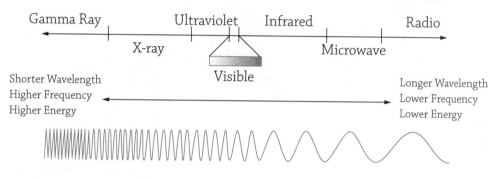

The electromagnetic spectrum is amazing. We use it for so many things. Our goal in this chapter is to understand how electromagnetic waves are formed and what their properties are.

The Electromagnetic Spectrum

Back in Chapter 10, we learned that charges create an electric field around themselves that extend infinitely far away. When a charge oscillates, it will create a disturbance in the very electric field it is producing. This jiggle in the electric field will move outward through the field and into space. This is a wave! Let's take a look:

An Electromagnetic (EM) Wave

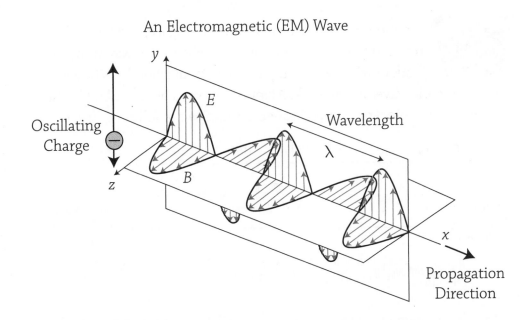

The really interesting thing is that this oscillating electric field wave induces an oscillating magnetic field wave at a right angle to itself, as seen in the previous figure. (Just like we learned in Chapter 14.) This re-creates the electric field wave, and the whole thing keeps oscillating back and forth and never stops. This electromagnetic, or EM, wave is self-propagating because the electric and magnetic fields are re-creating each other over and over. Therefore, EM waves can travel through a vacuum. This is a major departure from mechanical waves, like sound, that require a physical medium to move through. Another major difference between EM waves and mechanical waves is that EM waves travel at the **speed of light** in a vacuum, which is much faster than any mechanical wave.

EXAMPLE

▶ What makes sound waves different from microwaves?

▶ Sound waves are mechanical waves, which need a physical medium to transport through. Microwaves are EM waves, which can transport through the nonphysical medium of electric and magnetic fields. Therefore, EM waves can move through a vacuum. EM waves also travel much faster.

IRL We can see the Sun, but we can't *hear* the Sun because light travels through the vacuum of space, but sound waves can't.

All electromagnetic waves travel the same speed in a vacuum: $c = 3.0 \times 10^8$ m/s. The famous symbol for the speed of light is c. What makes one form of electromagnetic radiation different from another form is simply the frequency of the wave. AM radio waves have a very low frequency and a very long wavelength, whereas gamma rays have an extremely high frequency and an exceptionally short wavelength. But they're all just varying forms of the EM spectrum, which we simply call light waves.

BTW

The speed of light is fast. Really fast. But the universe is mind bogglingly **vast***. It takes 8.3 minutes for the light from the Sun to reach Earth. When you want to talk to your friend on Mars, the light wave will take anywhere from 3 to 22 minutes, depending on where both planets are in their orbits. It takes more than 4 years for light from the closest star system, Alpha Centauri, to reach Earth. So, when you look up at other stars, you are looking at light that may be thousands or millions of years old. This is why we say that a telescope is a time machine that lets astronomers study the history of the universe.*

EXAMPLE

The star Sirius is 8.6 light-years from Earth. How far is that in meters? (The conversion between light-years and meters is in the appendix.)

$$(8.6 \text{ ly})\left(\frac{9.46 \times 10^{15} \text{ m}}{1 \text{ ly}}\right) = 8.1 \times 10^{16} \text{ m}$$

EM waves are produced by any type of oscillating charge. If you charge up a balloon and shake it around, you are producing an EM wave. Remember from thermodynamics in Chapter 8, where we learned that atoms are in constant, random, *oscillating* motion. That means anything with a temperature above absolute zero, which is everything, will be radiating EM waves all the time. You are emitting infrared radiation right now. If we heat you up to about 1000°C, you will glow red!

BTW

The universe is so huge that astronomers use the distance that light travels in a year as a unit of measure for distance. One light-year: 1 ly = 9.46 × 10¹⁵ m.

EXAMPLE

> Which of these will produce an electromagnetic wave? An oscillating electron, a vibrating neutron, or a charged cat running in a circle.

> The electron and the cat. The neutron does not have a charge.

> The cat will be producing a wave because not only is it charged and running in a circle, but also it has a temperature above absolute zero, and its atoms will be radiating infrared EM waves due to their oscillation.

Look back at the diagram of the EM wave. Notice that EM waves are transverse, just like a wave in a string. Since light is a wave, it has properties just like other mechanical waves. In the next sections, I'll lead you through some of the wave properties that light exhibits and how we observe them and some of their consequences.

Polarization

One of the interesting differences between transverse and longitudinal waves is **polarization**. First off, this is not the same polarization that we talked about in static electricity in Chapter 10. That was charge polarization, where a neutral object can be induced to have one side more positive and the other side more negative. Here we are talking about wave polarization.

Look back again at the figure of an EM wave. Notice how the electric portion of the wave is oscillating up and down in the y-direction? That is because the charge that created the E-field wave was oscillating up and down as well. It is said to be polarized in the y-direction. If you want to polarize the E-field portion of the wave horizontally, just wiggle the charge back and forth in the horizontal (z) direction. Another way we get polarized light is when it reflects off of flat surfaces, like the surface of a lake or a mirror. It tends to be polarized in the plane of the surface. If you drive a car, you know about road glare. The bright reflection off the road tends to be horizontally polarized.

There are substances that will allow EM waves, polarized in the y-direction, to pass through them, but will absorb waves polarized in the z-direction. Visualize a picket fence with a rope passing through from one side to the other, like in the following figure. When a transverse wave is made in the rope, only the waves that align with the fence pass through, but waves that are perpendicular to the openings will be blocked.

When unpolarized light that is oscillating in all directions shines on a polarizing filter, only the light that is oscillating in the plane of the filter will get through. (See the next figure.) We use materials with atomic structures like the picket fence to create polarized sunglasses to block the horizontally polarized glare from a road. Longitudinal waves cannot be polarized, because they vibrate forward and backward.

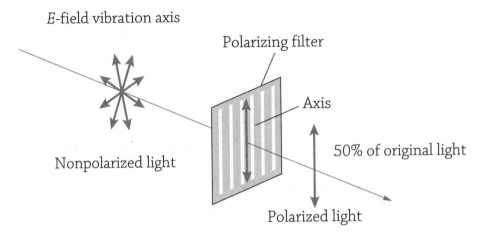

E-field vibration axis

Polarizing filter

Axis

50% of original light

Nonpolarized light

Polarized light

IRL 3D movies project two separate images simultaneously on the screen, each with a different polarization. Spectators use polarized glasses so that each eye sees only one of the two images projected onto the screen, and their brain turns the two images into a 3D image.

Here is a quick experiment you can perform:

EXAMPLE

▶ On a sunny day, put on polarizing sunglasses and look at the horizontal surface of a road or lake. Observe how the glasses block the horizontally polarized light reflecting from the surface. Now tilt your head sideways 90°. What will you see? Why do you now see so much glare from the surface?

▶ When you tilt your head 90°, you align the transmission axis of the lenses with the axis of the polarized light reflecting from the surface. All the glare goes right through the sunglasses. If you only turn your head 45°, a portion of the glare gets through the glasses.

▶ When your head is vertical, the transmission axis of the glasses, and the horizontally polarized light from the surface are at right angles, all of the glare is absorbed by the glasses and no glare gets through.

Diffraction and the Point-Source Model

In Chapter 15, we talked about how you can hear sounds around corners due to the property of diffraction. So, if sound waves bend around corners, why don't we see light bend around corners? Well, to understand this, let's review diffraction. Look at the wave fronts traveling upward toward the top of the page in the two figures provided next. Notice how the waves bend around the obstruction. Why does this occur?

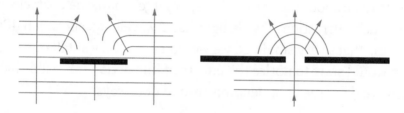

Huygens explained it this way: Each point on a wave is the starting spot for a new wave. This is called the **point-source model**. So, each point of the wave that goes past a boundary is the new starting spot for a wave. These waves don't just travel forward, they also travel outward to the sides, just like ripples in a pond produced by a dropped pebble, as seen in the next figure. Notice how the little point sources that travel through the opening past the barrier each produce their own wave, and these overlapping individual waves produce a new combined wave front that bends around corners. As a result, we can hear things around corners. If you have ever played in the band, you know that the mouths of woodwind and brass instruments curve outward to enhance this diffraction, helping the sound to bend outward and fill an entire auditorium.

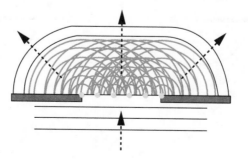

So we can hear things around corners, but why can't we see things around corners? Does this mean that light is not a wave? For years that was the argument, but it turns out that the smaller the wave is, compared to the boundary it is passing by, the less we see the effects of diffraction. Look at the following figure. When the opening in the boundary is smaller than the wavelength, the diffraction is very evident. But when the opening is large compared to the wavelength, we only see the effects of diffraction around the edges of the wave passing through. In general, the wavelength of the wave needs to be about the same size as, or larger than, the obstacle, or we won't get much diffraction. Visible light has a wavelength between 400 nm and 700 nm. That's very small. So we would need a very small opening before we would start to notice the effects of diffraction for visible light. That's why we don't see light bending around doorways.

Opening smaller than wavelength (λ)

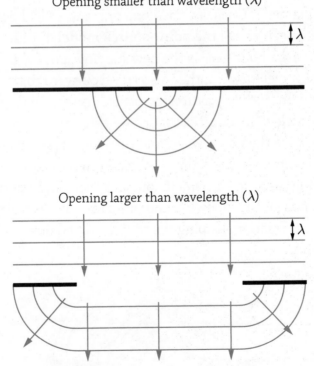

Opening larger than wavelength (λ)

Double-Slit Experiment

So there was this huge fight behind the physics building about light. One gang, commanded by Newton, of the "Call me Sir Isaac because I invented gravity and three laws" fame, championed the corpuscular theory of light, which claimed light was made up of tiny particles called corpuscles. The rival gang was headed by Christiaan "I think waves are groovy" Huygens, who put forth the outlandish claim that light was just a bunch of waves. These two gangs fought and argued over the nature of light for over 100 years until a dude by the name of Thomas Young yells, "Stop fighting! I've got an illuminating experiment to show you." (Illuminating...get it?)

Consider light shining through two tiny openings, or slits, located very close together—slits separated by tenths or hundredths of millimeters. The light shines through each slit and then hits a screen. Here is the genius of Young's experiment: If light is a particle, we should see two lines of light on the screen coming from the light passing through the two slits. But if light is a wave, we should see the effects of diffraction and interference. So, what did they see? Well rather than seeing two bright patches on the screen (which would be expected if light was made of particles), the physicists saw lots of alternating bright and dark patches. The only way to explain this phenomenon was to conclude that light behaves like a wave.

Look at the left side of the figure on the next page. When the light waves go through each slit, they are diffracted and bend around the corner. As a result, the waves that come through the left slit overlap with the waves that came through the right slit. Everywhere that peaks or troughs cross paths, either constructive or destructive interference occurs. So when the light waves hit the screen, some places they constructively interfered with one another and produced bright regions, and in other places they destructively interfered with one another and produced dark regions. That explains why the screen looks like the figure on the right.

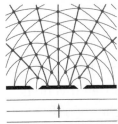
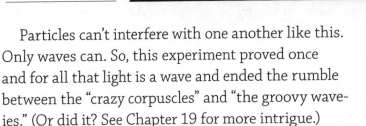

Particles can't interfere with one another like this. Only waves can. So, this experiment proved once and for all that light is a wave and ended the rumble between the "crazy corpuscles" and "the groovy wave-ies." (Or did it? See Chapter 19 for more intrigue.)

When light passes through slits to reach a screen, the equation to find the location of bright spots is as follows: $d \sin \theta = m\lambda$. Here, d is the distance between slits, λ is the wavelength of the light, and m is the "order" of the bright spot; I'll discuss m next. θ is the angle at which an observer has to look to see the bright spot.

BTW

This interference pattern can be created in all types of waves, like water waves in a wave pool at a water park, or with sound waves using two audio speakers. Acoustic engineers work hard to eliminate constructive interference—loud spots—and destructive interference—soft spots— at concert venues.

IRL Scientists use this diffraction equation to measure very tiny objects that are too small to measure directly. The smaller the wavelength used, the smaller the object we can measure. In fact, by shining a beam of X-rays on a sample, scientists can produce an interference pattern from the atoms themselves and determine the atomic structure of a molecule. This is what Rosalind Franklin did when she took the first picture of DNA.

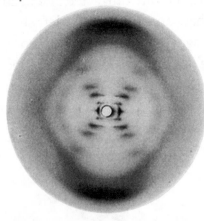

Photo 51: DNA X-ray diffraction image
Source: King's College London

IRL "Photo 51" is an iconic image of DNA, taken by Franklin and student Raymond Gosling. Franklin's work on X-ray diffraction images of DNA was essential to the discovery that DNA has a double-helix structure.

The variable m represents the "order" of the bright or dark spot, measured from the **central maximum** as shown in the following figure. Bright spots have integer values of m; dark spots get half-integer values of m. The central bright spot, or maximum, is represented by $m = 0$. The first constructive interference location to the side of the **central maxima** is called the **first-order maxima** and would be represented by $m = 1$. The second-order maxima would be $m = 2$, etc. The first destructive interference location at $m = \frac{1}{2}$ would be the **first-order minima**. So, for example, if you wanted to find how far it is from the center of the pattern, the first bright spot labeled $m = 1$ in the figure, you would plug in "1" for m. If you wanted to find the dark region closest to the center of the screen, you would plug in "½" for m.

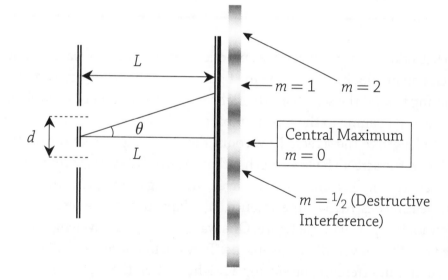

EXAMPLE

▶ You have a pair of rainbow diffraction glasses, which have thousands of parallel lines etched 4.9 µm apart into the plastic of the lenses. This creates a rainbow of color when you look through them. When you shine a green laser pointer with a wavelength of 532 nm through the glasses, it produces a diffraction pattern of dots. Calculate the angle to the first dot in the diffraction pattern.

▶ Use the double-slit interference equation with $m = 1$:

$$d \sin \theta = m\lambda$$
$$\left(4.9 \times 10^{-6}\,\text{m}\right) \sin \theta = (1)\left(532 \times 10^{-9}\,\text{m}\right)$$
$$\sin \theta = 0.11$$
$$\theta = \sin^{-1}(0.11) = 6.3°$$

▶ If you shine a red laser through the glasses, how will the pattern differ?

▶ Looking at the interference equation, we see that $\sin \theta \propto \lambda$. The red laser has a longer wavelength. Therefore, $\sin \theta$ must also increase. This will increase the angle. Thus, the red laser will produce an interference pattern that is more spread out than the green laser.

To better understand why waves produce this interference pattern but particles do not, look at the following figure where the waves are shown oscillating toward the screen on the right. Count the number of wavelengths from the top opening to the screen in part a: 4½ wavelengths. This is called the path length to the screen. Now count the path length from the bottom opening to the screen: 5 wavelengths. That means the waves will arrive ½ a wavelength off. Or, a crest of one wave is meeting a trough for the other wave, creating destructive interference, or a dark spot on the screen.

Now look at part b in the figure. One wave travels 4½ wavelengths, while the other travels 5½ wavelengths. They are in phase and will have constructive interference, producing a bright spot on the screen. The path length difference equation, $\Delta L = m\lambda$, shows us that whenever the difference in the length to the screen ΔL is a whole multiple ($m = 1, 2, 3, 4, 5$, etc.) of the wavelength, we will get constructive interference. When the waves are off by a ½ wavelength ($m = 1/2, 3/2, 5/2$, etc.), there will be destructive interference.

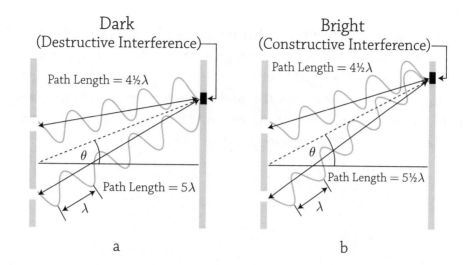

Dark
(Destructive Interference)

Path Length = 4½λ

θ

Path Length = 5λ

λ

a

Bright
(Constructive Interference)

Path Length = 4½λ

θ

Path Length = 5½λ

λ

b

IRL What do Morpho butterfly wings, CDs, iridescent beetle shells, and DVDs all have in common? They all have structural color. This is color created due to the structure of the surface, not due to pigment. CDs and DVDs have tiny lines of data that act as reflective grooves, creating a diffraction pattern and creating a rainbow of color. Morpho butterflies have amazing blue wings created by interference. Many beetles have iridescent green, blue, violet, and even red and gold shells. These vibrant colors also come from microscopic bumps and ridges that create constructive interference in a particular color.

Light at Boundaries

When light encounters a new medium, three things can happen:

- The light can reflect.
- The light can be absorbed.
- The light can transmit through the new medium.

Mirrors mostly reflect light in a uniform, *regular* way that allows you to see where the light came from and what object gave off the light. (See the following image.) Smooth surfaces behave like mirrors. Objects with bumpy or rough surfaces reflect light in a **diffuse** way so that we can't tell where the light originated. A white sheet of paper gives off a diffuse reflection, while a sheet of glass gives off a regular reflection. But isn't a sheet of paper smooth? To the touch it is, but if you look at it under a microscope it's made up tiny fibers that reflect light all over the place. So, the surface needs to be smooth on a microscopic scale to have a regular reflection.

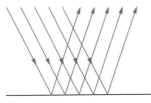
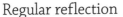
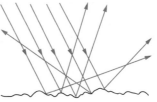

Regular reflection   Diffuse reflection

Opaque objects that are white also reflect most of the light that hits them, but a surface that is black or dark will absorb most of the light that lands on it. Absorbed light transfers its energy to the object, making the atoms vibrate faster and warming up the object. I live in Texas, where it is hot and sunny in the summer. It's definitely better to wear light-colored clothes that reflect light in the summer. Dark clothes absorb more of the sun's light and make it unpleasantly hot.

Transparent objects like water or glass allow most of the light that strikes them to pass through. When the light enters the new medium, it will travel at a slower velocity than in a vacuum. The speed of light in medium depends on the optical density of the material. We give optical density a number called the **index of refraction**, n, and calculate it using this equation: $n = \dfrac{c}{v}$, where n is the index of refraction, c is the speed of light in a vacuum which is $c = 3.0 \times 10^8 \, \text{m/s}$, and v is the speed of light in the medium that the light is traveling through. Notice that $\dfrac{c}{v}$ is a ratio of velocities. This means that the index of refraction does not have any units. The larger the index of refraction, the slower the speed of light will be in the substance. The following table shows a list of common indices of refraction.

Material	Index of Refraction (*n*)
vacuum	1.00
air	1.0003
water	1.33
ethanol	1.36
acrylic	1.49
plate glass (window glass)	1.52
polycarbonate	1.60
flint glass	1.61
cubic zirconia	2.17
diamond	2.42
silicon	3.45

When light transmits into a new medium, its frequency remains the same. This means that if the light slows down, the wavelength must also get shorter: $v_{wave} \propto \lambda$. So, when light enters a window, the light slows down and the wavelength gets shorter but the frequency remains the same. When the light passes out the other side of the window, the light returns to its original speed and wavelength. We will discuss this in more detail in Chapter 18.

One last thing about light when it encounters a boundary. Conservation of energy must be obeyed! The amount of light that is reflected, absorbed, and transmitted must all add up to the original amount of light. So, when light hits a window, 80% of the light might transmit and 15% is reflected. That means that 5% of the light must have been absorbed.

BTW

Glass windows are transparent to visible light, but not to all wavelengths of light. Glass is mostly opaque to UV light, which is why you don't have to worry about getting a sunburn while driving a car unless you roll the window down. Glass is also mostly opaque to IR light as well. This is why your car gets so hot in the summer time. Visible light comes through the windows and is absorbed by the seats and dashboard, which in turn heat up. The IR light that is emitted by the seats is trapped inside the car. This heats up your car to an astounding temperature. This is an example of the greenhouse effect. This is the same thing that happens in the Earth's atmosphere due to carbon dioxide, water vapor, and methane gas.

Thin Film Interference

When you were a kid, did you ever blow bubbles? Did you notice the amazing purple, blue, yellow, and green colors in the bubble? These colors are a beautiful example of light interference as light reflects off the front and back surface of the bubbles' skin. You have also likely seen this effect if you've ever noticed a puddle in a parking lot. If a little bit of oil happens to drop on the puddle, the oil forms a very thin film on top of the water. White light from the sun reflecting off of the oil undergoes interference, and you see some of the component colors of the light enhanced.

How does **thin film interference** work? Take a look at the following figure. At the top surface, some light will be reflected and some will transmit through the film. The same thing will happen at the bottom surface: some light will be reflected back up through the film and some will keep on traveling out through the bottom surface. Notice that the two reflected light waves overlap; the wave that reflected off the top surface and the wave that traveled through the film and reflected off the bottom surface will interfere. Depending on the thickness of the film, the overlapping waves will line up to constructively or destructively interfere, enhancing or diminishing a particular color from the visible spectrum. This creates the vibrant colors that we see.

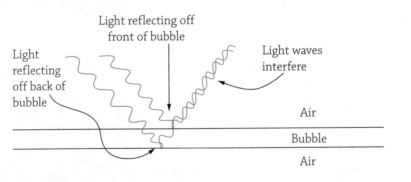

Light reflecting off front of bubble

Light reflecting off back of bubble

Light waves interfere

Air

Bubble

Air

 IRL　Most prescription glasses come with an anti-reflective coating. This is a thin layer of film on the surface that creates destructive interference for the reflected light from the surface of the lens. You should also make sure that your glasses have a UV filter that absorbs the UV light and protects your eyes from sunburn.

REVIEW QUESTIONS

Let's practice what we have learned about light by answering the following questions.

1. What type of wave is light?

2. How fast does light travel?

3. How do we know that light is a transverse wave?

4. How are EM waves produced?

5. Explain why you give off EM radiation.

Questions 6 to 7: the North Star is 323 light years from Earth.

6. How long does it take light from the North Star to get to Earth?

7. How far away is the North Star from Earth in meters?

8. Rank these in order of highest to lowest frequency: visible light, gamma ray, radio wave.

9. Rank these in order of longest to shortest wavelength: ultraviolet, X-ray, microwave.

10. Rank these in order of highest to lowest energy: infrared, visible light, ultraviolet.

11. How much light transmits through a polarizer, and what happens to the light that does not make it through the polarizer?

12. Why do waves bend around corners?

13. Why do we hear sound around corner of a doorway but can't see around the doorway?

14. How do we know that light has wave properties?

15. To measure the width of a human hair, you pull the hair tight and shine a 650 nm red laser past it. The hair acts like a double slit because it blocks the laser in the middle and lets the light on the sides pass by. This creates a diffraction pattern on a screen with the angle between the central maxima and the first-order maxima of 7.5°. Calculate the width of the hair.

16. A pair of sunglasses absorbs 55% of the light that hits them and transmits 40% of the light. How much of the incident light is reflected?

17. You can see your reflection in a freshly polished car. What does that tell you about the surface of the car?

18. In which substance does light travel fastest: water or glass?

19. What is the speed of light in polycarbonate?

20. What happens to the wavelength, frequency, and speed of light when it travels from air into plate glass?

21. Why do we see vibrant colors in soap bubbles?

17 Reflection and Mirrors

MUST ⚡ KNOW

⚡ Mirrors reflect light and can form images.

⚡ When light reflects off a surface, it always reflects off at the same angle it hit the surface.

⚡ The curvature and focal length of a mirror determine what type of image a mirror will form.

ou are so familiar with mirrors, you probably don't even think about them. Light reflects off a window or a pond. You see yourself every day in the bathroom mirror. You use them when you drive . . . or at least I hope you check your mirrors when you drive. Mirrors are all around us.

Remember from Chapter 16 that when light encounters a boundary, or change in medium, that three things can happen: reflection, transmission, and absorption. In this chapter, our goal is to understand reflection in more detail. We'll talk about how a mirror reflects light to form an image that our eyes can see and how these images change depending on the contours of the mirror. Note that we will be concentrating on visible light in this chapter, but everything we learn also applies to other waves: radio waves, microwaves, infrared, even sound waves reflect just like visible light does. In Chapter 18, we'll discuss what happens to the light that transmits through the new medium.

The Law of Reflection

Okay, physics fans! Here comes the simplest physics equation in the entire book. The law of reflection is $\theta_{incidence} = \theta_{reflection}$. (It's nice to have an easy one once in a while!) The law of reflection simply says that light rays always reflect off at the same angle they strike the surface. In the figure on the next page, we see a light wave hitting a flat surface and reflecting off. In physics, we always measure our angles from the dashed line, which is called a **normal line**. A normal line is just a perpendicular line to the surface. The angle between the incoming ray, called the **incident ray**, and the normal line is the **angle of incidence**. The angle between the normal line and the outgoing ray, called the **reflected ray**, is known as the **angle of reflection**.

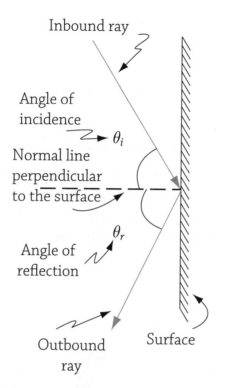

Inbound ray

Angle of
incidence
θ_i

Normal line
perpendicular
to the surface

θ_r

Angle of
reflection

Outbound
ray

Surface

BTW

*Why measure angles
from the normal line?
Why not just measure
between the ray and
the surface? Well, you
will see in just a bit that
not all surfaces are flat,
so we need a reliable
reference line to
measure angles from.
The normal line is an
easy reference line to
draw because it is the
same as the radius line
of a curved surface.*

BTW

*The terminology of
incident and reflected
rays and how the
angles are measured
are actually much
harder to learn than the
actual concept. But it is
important to learn the
language so that there
is no confusion.*

In the last chapter, we talked about two types of
reflection: regular and diffuse:

Regular reflection   Diffuse reflection

This time I have drawn in the normal line for each ray so that you can
see why a bumpy surface produces a diffuse reflection. See how the normal
lines are all parallel for a regular/smooth surface? This will produce a nice

regular parallel ray reflection. Diffuse reflections occur because the normal lines for the surface are pointing in different directions, causing the reflected light to scatter from the surface in nonparallel directions. For the rest of this chapter, we are going to assume that our mirrors are nice and smooth with regular reflections.

▶ In the following figure, you see two mirrors situated so that they create a 90° angled corner. Complete the path of the light ray as it strikes both mirrors. Be sure to draw a normal line when the light strikes the mirror and label the angle of incidence and reflection.

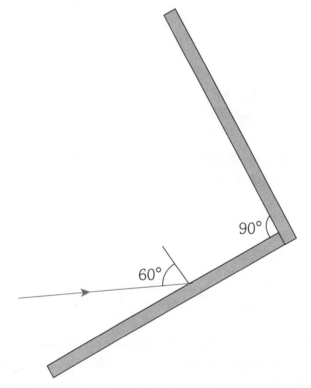

▶ The angle of incidence equals the angle of reflection. Remember from geometry class that complementary angles add up to equal 90° and that the angles of a triangle always add up to 180° to help you determine the path of the light ray. (See the next figure.) I have labeled the light ray angles and the angles in the triangle that is formed in the corner of the mirror for reference.

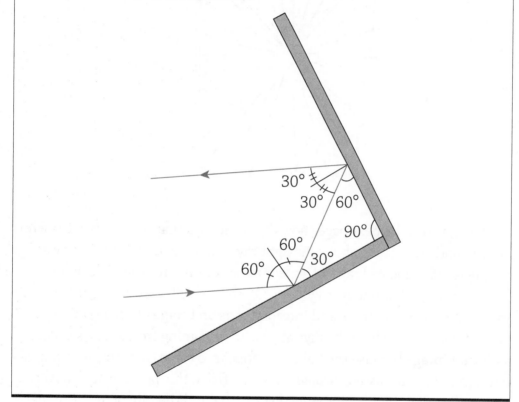

Plane Mirrors

The simplest mirror is a flat mirror, like the one on your bathroom wall, called a **plane mirror**. In the following figure, you see a lightbulb in front of a plane mirror. The light rays from the bulb travel toward the mirror. Some

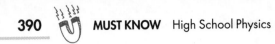

hit the mirror and some don't. The rays that strike the mirror reflect off with the same angle that they hit the mirror.

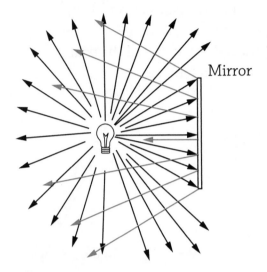

The figure on the next page shows only the rays that reflect off the mirror. Do you notice anything about the light rays? Imagine that the mirror was not there. The reflected rays all appear to be coming from a single point. That is exactly what our eyes will see. Our brains will take the light ray information from our eyes and interpret it as an **image** of the lightbulb behind the mirror where the rays appear to be coming from. This is called a **virtual image** because the light rays appear to come from this location, but never actually passed through, or came from, this point. Notice how the image of the lightbulb is the same distance behind the mirror as the actual lightbulb is in front of the mirror.

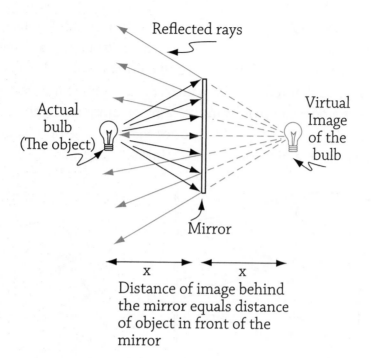

Reflected rays

Actual
bulb
(The object)

Virtual
Image
of the
bulb

Mirror

x   x

Distance of image behind
the mirror equals distance
of object in front of the
mirror

BTW

If we cover up half the mirror, there will still be light from the bulb hitting the uncovered part of the mirror. So, the image will still form—it just won't be as bright because fewer rays bounce off the mirror to form the image.

 IRL Stand in front of your bathroom mirror, reaching out to touch the mirror's surface with one of your hands. Now, look at your mirror-image self. It's reaching forward in the opposite direction to touch the mirror because it is exactly the same distance behind the mirror that you are in front of the mirror.

BTW

Virtual images cannot be projected onto a screen, because the light that forms the image does not actually pass through the image location. Therefore, a screen placed at the image location won't have any light falling on it.

Now let's take a look at a more complex object in front of a plane mirror. (See the figure below.) Light is coming from all parts of the face and striking the mirror. Let's look at the hair, nose, and chin of the face. I've drawn the light rays for each that reflect off the mirror and enter your eyes. Notice how the image is reproduced **upright** in the same **orientation** as the object and has the same size as the object. Also notice that the nose of the person is closer to the mirror, and the same is true in the image.

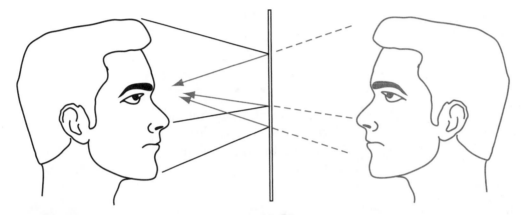

 **IRL** Most people think that their bathroom mirror flips their image left to right. This is not true. When you raise your right hand to wave at your image, your image also raises its hand that is on the right side. (If the mirror actually flipped your image, the hand on the opposite side of the image would wave back at you. Not only that, you would be looking at your feet instead of your face!) Your nose is the closest thing to the mirror, and your ears are farther away. The mirror reproduces this the same way, but on the opposite side of the mirror. So, your reflection is inverted front to back.

So, to recap: Plane mirrors produce virtual, upright images of the same size as the object that are located the same distance behind the mirror that the object is in front of the mirror.

EXAMPLE

▶ You stand 2.5 m in front of your bathroom mirror: How far is your image away from you?

▶ Remember that the image is located the same distance behind the mirror that the object is in front of the mirror. Your image will be a total of 5.0 m away from you.

▶ How tall is your image? The same height as you are.

▶ Is your image virtual or real? Virtual.

▶ What is the orientation of your image: upright or inverted? Upright.

Now let's try drawing the image produced by a plane mirror.

EXAMPLE

▶ The following figure shows the letter "F" in front of a mirror. Sketch the image of the letter.

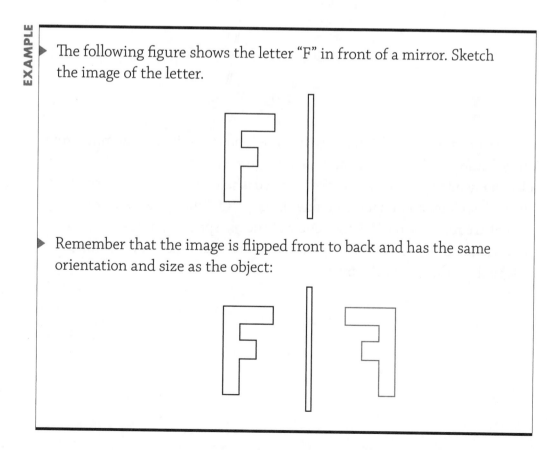

▶ Remember that the image is flipped front to back and has the same orientation and size as the object:

Convex Mirrors

Not all mirrors are flat. Many are curved. We are going to take a look at spherically curved mirrors. Imagine a basketball with a section sliced off to form a smooth bowl shape, like a contact lens. Now take this bowl shape and flip it over so that it bends out toward you, and you have the shape of a **convex mirror**. The following figure shows a convex mirror with an object—a flower—in front of it. Notice how the mirror bends out toward the object/flower.

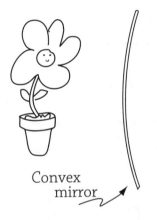

Convex mirror

Now imagine parallel light rays, like the kind from the Sun, coming from the left and striking the mirror. (See the figure on the next page.) Every place a ray hits the mirror, I have drawn a dashed normal line. Notice how the normal lines all radiate from one single point. This is the **center of curvature** for the lens. It is the center of the big sphere that the lens is a part of. Therefore, a line drawn from the center of curvature to the mirror will be the radius, R, of the mirror.

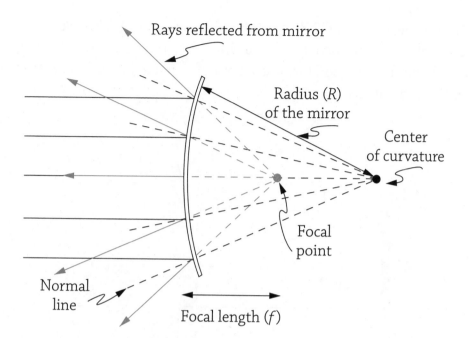

Rays reflected from mirror

Radius (R) of the mirror

Center of curvature

Focal point

Normal line

Focal length (f)

Now look at how parallel light rays coming from the left reflect back off the mirror. Notice how they all appear to radiate from a point halfway between the center of curvature and the mirror. This spot is called the **focus**, *f*. This is called a **virtual focus** because the light rays that reflect off the mirror appear to come from this point but do not actually pass through the point. The distance from the mirror to the **focal point** is called the **focal length**, and it will always be half of the radius: $f = \dfrac{R}{2}$. A convex mirror is also called a **diverging mirror** because the light that falls on it will be spread out, or diverges, as if it came from the virtual focus.

Let's find out what kind of image a convex mirror will form. The following figure shows an arrow as the object in front of the mirror. Two easy rays can be drawn to find out where the image will form:

Ray 1   A light ray traveling horizontally toward the mirror and reflecting off as if it came from the focal point, just as in the previous figure.

Ray 2   A light ray drawn toward the focus and reflecting off horizontally from the mirror. This light ray is the opposite of ray 1.

Notice how rays 1 and 2 diverge off the mirror. If we backtrack to see where they appear to come from, you will see that the two rays diverge from a point behind the mirror, forming an upright virtual image that is smaller than the original object.

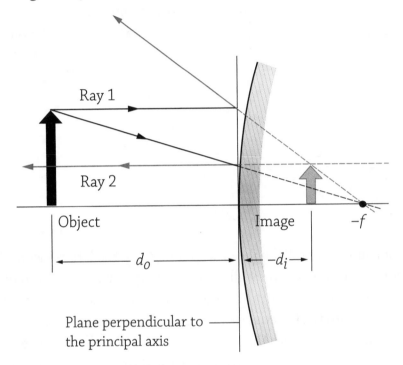

Convex, or diverging, mirrors always form the same type of image: upright, virtual, and smaller than the original object.

 **IRL** Convex mirrors are used in stores to allow customers and employees to see around the corners of the aisles to avoid collisions. We also have them on the outside of our cars on the passenger side of the vehicle. It's the one that says "Objects are closer than they appear," because images in convex mirrors are always smaller than the object.

Mirror and Magnification Equations: Part 1

The location and size of images formed by mirrors can be found using two equations. **The mirror equation** helps us find the location of the image: $\frac{1}{f} = \frac{1}{d_i} + \frac{1}{d_o}$. f is the focal length of the mirror, d_o is the distance from the mirror to the object, and d_i is the distance from the mirror to the image.

There are some sign conventions we need to talk about in order to use this equation correctly:

- $+f$ is for a real focal length. **Concave** mirrors, which we will discuss next, have a real focal point.

- $-f$ is for a virtual focal length. **Convex** mirrors have a virtual focus because light diverges from this point but does not actually pass through the focus, as seen earlier.

- $+d$ designates real distances. For instance, objects always have real distances because light actually comes from the object toward the mirror. (We haven't seen any real images yet, but they are coming soon.)

- $-d$ designates virtual distances. Look back at the previous figure. See how the image is behind the mirror. The light rays that form the image diverge as if they came from the image location, but actually don't. Therefore, this is a virtual image.

The magnification equation helps us determine the size of the image: $M = \frac{h_i}{h_o} = -\frac{d_i}{d_o}$. M is the magnification of the image, h_i is the height of the image, and h_o is the height of the object.

The sign conventions for the magnification are:

- $+M$ indicates that the image is upright.

- $-M$ indicates that the image is inverted.

- When M is greater than 1, the image is enlarged. When M is less than 1, the image is smaller than the object. If M is equal to 1, the image is the same size as the object.

- h_o is always positive because we always assume that the object is upright.

- $+h_i$ means the image is upright, and $-h_i$ tells us the image is flipped over.

- $+d$ designates real image distances, and $-d$ designates virtual image distances, just as with the mirror equation.

Here is an example.

EXAMPLE

The following figure shows a toy car placed in front of a convex mirror with a radius of 30 cm.

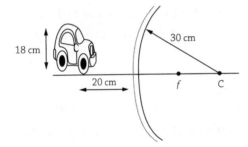

First let's calculate the focal length of the mirror. Remember that $f = \dfrac{R}{2}$.

$$f = \frac{R}{2} = \frac{30\,\text{cm}}{2} = 15\,\text{cm}$$

Remember that this is a virtual focus. Therefore,

$$f = -15\ \text{cm}$$

▶ Now use the mirror equation to find the location of the image:

$$\frac{1}{f} = \frac{1}{d_i} + \frac{1}{d_o}$$

$$\frac{1}{-15\,\text{cm}} = \frac{1}{d_i} + \frac{1}{20\,\text{cm}}$$

$$d_i = -8.6\,\text{cm}$$

▶ The negative image distance tells us that the image is a virtual image located behind the mirror. Now calculate the height and magnification of the image:

$$M = \frac{h_i}{h_o} = -\frac{d_i}{d_o}$$

$$M = \frac{h_i}{18\,\text{cm}} = -\frac{(-8.6\,\text{cm})}{20\,\text{cm}}$$

$$M = 0.43 \quad \text{and} \quad h_i = 7.7\,\text{cm}$$

▶ The image will be 7.7 cm tall, which is only 43 percent as tall as the object. Both magnification and image height are positive, indicating that the image will be upright.

Concave Mirrors

So far, all of our mirrors have only produced a single type of image. Not so with the *concave* mirror! This one produces a variety of images, depending on where the object is placed in relation to the mirror. Let's start with the basics first. Look at the following figure of a concave mirror with the person in front of it. Notice how the concave mirror bends away from the person.

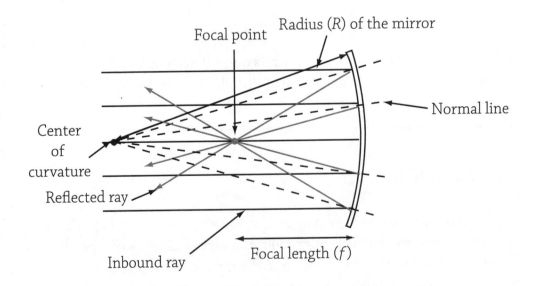

Again, imagine parallel light rays, like the ones from the Sun, coming from the left and striking the mirror. (See the next figure.) Every place a ray hits the mirror, I have drawn a dashed normal line. Notice again how the normal lines all radiate from one single point called the center of curvature of the lens. Remember that it is the center of the big sphere that the lens is a part of and that a line drawn from the center of curvature to the mirror will be the radius R of the mirror. But this time, the center of curvature is in front of the mirror on the reflective side.

IRL   Satellite dishes use concave metal mirrors to reflect radio waves from satellites way up in orbit to a receiver at the focal point. Concave mirrors can also be used to concentrate sound waves to a microphone at the focus to hear sounds from very far away, just like in spy movies.

Now look at the reflected rays that bounce off the mirror. Notice how they all converge on a single point halfway between the center of curvature and the mirror called the focus, f. This is a **real focus** because the light rays that reflect off the mirror actually pass through this point. The distance from the mirror to the focal point is called the focal length, and it will always be half of the radius: $f = \dfrac{R}{2}$. A concave mirror is also called a **converging mirror** because the light that falls on it will converge, or meet and cross paths, at the real focus.

Let's find out what kind of images a concave mirror produces. The following figure shows an arrow as the object. Two easy rays can be drawn to find out where the image will form:

Ray 1   A light ray traveling horizontally parallel to the **principal axis** toward the mirror and reflecting off and passing through the focal point, just like in the previous figure.

Ray 2   A light ray drawn toward and then through the focus will reflect off the mirror horizontally. This light ray is the opposite of ray 1.

Notice how rays 1 and 2 converge after reflecting off the mirror. This is the point where the image will form. Notice that the image is inverted and smaller than the original object. The image is also **real**. This means that the light reflecting off the mirror actually converges at and passes through the image location. If we place a piece of paper at the image location, the image will appear on the screen. Every time the object is out beyond the center of curvature of a concave mirror, the image will be inverted, smaller, and real.

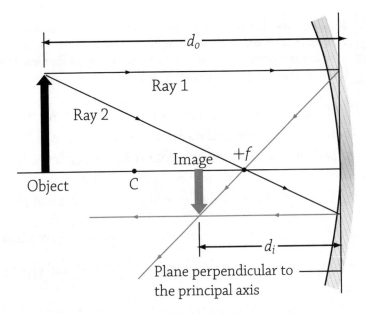

Ray 1

Ray 2

Image $+f$

Object C

d_o

d_i

Plane perpendicular to
the principal axis

IRL Several power plants around the world use hundreds of converging mirrors to focus the light of the sun onto a heat collector to boil water—creating steam—to turn a turbine that rotates magnets past wires to produce electricity by electromagnetic induction, like we learned in Chapter 14.

Now let's move the object a little closer to the mirror and place it exactly on the center of curvature. Notice that our singing stick figure's image is exactly the same size and location as the original object. (Look at the figure on the next page.) The only difference is that the image is inverted. Can you guess what type of image this is? Is it virtual or real? Well, the light rays that bounce off the mirror converge to the location of the image. If we put a screen at the image location, the light would actually fall on the screen to form an image. This must be a real image.

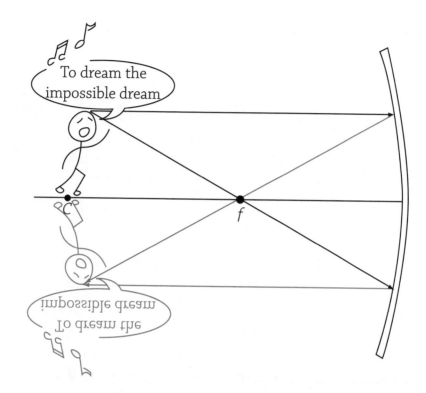

Now, let's move the object between the center of curvature and the focal point. (See the next figure.) This time the image is enlarged, inverted, and real. This type of mirror can be used to project large images on a screen.

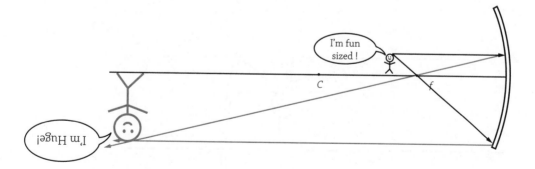

As we continue to move the object toward the mirror, something strange occurs. When the image is placed on the focal point, there is no image. All of the light rays that reflect off the mirror bounce off parallel and never converge to make an image. This is the only location in front of the mirror that images will not form:

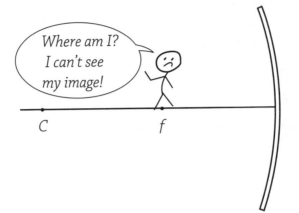

Finally, when the object is placed between the focal point and the mirror, the light rays that bounce off the mirror appear to be diverging from a point behind the mirror. (When you backtrack the reflected rays that bounce off the mirror, like I have done in the following figure, you will find where the image forms.) This is the hallmark of a virtual image. The virtual image is upright and larger than the object.

Some of you may be familiar with this type of mirror. It is commonly referred to as a makeup mirror. It is used to magnify your face for closeup detail.

 IRL Dentists can use this type of mirror to enlarge your teeth for a closer look at that cavity you have.

Mirror and Magnification Equations: Part 2

The equations and sign conventions we learned earlier are good for both concave and convex mirrors, so next I'll list the equations and sign conventions again for your convenience, and then we'll go through a couple of examples.

The **mirror equation** helps us find the location of the image:

$\frac{1}{f} = \frac{1}{d_i} + \frac{1}{d_o}$. f is the focal length of the mirror, d_o is the distance from the mirror to the object, and d_i is the distance from the mirror to the image.

Here are the sign conventions we need in order to use this equation correctly:

- $+f$ is for a real focal length. Concave mirrors have a real focal point because light actually converges to the focal point and passes through it after it bounces off the mirror.

- $-f$ is for a virtual focal length. Convex mirrors have a virtual focus because light diverges from this point but does not actually pass through the focus, as seen earlier.

- $+d$ designates real distances. For instance, objects always have real distances because light actually comes from the object toward the mirror. Real images form when light from the mirror converges to and passes through the image location.

- −d designates virtual distances. The light rays that form virtual images diverge, as if they came from the image location, but actually don't. Therefore, the image is virtual.

The **magnification equation** helps us determine the size of the image:

$M = \dfrac{h_i}{h_o} = -\dfrac{d_i}{d_o}$. M is the magnification of the image, h_i is the height of the image and h_o is the height of the object.

The sign conventions for the magnification are

- +M indicates that the image is upright.

- −M indicates that the image is inverted.

- When M is greater than 1, the image is enlarged. When M is less than 1, the image is smaller than the object. If M is equal to 1, the image is the same size as the object.

- h_o is always positive because we always assume that the object is upright.

- +h_i means the image is upright, and −h_i tells us the image is flipped over.

- +d designates real distances, and −d designates virtual distances, just as with the mirror equation.

Let's look at an example:

EXAMPLE

▶ You have not been brushing your teeth, so now a dentist is using a concave lens to examine a cavity in your teeth. The dentist holds the mirror 2.5 cm from your tooth, and the mirror produces an image that is 20 percent bigger than your tooth. What information did I give you, and what can we find out from this information?

▶ Your tooth is the object, and it is 2.5 cm from the mirror: $d_o = 2.5$ cm. The image is 20 percent bigger than the tooth, of a magnification: $M = 1.2$. Since magnification is positive, the image is upright.

▶ From this information we can determine the image distance using the magnification equation:

$$M = -\frac{d_i}{d_o}$$

$$1.2 = -\frac{d_i}{2.5\,\text{cm}}$$

$$d_i = -3.0\,\text{cm}$$

▶ Why is the image distance negative? This means the image is virtual and is located behind the mirror.

▶ Now we can use the mirror equation to find the focal length of the mirror. Remember that your image distance is negative:

$$\frac{1}{f} = \frac{1}{d_i} + \frac{1}{d_o} = \frac{1}{-3.0\,\text{cm}} + \frac{1}{2.5\,\text{cm}}$$

$$f = 15\,\text{cm}$$

▶ The focal length is positive. This tells us that this is a concave mirror.

Let's try another one:

▶ A 25-cm-tall puppy sits 2.0 m from a concave mirror with a radius of curvature of 1.0 m. Let's determine where the image shows up and what kind of image it is.

▶ First off, we need to take care of our units! Not all of the units are the same. I'm going to convert the dog's height to meters: 0.25 m. (You could have also converted everything to centimeters if you wanted to. It does not really matter—we just need all the units to be the same in the equation.) Second, we know that the focal length is half the radius of the mirror. Therefore, $f = 0.5$ m.

▶ Using the mirror equation:

$$\frac{1}{f} = \frac{1}{d_i} + \frac{1}{d_o}$$

$$\frac{1}{0.50\,\text{m}} = \frac{1}{d_i} + \frac{1}{2.0\,\text{m}}$$

$$d_i = 0.67\text{ m}$$

▶ Since the image distance is positive, the image will be real and located in front of the mirror. Now let's use the magnification equation to find the height of the puppy's image.

$$M = \frac{h_i}{h_o} = -\frac{d_i}{d_o}$$

$$M = \frac{h_i}{0.25\,\text{m}} = -\frac{0.67\,\text{m}}{2.0\,\text{m}}$$

$$h_i = -0.084\text{ m}$$

$$M = -0.33$$

▶ The image of the puppy is one-third the size and inverted.

REVIEW QUESTIONS

Let's demonstrate what we have learned about mirrors by answering the following questions.

1. Label the angle of incidence and angle of reflection in the figure provided.

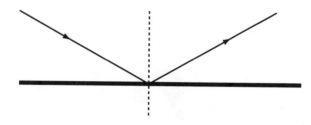

2. An object is placed in front of a mirror, as shown in the following figure. Draw light rays to show where the "disembodied eye of science" will see the image.

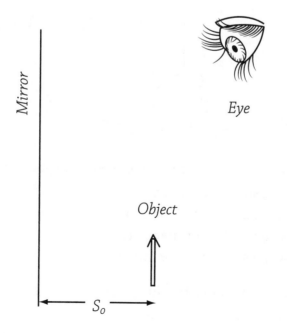

3. A cat is looking at the mouse in a mirror. Draw two rays showing where the image of the mouse would appear to the cat.

Plane mirror

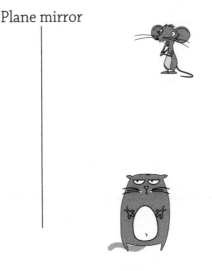

4. A mirror with a focal length of 25 cm is lying on a table, as shown in the following figure. An itsy-bitsy spider is suspended 10 cm above the mirror by a web.

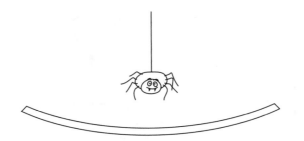

Looking down at the mirror from above, where will you see the image?

a. 7.1 cm above the surface of the mirror

b. 8.3 cm above the surface of the mirror

c. 12.5 cm below the surface of the mirror

d. 16.7 cm below the surface of the mirror

5. What type of mirror produces real images?

6. Which type of mirror can produce images that can be projected on a screen?

7. Which type of mirror produces upright images?

8. The next figure shows three mirrors with objects in front of them. Draw rays from the object to the mirrors to find out where the image will be located. Label each image with its orientation and size, and indicate if the image is real or virtual. Careful! I mixed up the types of mirrors and where the objects are located. Hint: Identify if the mirror is concave or convex before you begin.

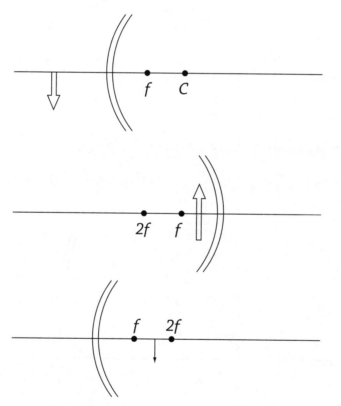

The following figure shows a flower in front of a convex mirror with a radius of 40 cm.

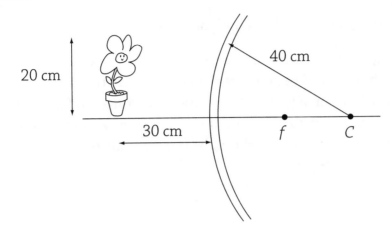

9. Calculate the focal length of the mirror.

10. Calculate the location of the image.

11. Calculate the height and magnification of the image.

A frightened bunny is 1.5 m from a mirror with a focal length of 1.0 m, as seen in the next figure.

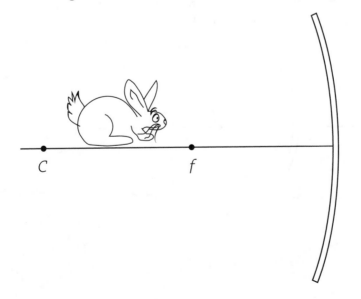

12. Calculate the location of the bunny's image.

13. How does the size of the image compare to the bunny?

14. Describe the image.

Refraction and Lenses

 Light changes speed when moving from one transparent medium to another.

 When light changes speed, it also refracts—changes direction—depending on whether the light speeds up or slows down.

 Lenses refract light to form images, just like mirrors do.

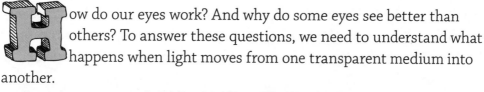ow do our eyes work? And why do some eyes see better than others? To answer these questions, we need to understand what happens when light moves from one transparent medium into another.

Once again, remember from Chapter 16 that when light encounters a boundary, or change in medium, that *three things can happen:* reflection, transmission, and/or absorption. In the last chapter, we concentrated on what happens to the reflected part of the light waves. In this chapter, our goal is to investigate what happens to the portion of the light that transmits into the new medium. We'll talk about how light changes direction as it changes speed. We'll also see how images can change depending on the shape of a lens. There will be a lot of similarities between the images a mirror forms and the ones produced by lenses.

Index of Refraction

Light, or any electromagnetic wave, travels at a speed of $c = 3.0 \times 10^8$ m/s. But it only travels at this speed through a vacuum, when there aren't any pesky molecules to get in the way. When it travels through anything other than a vacuum, light slows down due to the interactions it has with the molecules that are in the way. The amount by which light slows down in a substance is called the material's **index of refraction** (n). The index of refraction for any material can be calculated using this equation: $n = \dfrac{c}{v}$.

This equation says that the index of refraction of a certain material, n, equals the speed of light in a vacuum, c, divided by the speed of light through that material, v. Notice that the index of refraction does not have any units because it is a ratio.

For example, the index of refraction of window glass is about 1.5. This means that light travels 1.5 times faster through a vacuum than it does through glass. The index of refraction of air is approximately 1. Light travels through air at just about the same speed as it travels through a vacuum. The following table gives a list of indices of refraction for common materials.

Material	n
vacuum	1.00
air	1.0003
water	1.33
ethanol	1.36
acrylic	1.49
window glass	1.52
polycarbonate	1.60
flint glass	1.61 avg
cubic zirconia	2.17 avg
diamond	2.42

When the speed of light changes in the new medium, the wavelength, λ, also changes, but the frequency, f, of the wave remains the same. When light waves go from a medium with a low index of refraction to one with a high index of refraction, they get squished together. So, when light with a wavelength of 500 nm traveling through air ($n_{air} = 1.0003$) enters water ($n_{water} = 1.33$) and then emerges back out into air again, the waves would look like the following figure.

BTW

Air has an index of refraction of 1.0003. This means that light travels almost as fast in air as in a vacuum. So, we usually just assume that the speed of light in air is 3.0×10^8 m/s.

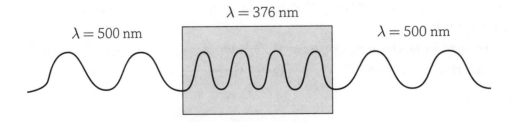

$\lambda = 500$ nm $\qquad \lambda = 376$ nm $\qquad \lambda = 500$ nm

The equation that goes along with this change in wavelength is $n = \dfrac{\lambda}{\lambda_n}$. In this equation, λ_n is the wavelength of the light traveling through the transparent medium (like water, in the figure shown earlier), λ is the wavelength in a vacuum, and n is the index of refraction of the transparent medium. It is important to note once again that even though the wavelength of light changes as it goes from one material to another, its frequency remains constant. The frequency of light is a property of the photons that comprise it (more about that in Chapter 19), and the frequency doesn't change when light slows down or speeds up.

BTW

Notice in the figure from the previous page how the light speeds back up once it leaves the water! Remember that the medium determines the speed of the wave. So, once the light emerges out of the water, the wavelength returns to 500 nm, and the light *instantaneously* speeds back up to its original velocity. This is a common property of all waves.

EXAMPLE

▶ You shine a 650 nm laser pointer at a window. What are the speed and wavelength of the laser before it strikes the glass?

▶ The wavelength of the laser was given: 650 nm. Now let's make the reasonable assumption that the laser is traveling through air before it strikes the glass and calculate the speed of the laser in the air.

$$n = \frac{c}{v}$$

$$v = \frac{c}{n} = \frac{3.0 \times 10^8 \, \text{m/s}}{1.0003} = 2.999 \times 10^8 \, \text{m/s} \approx 3.0 \times 10^8 \, \text{m/s}$$

▶ Notice that the speed of light in air is nearly the same as the speed of light in a vacuum. For this reason, we usually just assume that the speed of light in air is the same as in a vacuum.

▶ What is the speed of the laser light inside the glass window?

▶ Use the index of refraction equation:

$$n = \frac{c}{v}$$

$$v = \frac{c}{n} = \frac{3.0 \times 10^8 \,\text{m/s}}{1.52} = 1.97 \times 10^8 \,\text{m/s}$$

▶ What is the wavelength of the light inside the window?

▶ Use the wavelength equation:

$$n = \frac{\lambda}{\lambda_n}$$

$$\lambda_n = \frac{\lambda}{n} = \frac{650 \,\text{nm}}{1.52} = 428 \,\text{nm}$$

▶ What are the wavelength and speed of the laser light when it exits out the other side of the window?

▶ Since the laser has traveled back into air, both the wavelength and speed of light return to their original values of 650 nm and $c = 3.0 \times 10^8 \,\text{m/s}$.

Refraction

In addition to changing speed and its wavelength, light can change its direction when it travels from one medium to another. The way in which light changes its direction is described by Snell's law: $n_1 \sin \theta_1 = n_2 \sin \theta_2$, where n_1 and n_2 are the indices of refraction of the materials from which the light is coming from and going into. θ_1 and θ_2 are the angles of the light in the material the light is coming from and going into. To understand Snell's law, it's easiest to see it in action. The next example should help.

▶ In the following figure, a ray of light is going from air and moving into water. The dotted line perpendicular to the surface is the normal line and is the reference line we use to measure our angles for Snell's law.

▶ As the light ray enters the water, it is bent, or changes direction, toward the normal. The angles θ_1 and θ_2 are marked on the figure, and the index of refraction of each material, n_1 and n_2, is also indicated.

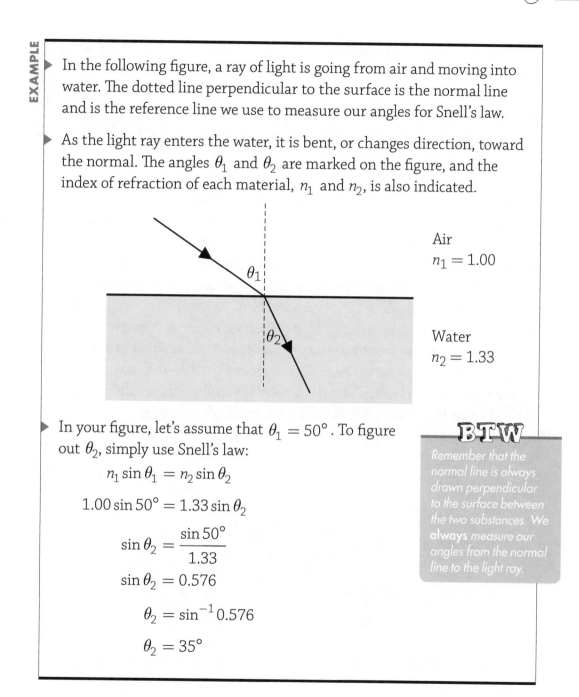

Air
$n_1 = 1.00$

Water
$n_2 = 1.33$

▶ In your figure, let's assume that $\theta_1 = 50°$. To figure out θ_2, simply use Snell's law:

$$n_1 \sin \theta_1 = n_2 \sin \theta_2$$

$$1.00 \sin 50° = 1.33 \sin \theta_2$$

$$\sin \theta_2 = \frac{\sin 50°}{1.33}$$

$$\sin \theta_2 = 0.576$$

$$\theta_2 = \sin^{-1} 0.576$$

$$\theta_2 = 35°$$

BTW

*Remember that the normal line is always drawn perpendicular to the surface between the two substances. We **always** measure our angles from the normal line to the light ray.*

BTW

Since we are using degrees, you need to make sure that your calculator is in degree mode.

Whenever light goes from a medium with a lower index of refraction to one with a higher index of refraction, as in the example provided earlier, the ray bends toward the normal. Whenever light goes in the opposite direction from a medium with a larger index of refraction to one with a smaller index of refraction, say, from water into air, the ray bends away from the normal.

But why, you may be asking, does light change directions just because it moves into a new material? The point source model of waves (Huygens' principle) tells us why. In the next figure, the waves moving through medium 1 are traveling faster than the waves traveling in medium 2.

Using the endpoints of the wave front as point sources, we can draw the two waves that are produced. Notice how the wave in medium 1 has a longer wavelength because it is moving faster. When we connect these two point sources, the new wave front has changed direction. It's heading in a new direction closer to the normal line. This happens because the wave in the second medium slows down and has a shorter wavelength, resulting in an angle of refraction that is smaller than the angle of incidence.

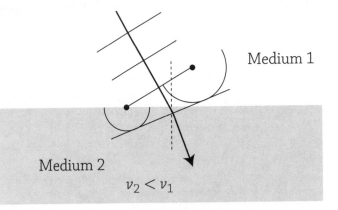

Medium 1

Medium 2

$v_2 < v_1$

Now let's reverse the situation. In the figure on the next page, medium 1 is the slow material and medium 2 is the fast material. The effect is reversed. When the wave speeds up, it will turn away from the normal. It has to.

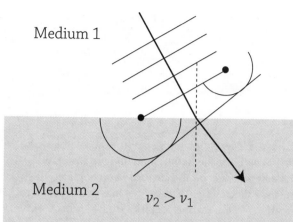

Medium 1

Medium 2 $v_2 > v_1$

Here is another way to remember the "bending" rule. A bigger velocity of light means a bigger angle with the normal. A smaller velocity of light means a smaller angle with the normal. So, just by looking at the figure, we can tell which medium is the slower medium. It is always the one where the light is closer to the normal. There is only one exception to this rule. If light hits the new medium head on (angle of incidence of 0°), it won't change direction. It just plows straight into the new medium with no turning.

Let's practice with another example.

EXAMPLE

▶ A green laser pointer sends a beam of light from air to ethanol. Which way does the green beam bend?

▶ The light is going from fast ($n_{air} = 1$) to slow ($n_{ethanol} = 1.36$), so it will bend toward the normal line.

▶ If the angle of incidence is 60°, what is the angle of refraction?

▶ Use Snell's law:

$$n_1 \sin \theta_1 = n_2 \sin \theta_2$$

$$1.00 \sin 60° = 1.36 \sin \theta_2$$

$$\theta_2 = 40°$$

BTW

The bigger the change in speed when moving from one medium to the next, the bigger the refractive bending. This means that the bigger the difference in index of refraction there is, the more the light will bend and change direction when entering a new medium.

IRL   If you have a laser pointer, try shining it into some water with just a tiny bit of milk in it. You'll see the beam bend into the water. But you'll also see a little bit of the light reflect off the surface at an angle equal to the initial angle. (Be careful the reflected light doesn't get into your eye!) In fact, at a surface, if light is refracted into the second material, some light must always be reflected.

Total Internal Reflection

Sometimes, when light goes from a medium with a high index of refraction to one with a low index of refraction, we can get **total internal reflection**. This means that none of the light escapes the first medium. All the light is trapped, and 100 percent of the light simply reflects off the surface without any of it passing into the second medium. The surface acts like a perfect mirror. For total internal reflection to occur, the light ray must be traveling in a slower substance toward a faster substance and be at an angle beyond the **critical angle**. The critical angle $\theta_{critical}$ is the angle past which rays cannot be transmitted from one material to another: $\sin \theta_{critical} = \dfrac{n_2}{n_1}$, where n_1 and n_2 are the indices of refraction of the first and second mediums.

EXAMPLE

> The following figure shows a block of window glass surrounded by air. Let's calculate the critical angle of the glass:
>
> $$\sin \theta_{critical} = \frac{n_2}{n_1} = \frac{1.00}{1.52} = 0.658$$

$$\theta_{critical} = \sin^{-1} 0.658 = 41°$$

▶ In this figure, a ray of light is shining up through a window glass block, attempting to exit out the top surface of the glass. Since the angle of the incident ray is greater than the critical angle, the light cannot be transmitted into the air. Instead, all of it reflects inside the glass. Total internal reflection occurs anytime light strikes a surface with an angle of incidence greater than the critical angle.

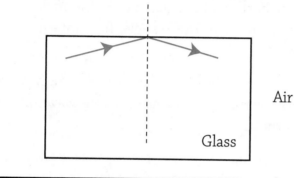

The next diagram shows a laser beam coming from the upper left, shining on a glass block.

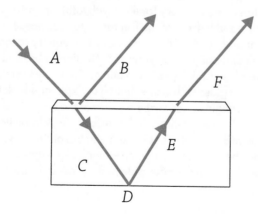

BTW

We only have a critical angle when light goes from a slower substance into a faster substance. When going from a fast to a slow medium, the light bends toward the normal, and some of the light always enters into the second medium.

This diagram lets us see all the behaviors of light in one image:

- First, the light strikes the surface at point *A*. Part of the light reflects, *B*, and part transmits, *C*. The reflected light takes off at an angle equal to the angle of incidence.

- The transmitted wave, *C*, refracts toward the normal because it is slowing down.

- None of the wave exits the bottom of the glass block. The angle of incidence must be larger than the critical angle. Total internal reflection occurs at point *D*.

- After reflecting off the bottom of the glass, the wave, *E*, strikes the top of the glass and refracts away from the normal, *F*, as the wave speeds back up when it reenters the air.

Note that when ray *E* strikes the upper surface, part of the light would probably reflect again. However, for each successive interface the light strikes, there is less and less of the original light left, and the ray gets dimmer and dimmer.

IRL If you have a fish tank, you are very familiar with total internal reflection because, when you stand in front of the fish tank, you can see straight through the glass and water and out the other side. But when you try to look through the front of the tank and out the side you can't. The light is beyond the critical angle, and the side of the tank looks like a mirror. For those that don't have a fish tank, fill up a clear glass with water, hold it up high, and look through the bottom of the glass straight up toward the water at the top. You should have no trouble seeing out the water through the surface. Now look through the side of the glass toward the water's surface at the top of the glass. You should see a silvery surface because the top of the water has become a mirror. You can even place your hand on the other side of the glass and see its reflection.

> **IRL** | One of the uses for total internal reflection is fiber optics. A light beam can be shot into one end of a narrow tube of plastic called an optical fiber. When the light hits the sides of the tube, it is beyond the critical angle and continues to bounce down the tube until it reaches the other end. We can use this to transport massive amounts of information at great speeds.

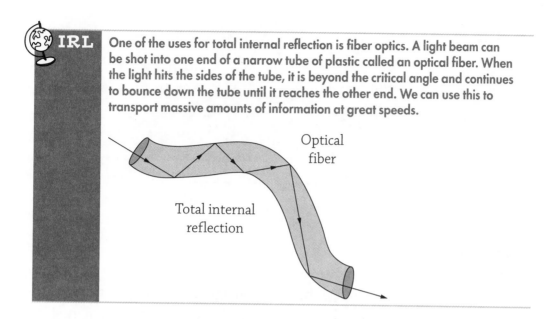

Optical fiber

Total internal reflection

Dispersion

Okay, I lied to you—but just a little! The indices of refraction that I gave you in the table earlier in the chapter are actually averages for visible light. It turns out that the index of refraction is a little different for each frequency of light. For instance, in window glass, the real index of refraction for red light with a wavelength of 700 nm is more like $n_{red} = 1.51$, and for violet light with a wavelength of 400 nm it is about $n_{violet} = 1.53$. In the index of refraction table you will see an average for visible light of $n = 1.52$. So, what's the big deal? Well, different indices of refraction mean different speeds of light, and thus not all light bends at the same angle when it passes through a transparent medium. This is why light that passes through a **prism** can form a rainbow of color, as seen in the figure on the next page.

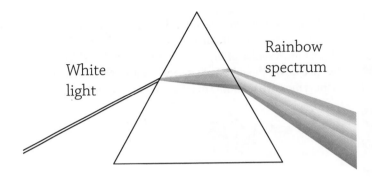

Lenses

Just like a mirror, lenses change the direction light is traveling to form images that we can see. How do lenses work? Look at the glass prism in the following figure. The ray coming from air on the left turns toward the normal as it slows down in the glass: a smaller speed means a smaller angle. (There is a partial reflection as well.) Once inside the prism, the light hits the right side of the prism. Part of the ray reflects again. The rest refracts away from the normal as it speeds back up in air: a bigger speed means a bigger angle. After leaving the prism, the ray has a new downward direction.

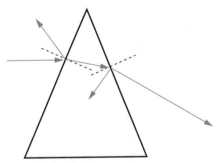

Let's concentrate only on the rays that transmit through the prism and ignore the reflected rays and make some lenses! Place two prisms together flat side to flat side, as seen in the following figure. Notice how the parallel rays change direction inside the two prisms and exit heading toward each other.

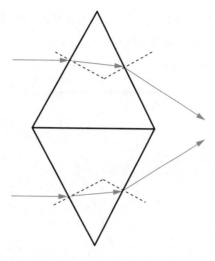

Now, if we smooth the two prisms together—Shazam!—we have a **converging** or **convex lens**. Convex lenses have a **positive focal point** because parallel light converges to this location and actually passes through the **focus**.

Refraction by a Converging Lens

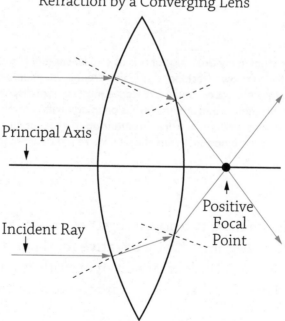

Incident rays that travel parallel to the principal axis
will refract through the lens and converge to the focal point.
The focal length will be positive for converging lenses.

Now let's flip over the two prisms so that we can place them point to point. This time parallel rays change direction inside the two prisms, causing them to exit heading away from each other.

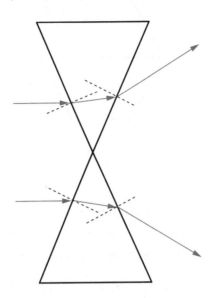

When we smooth these two prisms together, we get a **diverging** or **concave lens**. Concave lenses have a **negative focal point** because incoming parallel rays of light exit the lens on a diverging path as if they came from the focal point, but they never really passed through the focus.

Refraction by a Diverging Lens

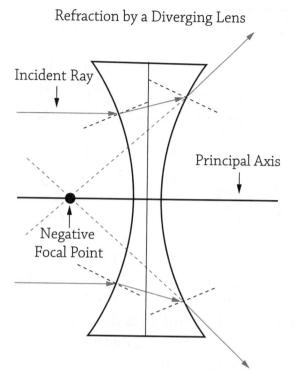

Incident Ray

Principal Axis

Negative
Focal Point

Incident rays that travel parallel to the principal axis will refract through
the lens and diverge as if coming from the focal point, never intersecting.
The focal length will be negative for diverging lenses.

EXAMPLE

▶ The following figure shows a concave and convex lens with
horizontal light rays entering the lens from the right. Complete the
paths of the rays, and use the rays to locate the focal points. Be sure
to draw the normal lines wherever the rays pass through a lens's
surface.

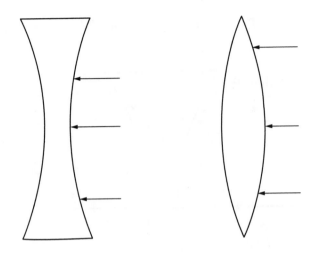

In the next figure notice that the ray in the middle passes straight through the lens because it is striking the lens along the normal line. The rays are closer to the normal line inside the lens because the speed of light is slower inside the lens than in the air outside the lens. The light rays will diverge when passing through the concave lens and converge when passing through the convex lens.

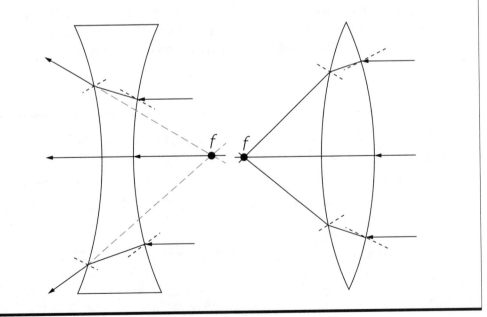

IRL People who are nearsighted, which means they can see up close okay but things that are far away look blurry, wear diverging/concave corrective lenses. People who are farsighted, which means they can see things far away but up-close things are fuzzy, wear converging/convex corrective lenses.

So, lenses work by bending, or refracting, light. The shape of the lens is very important, as you can see. Any lens that is curved so that it is thicker in the middle will behave like a convex/converging lens. Lenses that are thinner in the middle will behave like a concave/diverging lens.

EXAMPLE

▶ The following figure shows several lenses.

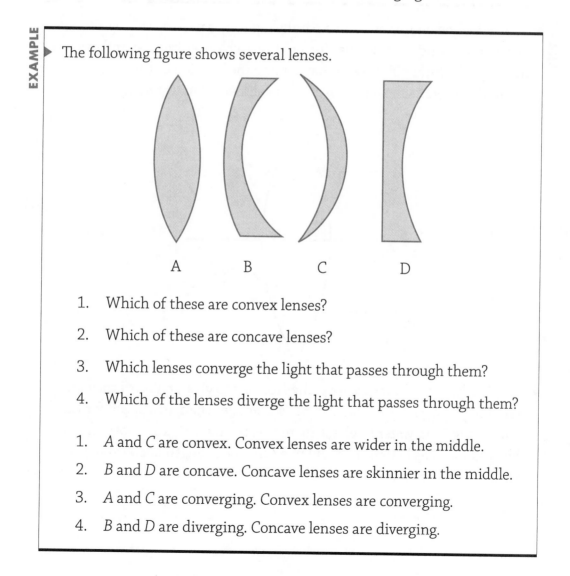

1. Which of these are convex lenses?

2. Which of these are concave lenses?

3. Which lenses converge the light that passes through them?

4. Which of the lenses diverge the light that passes through them?

1. *A* and *C* are convex. Convex lenses are wider in the middle.

2. *B* and *D* are concave. Concave lenses are skinnier in the middle.

3. *A* and *C* are converging. Convex lenses are converging.

4. *B* and *D* are diverging. Concave lenses are diverging.

Equally important to a lens is how much the light changes speed when entering a lens. The greater the difference there is between the index of refraction of the lens and air, the greater the bending (refracting). In fact, if you place the lens in a fluid that has the same index of refraction as the lens itself, the lens won't bend the light at all, because there is no change in velocity as the wave moves from the fluid to the lens. No change in velocity equals no refraction.

▶ The following figure shows two parallel light rays passing through a glass converging/convex lens. The light rays change direction and pass through the focal point. What happens to the location of the focal point if the lens is moved from air and submerged in water?

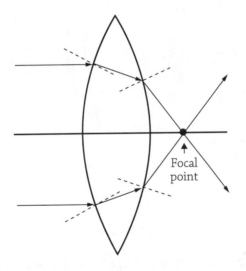

Focal point

▶ Originally, the light rays were going from air to glass, but now they are traveling from water to glass. There is a smaller change in speed for the light going from water to glass. Therefore, the light rays are not going to bend as much as before. This will push the focal length outward to the right, away from the lens:

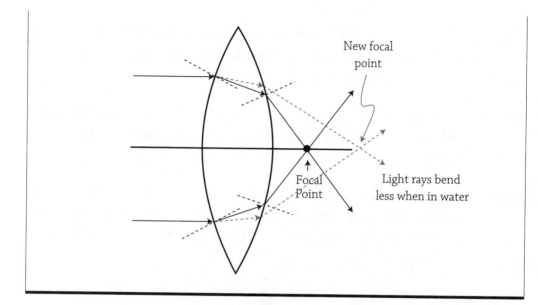

Now, let's take a look at what these two types of lenses can do for us.

BTW

Lenses and mirrors form the same types of images. The only difference between the two is how they form the images. Mirrors reflect light to form images and have only one reflective side. Lenses refract light that transmits through the lens and therefore can operate with light passing through from either side.

Ray Diagrams

Just like for mirrors, we can draw light rays from the object to find where the lens will form an image.

BTW

There are two important locations to know for a lens. The focal point f, and twice the focal length, 2f. The size and type of image produced will depend on where the object is placed in relationship to these two points. Notice that a lens has a focal point on both sides because light can pass through the lens in either direction.

Let's find out what kind of images a convex/converging lens will form. The next figure shows an arrow to represent an object located out beyond $2f$. Two easy rays can be drawn to find out where the image will form:

Ray 1 A light ray traveling horizontally toward the lens will pass through the lens and bend toward and through the focal point on the opposite side of the lens.

Ray 2 A light ray drawn toward and through the focus on the front side of the lens will pass through the lens and bend to exit horizontally out the opposite side of the lens. This light ray is the opposite of ray 1.

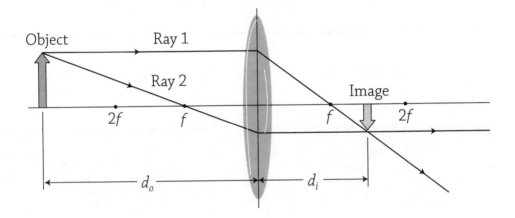

Notice how rays 1 and 2 converge and cross on the opposite side of the lens, between f and $2f$. This is where the image will form. You can see that the image is inverted and smaller than the object. The image is real because the light rays from the lens actually pass through the image location. This means that the image can be projected on a screen.

IRL The figure above, of a convex lens with the object out beyond $2f$, is an example of what happens in a camera. For example, the light from a person standing in front of the camera passes through the lens and is shrunk down so that it fits on the camera sensor. The image on the sensor is inverted, but the camera's electronics flip the image back over to make it upright again.

Now let's replace our object with a girl and get her to move inward to see what happens to the image (see the following figure). When the girl is exactly a distance 2*f* from the lens, the image forms exactly at 2*f* on the other side of the lens. The image will be identical in size, but inverted. Since the light rays converge at the image location, the image is real and can be projected on a screen.

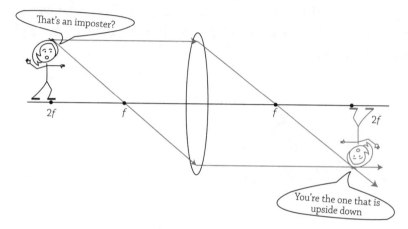

With the girl continuing to move inward toward the lens, we see that she is now between 2*f* and *f*. Using our two rays, we see that the image is now formed out beyond 2*f* on the other side of the lens, creating an inverted, larger, and real image:

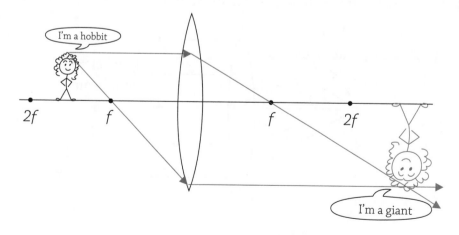

When we place the object on the focus, an interesting thing occurs. The light rays that pass through the lens come out parallel and never cross. (See the figure below.) This is the one location where an object can be placed and an image will not form.

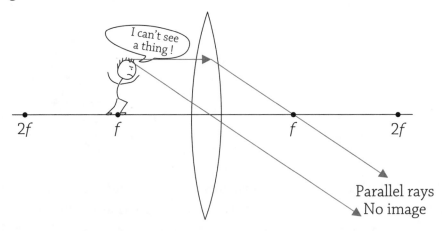

If we place a light at the focus of a lens, the light that exits the lens will form a parallel beam as shown in the next figure.

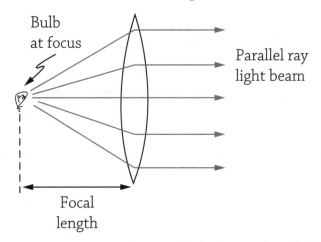

This is one way you can create a beam of light for car headlights.

Our convex lens has one more image to show us. When the object is placed between the focal point and the lens, the rays that pass through the lens diverge and will not form an image on the right side of the lens. The diverging rays appear to come from a spot on the left side of the lens. This creates a virtual, upright image that is larger than the object itself:

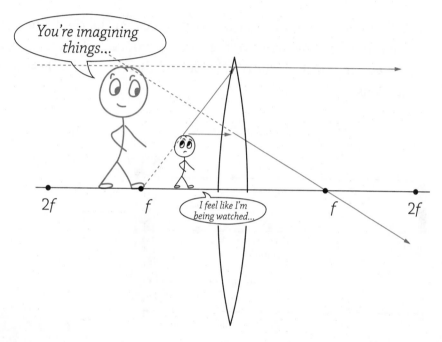

 IRL I am sure you have used a convex lens to produce this type of image—a magnifying glass. Holding the lens close to the object produces an upright, enlarged image.

Wow! The convex/converging lens produced a lot of different images. The concave/diverging lens is, in contrast, a little boring. It only produces one single type of image. In the figure below, an arrow represents

the object. Once again, two rays can be drawn to find out where the image will form:

Ray 1 A light ray traveling horizontally toward and passing through the lens. The ray travels as if it came from the virtual focal point on the left side of the lens. To recap: horizontally to the lens and then away from the virtual focus.

Ray 2 A light ray drawn toward the virtual focus on the right side of the lens. This light ray will exit the lens horizontally. To recap: toward the virtual focus on the other side and then horizontally away from the lens.

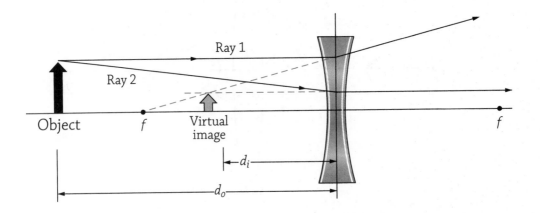

Notice how rays 1 and 2 diverge after passing through the lens. This means they will form a virtual image. When we backtrack both rays, we see that they diverge from a point behind the lens, forming an upright virtual image that is smaller than the original object. This is the only type of image the concave/diverging lens produces.

EXAMPLE

▶ The figures provided next show objects placed near a lens.

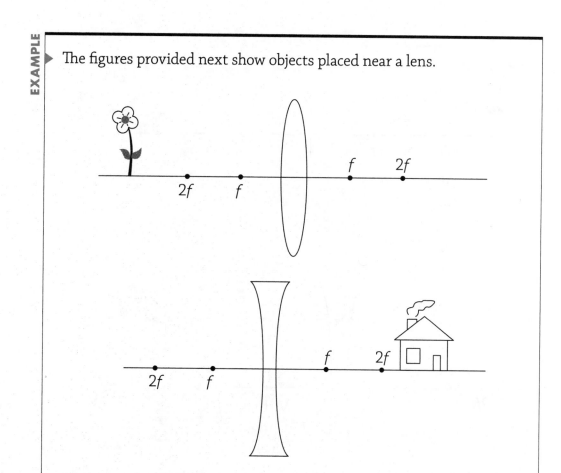

Draw a ray diagram to find the location of both images. Also, label each image with its orientation and size, and indicate if the image is real or virtual.

▶ Let's review our rules for drawing light rays to locate the image. Notice that I have placed the house on the right side of the lens so our rays from the house will travel to the left toward and through the lens:

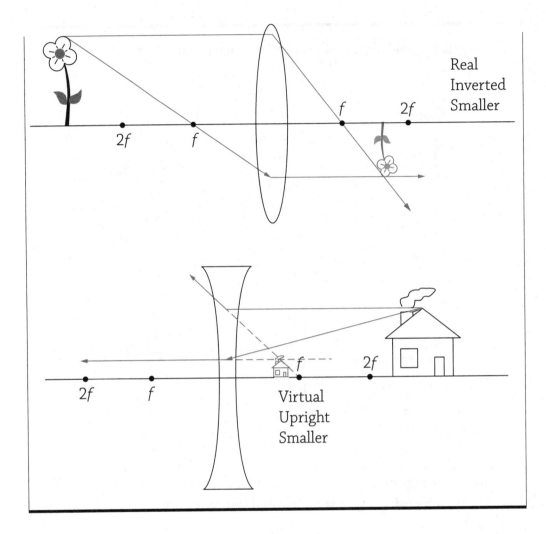

Lenses and Magnification Equations

The location and size of images formed by lenses can be found using two equations that will look very familiar—surprise! The equations for lenses are exactly the same as the ones for mirrors. I'm going to repeat the equations for you. Once again, pay close attention to the sign conventions, because they are important and just a little different from the ones in the last chapter.

The **lens equation** helps us find the location of the image: $\dfrac{1}{f} = \dfrac{1}{d_i} + \dfrac{1}{d_o}$.

f is the focal length of the lens, d_o is the distance from the lens to the object, and d_i is the distance from the lens to the image.

Here are the sign conventions we need to talk about in order to use this equation correctly:

- $+f$ is for a real focal length. Convex/converging lenses have a real focal point because light actually converges to this point.

- $-f$ is for a virtual focal length. Concave/diverging lenses have a virtual focus because light diverges from this point, but does not actually pass through the focus.

- $+d$ designates real distances. For instance, objects always have real distances because light actually comes from the object toward the lens. Images are real when light actually converges and passes through the image.

- $-d$ designates virtual distances. The light rays that form virtual images diverge as if they came from the image location, but actually don't. Therefore, the image is virtual.

The **magnification equation** helps us determine the size of the image: $M = \dfrac{h_i}{h_o} = -\dfrac{d_i}{d_o}$. M is the magnification of the image, and h_i and h_o are the height of the image and the object respectively.

The sign conventions for the magnification are:

- $+M$ indicates that image is upright.

- $-M$ indicates that the image is inverted.

When M is greater than 1, the image is enlarged. When M is less than 1, the image is smaller than the object. If M is equal to 1, the image is the same size as the object.

- h_o is always positive because we always assume that the object is upright.

- $+h_i$ means the image is upright, and $-h_i$ tells us the image is flipped over.

- $+d$ designates real distances, and $-d$ designates virtual distances, just like for the lens equation.

Here is an example for a concave/diverging lens.

A 15-cm-tall stick figure stands 25 cm in front of a concave lens with a focal length of 10 cm. Since this is a diverging optical device, the focal length will be negative.

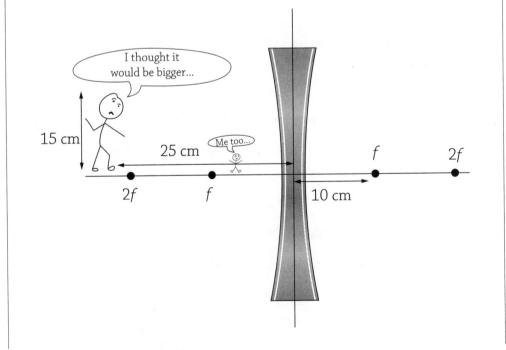

▶ Use the lens equation to find the location of the image:

$$\frac{1}{f} = \frac{1}{d_i} + \frac{1}{d_o}$$

$$\frac{1}{-10\,\text{cm}} = \frac{1}{d_i} + \frac{1}{25\,\text{cm}}$$

$$d_i = -7.1\,\text{cm}$$

▶ The negative image distance tells us that the image is a virtual image located on the left side of the lens.

▶ Now calculate the height and magnification of the image:

$$M = \frac{h_i}{h_o} = -\frac{d_i}{d_o}$$

$$M = \frac{h_i}{15\,\text{cm}} = -\frac{\left(-7.1\,\text{cm}\right)}{25\,\text{cm}}$$

$$M = 0.28 \text{ and } h_i = 4.3\,\text{cm}$$

▶ The image will be 4.3 cm tall, which is only 28 percent as tall as the object. Both magnification and image height are positive, indicating that the image will be upright.

Let's look at another example.

EXAMPLE

▶ A projector is used to display a movie on a wall. The glowing picture inside the projector is 17 cm from the lens. The wall where the movie will be projected is 3.0 m from the lens. What kind of lens will we need to accomplish this?

▶ Well, our lens will have to be a converging lens because those are the only ones that create real images. Let's find the focal length of the lens using the lens equation. Be careful! We have different units. I'm going to convert all the units to centimeters. We also know that our image distance will be positive, because the image is real since it is being projected on a wall.

$$\frac{1}{f} = \frac{1}{d_i} + \frac{1}{d_o}$$

$$\frac{1}{f} = \frac{1}{300\,\text{cm}} + \frac{1}{17\,\text{cm}}$$

$$f = 16\,\text{cm}$$

▶ So now we know that the projector has a convex lens with a focal length of 16 cm. Now, let's use the magnification equation to find out how much our movie has been magnified:

$$M = \frac{h_i}{h_o} = -\frac{d_i}{d_o}$$

$$M = -\frac{300\,\text{cm}}{17\,\text{cm}}$$

$$M = -17.6$$

▶ So, our movie will be almost 18 times bigger than the original inside the projector. The image is also inverted, meaning our projectionist will have to put the original in upside down so that the movie on the wall is right side up!

Our Eyes

Human eyes are amazing pieces of physics. Eyes are actually **compound lenses,** meaning they have two refractive surfaces. (See the figure below.) Light from objects first enters the **cornea,** which acts as a smooth hard outer lens to protect the inner eye. The cornea also bends the incoming light. In fact, the cornea does as much as 75 percent of the refracting for the eye. Next, the light enters the lens, which is flexible. Muscles and ligaments in our eye bend and shape the lens. This changes the focal length of the lens for fine-tuning the image location. When our eyes are correctly focusing on an object, a clear image will form on the **retina**. Nervous system sensors pick up color and brightness of the image and send this to the brain. Since the image on the retina must be a real image, it is inverted. Your brain has learned to flip the image over so that you see the world right side up. Your brain learns to flip the image over when you are a baby.

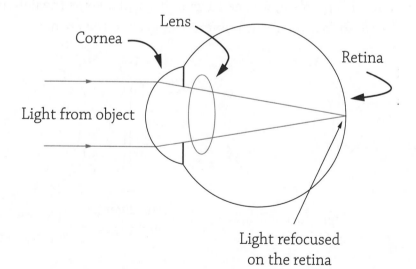

Not all eyes can focus properly. Some people are **farsighted** (**hyperopia**), which means they can see objects far away but up-close images are blurry.

This is because the image formed by the lens is behind the retina. To fix this problem, we need a lens that converges the light before it enters the eye. Therefore, the corrective lens is a convex lens:

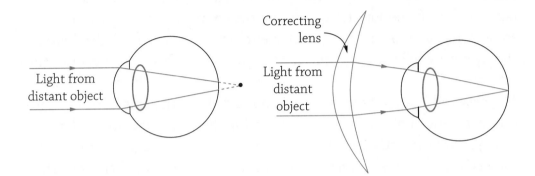

A more common problem for people is **nearsightedness (myopia)**, where a person can see things up close just fine but objects far away are all fuzzy. This is caused because the eye is focusing the light too soon in front of the retina. To fix this problem, we need a lens that will diverge the light before it enters the eye so that the focus point moves back to the retina:

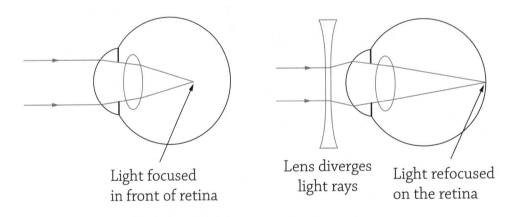

REVIEW QUESTIONS

I hope you have enjoyed refraction and lenses. Let's show off our skills by answering the following questions.

The following figure shows light traveling from the right to the left. Use it for questions 1 to 4.

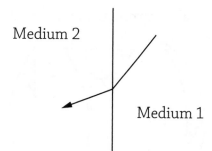

Medium 2

Medium 1

1. In which medium is light traveling the fastest?

2. Which medium has the largest index of refraction?

3. In which medium is the wavelength of the light shorter?

4. In which medium is the frequency the highest?

A violet laser with a wavelength of 405 nm is passing through flint glass. Use this information for questions 5 and 6.

5. Calculate the speed of light in flint glass.

6. Calculate the wavelength in the flint glass.

7. Light travels from air at an angle of 33.0° into an unknown substance at an angle of 19.9°. Determine what the unknown is.

8. A light beam traveling in water strikes the surface at an angle of incidence of 25° and exits into air. What is the exit angle into air?

9. Beyond what angle will light not exit from water into air?

Use the following lenses and mirrors to answer questions 10 to 13.

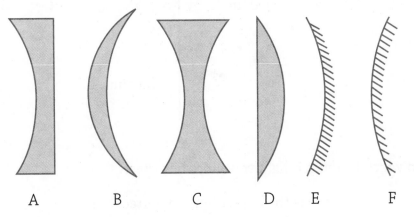

A   B   C   D   E   F

10. Which of these are convex?

11. Which of these are concave?

12. Which are converging?

13. Which are diverging?

14. A convex lens has a focal length of 45 cm. Is the focal length positive or negative?

15. Where would you stand in front of a convex lens so that you produce an enlarged upright image of yourself?

16. An apple is placed 40 cm in front of a convex lens with a focal length of 10 cm. What kind of image is formed?

The figures on the next page shows two lenses with objects. Use them to answer questions 17 and 18. I have drawn the objects upside down to see if you can still correctly draw the ray diagram. I know you can!

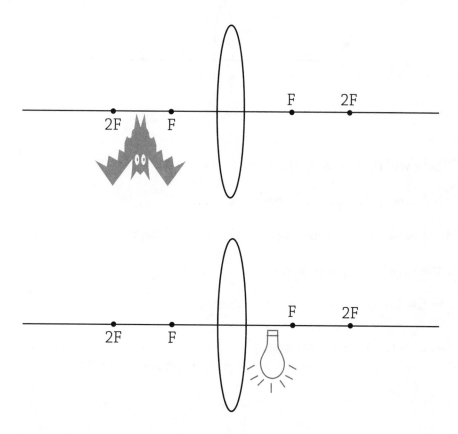

17. Draw a ray diagram to find the location of both images.

18. Label each image with its orientation and size, and indicate if the image is real or virtual.

The following figure is of a 26-cm-tall flower sitting 60 cm in front of a converging lens with a focal length of 20 cm, Refer to it to answer questions 19 to 23.

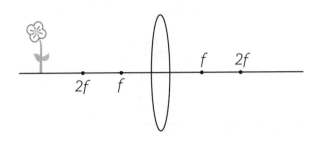

19. Where will the image of the flower appear?

20. What kind of image is it? Real or virtual?

21. What is the height and magnification of the image?

22. Is the image upright or inverted?

23. Can the image be projected on a screen?

The figure below is of a toy house placed 23 cm from a diverging lens with a focal length of 10 cm, Refer to it to answer questions 24 to 28.

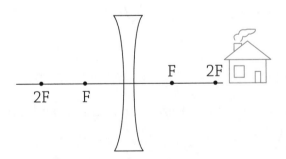

24. Where is the image of the toy house?

25. What kind of image is it? Real or virtual?

26. Is the image upright or inverted?

27. Is the image larger or smaller?

28. Can the image be projected on a screen?

A $100 bill is located 30 cm in front of a lens with a focal length of 20 cm. Use this information to answer questions 29 to 31.

29. Where will the image of the $100 bill be located?

30. Will the image be bigger or smaller?

31. Will the image be upright or inverted?

Flashcard App

PART SIX
Modern Physics

The high-speed world and the nano-size world are really different and kind of strange. The classical physics we use to understand how our everyday world operates does not accurately describe what happens when objects move really fast or get really small. A new set of physics rules is needed to understand this wonderfully bizarre world at the extremes. In this section of the book, we'll be introducing a more modern understanding of how physicists believe the universe works. We'll learn some relativity, introduce the ideas of quantum behavior, and end with a study of how atoms transmute from one element to another.

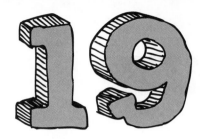 

Relativity and Quantum Behavior

MUST KNOW

⚡ Einstein gives us a framework, called relativity, that describes time and space as changeable and asserts that mass and energy are actually two different aspects of the same thing.

⚡ Light has both wave and particle properties.

⚡ Nano-sized objects about the size of an atom and smaller also behave as if they are both a wave and a particle.

Our modern understanding of physics shows us that time, space, and matter are not exactly as they appear. The nano world of atoms and subatomic particles can seem strange and does not always follow the same rules as the world we live in. But the weird and wonderful world of the ultra-small comprises the building blocks that make up everything we know and experience in the human-sized world we live in.

Our goal in this chapter is to understand the foundations of modern physics, which takes us into the bizarre world of the uber-fast and the nano-small. This is the realm where the distinction between waves and particles gets fuzzy, where things without mass can have momentum, where elements spontaneously break apart, and where time and space are no longer constant. Modern physics refers to all of the physics developed from about 1900 on, and that is when things really got weird. In fact, things got so bizarre that Albert Einstein refused to believe some of it, even though he was one of the pioneers of this new "modern physics." But quantum theory, a big part of modern physics, has proven over and over again to be correct, so much so that practically all of our modern electronic toys are based upon it. So, let's dive into modern physics.

Space, Time, and the Speed of Light

Crazy-haired Einstein changed everything when he made these two simple **postulates of relativity** (a postulate being just a fancy word for a statement that is assumed to be true based on reasoning):

1. All the laws of physics are exactly the same in every uniformly moving frame of reference.

2. The speed of light in a vacuum is always a constant, c, no matter the motion of the source of the light or the motion of the receiver of the light.

The first postulate means all the physics we have learned earlier in this book behave exactly the same when we are sitting still or moving at a constant velocity. You already take this for granted. For instance, when you are flying on a jet plane, you can walk up and down the aisle just like you walk across a room on the ground. You don't have to do anything special like lean forward when you walk toward the cockpit. You can just walk normally. And when you get a drink from the flight attendant, you don't have to do anything special when you pour the drink into your cup of ice. You don't have to worry that the plane is flying 500 knots and that the drink might fly backwards because you are traveling so fast. Everything is "normal." But if the plane flies through turbulence or the pilot decides to dive the plane, you might get wet, and you need to strap in or get thrown about. As long as the plane isn't accelerating, everything is the same as if the plane was just sitting on the runway. This fact has an odd consequence; it is impossible to know if you are sitting still or moving at a constant velocity.

Imagine that you are in a cabin on a train with no windows to the outside world. Now imagine that you take a baseball and drop it to the floor. The ball falls straight to the floor. You measure the distance the ball falls and time how long it takes the ball to hit the ground. You calculate that the ball accelerates downward to the ground at 9.8 m/s^2. Based on this observation, what can you determine? Well, you can tell that the cabin is not accelerating forward or backward because the ball fell straight downward, not to the side. You can also tell that the cabin is not accelerating up or down because the ball accelerated at 9.8 m/s^2, just like you would expect. But what you can't possibly know is if the cabin is stationary or moving at a constant velocity because the pull of gravity downward on the ball is the same in both cases. In fact, there isn't any experiment you can do to determine how fast you are going without measuring your motion relative to something else.

Now imagine that you are in a spaceship with a window. As you look out the window, you see an asteroid moving past your window at 100 m/s:

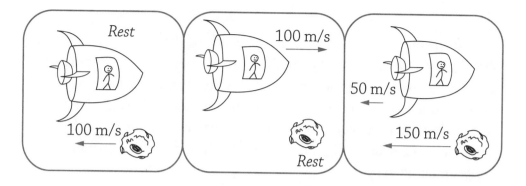

Now think about it...which one is moving? Your spaceship or the asteroid? Are you stationary and the asteroid is moving at 100 m/s? Or is the asteroid stationary and you are moving 100 m/s? Or, maybe, the asteroid is moving 150 m/s and you are moving 50 m/s in the same direction, so that the difference is still 100 m/s. In each case, there is a 100 m/s relative motion between you and the asteroid. It turns out that there isn't any way for you to know which is correct. All you can tell is that you are moving at a rate of 100 m/s relative to the asteroid. This is why Einstein's idea is called the **theory of relativity,** because he showed us that all motion is actually relative to how you decide to measure it.

IRL   We don't have a good intuitive feel for relativity, because on Earth we measure everything relative to the Earth itself. As I type this sentence, I assume that I am not moving because I am stationary relative to my computer, and desk, and the Earth. But the Earth is spinning and orbiting around the Sun. The Sun and solar system are in the Orion arm of the galaxy, traveling around the galactic center of the Milky Way, which is hurtling through space toward Andromeda, and...the universe is expanding...Gaahhhh! I'm just going to say that I'm at rest...relative to the Earth.

EXAMPLE

▶ Imagine you have a truck that you are driving 100 miles per hour. Now imagine that you put a professional baseball pitcher in the back who can throw a 100 mph fastball.

▶ When the pitcher throws the baseball forward while you are driving forward, the total speed of the ball will be 200 mph because the ball is being thrown forward relative to the truck and the velocities will add. If the pitcher throws the ball backward, the ball leaves their hand at exactly zero velocity relative to the Earth because the pitcher is actually just slowing the ball down to zero. The 100-mph vector forward due to the truck plus the 100-mph vector backwards due to the pitcher add up to equal zero. After the ball leaves the pitcher's hand, it would drop straight downward to the ground. It's actually quite amazing to see:

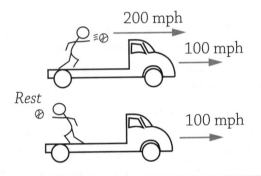

Einstein says that's not how things work for light. The second postulate says that the speed of light in a vacuum is always the same no matter how you measure it.

▶ Imagine a car sitting still with its headlights on. The light leaves the car at $c = 3 \times 10^8$ m/s as observed by the passengers in the car (see the following figure).

At rest

▶ But what if the car is moving forward at $0.5c$?

▶ We would expect that the people inside the car would see the light leave at c, but anyone watching from the side of the road would see the light traveling at $1.5c$, but that's not what happens.

▶ Both the people in the car and observers on the side of the road see the light traveling at c. That means the passengers in the car see the people on the side of the road moving by at $-0.5c$ and the light leaving them at c, as shown in the following figure.

▶ The observers on the side of the road see the car zipping by at $0.5c$ and the light beam traveling at c so that the light beam is only leaving the car at $0.5c$.

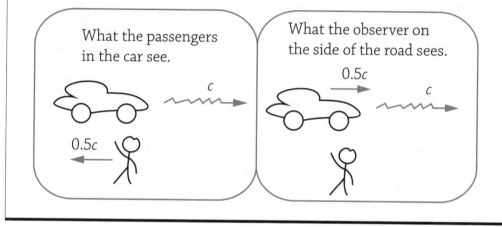

This is really weird. How can that possibly happen? Einstein realized that in order for the speed of light to always equal c, time and space cannot be constant. The faster you go, the slower time moves and the shorter distances become in the direction you are going. The really odd thing is that you will not notice any of these changes happening to you because of postulate 1. All the laws of physics will be exactly the same for you in your frame of reference no matter what speed you travel. All the effects of relativity are observed to happen to the "other guy," outside of your frame of reference. So for the passengers of the car, life seems normal, but the outside world seems to shrink in the direction they are traveling. This allows them to measure the speed of light to be c leaving the car. For the bystander on the side of the road, life seems normal, but the car seems to

BTW

High-speed relativity effects are called **time dilation** and **length contraction**.

shrink in the direction of travel, and time seems to be traveling slower for the moving car. This allows the bystander to measure the speed of the car to be 0.5c and also measure the speed of light to be c.

This is very confusing because we never see these effects in our everyday life, but they are real. Here is an example to help.

▶ Let's say I get on a spaceship and travel to a distant exoplanet to look for intelligent life. The exoplanet is 10 light-years away, which means that it takes 10 years for light to travel between Earth and the exoplanet. The fancy rocket transporting me can travel at 0.95c.

▶ As you watch me through a telescope, you see that my ship seems to shrink in the direction it is traveling to only about 30 percent of its original length. You also observe that I seem to be moving really slowly and my heart rate is only about 30 percent its normal speed. It's as if time has slowed for me, and you predict that I will only age about 30 percent as much as I should on the 10-year trip because time has slowed for me.

▶ I, on the other hand, don't notice anything strange happening to me at all. But when I look outside, the distance in the direction I'm traveling to the exoplanet is shrinking to only 30 percent of the original distance. This means I'm getting to the exoplanet much faster than expected.

▶ By my calculations I predict I will only age about 3 years on the 10-year trip because the distance has become shorter. Notice how both you—the observer—and I—the traveler—calculate that I will only age about 3 years on the trip, but for different reasons. I observe the distance diminish. You observe me moving through time slower.

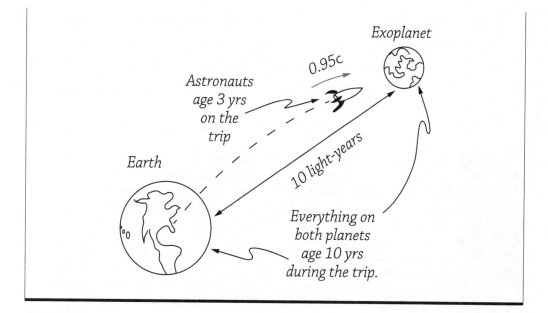

Exoplanet

0.95c

Astronauts age 3 yrs on the trip

Earth

10 light-years

Everything on both planets age 10 yrs during the trip.

IRL Time dilation will allow future humans to travel at high speeds to planets far away, because time will slow down for them. Unfortunately, everyone they leave behind on Earth will age normally. High-speed travel will be like traveling in a time machine into the future with no way to return to your original time frame. So, it is likely that these future journeys to far-away planets will be a one-way trip, because everyone the astronauts left behind on Earth will be gone by the time they return. Kind of sad....

Remember that relativity effects seem to happen to the other guy? So why it is that I only age 3 years on the trip and you age 10 years? Why not the other way around? Well, it's because I have changed from your slow-moving frame of reference to a high-speed moving frame of reference during the trip through space and then back to your slow frame of reference. Your slow speed frame of reference is the common reference between us. So, the

BTW

Have you ever wondered how light from distant stars makes it all the way to Earth from hundreds of millions of light-years away? Why doesn't the light fade away? Light always travels at the speed of light. At that speed, time stops. Light does not experience time at all. Light from a distance star arrives instantaneously from its own point of view.

relativity effects end up being measured from your time frame because I have been jumping between different time frames. This is really mind-bending, because we never see this kind of stuff in everyday life. The effects of relativity don't become easily visible unless traveling at a good fraction of the speed of light.

IRL Muons, subatomic particles similar to electrons, are created in the upper atmosphere by cosmic radiation from space. These muons have a very short life span. Almost all should disappear before they have a chance to hit Earth's surface, but somehow large numbers make the trip. This presented a dilemma for scientists to explain. Relativity supplied the answer. The muons are traveling at a good fraction of the speed of light. From our perspective on Earth, the muons are able to make the trip to the ground, because time slows down for the muon and they live long enough to make it to the ground. From the perspective of the muon, they are able to reach the ground because the space around them shrinks, and so the trip to the ground is shorter.

IRL Scientists have measured time dilation by putting a very accurate "atomic clock" on a high-speed jet and flying it around for a while. When the plane lands, the clock on the plane is slightly slower than the clocks on the ground. The difference in the clocks is so small that we never notice the effect in real life.

Another gift from Einstein is the most famous physics equation of all time: $E = mc^2$. It is a matter-shattering idea–literally. The equation presents the idea that energy and mass are just two aspects of the same thing. Just like we learned way back in Chapter 4, different forms of energy can transform from one type to another, like kinetic into potential. Mass is like a solid form of energy, and it can be converted into energy. (Lots and lots of energy because c^2 is a huge number!) Energy can also be converted into mass. You heard that right. *You can create matter out of energy.* These things happen all the time in the nano-world of subatomic particles. We will cover this idea in greater detail in Chapter 20.

The Electron-Volt

We are going to start delving into the very small. At nano-sizes, some of the units we have been using are just too big. The **electron-volt**, abbreviated **eV**, is a unit of energy that's particularly useful for problems involving atomic-sized particles. One eV is equal to the amount of energy needed to change the potential of an electron by 1 volt. For example, imagine an electron nearby a positively charged particle such that the electron's potential is 4 V. If you were to push the electron away from the positively charged particle until its potential was 5 V, you would need to use 1 eV of energy to accomplish this.

$$1 \text{ eV} = 1.6 \times 10^{-19} \text{ J}$$

The conversion to joules shows that an eV is an itty-bitty unit of energy. However, such things as electrons and protons *are* itty-bitty particles. So, the electron-volt is actually a perfectly sized unit to use when talking about the tiny bits of energy.

Light: Wave *and* Particle

Back in Chapter 16 we saw how the double-slit experiment proved once and for all that light had to be a wave, because light passing through two narrow slits produced an interference pattern that can only be explained if light has wave properties. Case closed. Huygens wins! Light is a wave. But the ghost of Newton is restless and demands to be heard. It turns out that there is still more to the story of light.

This new chapter to the story begins with glowing objects and something called **black body radiation**. When an object is heated up, it begins to glow. First a dark red, then a brighter red, turning into an orangey-red as it gets hotter. You can see this when you turn on your oven or cook top in the kitchen or when you burn wood in a fire. As the temperature of the object goes up, the color it emits becomes yellow, then white, and then a blueish white. Scientists wanted to know why this occurred, so they measured the light intensity radiating from the hot objects and built the graph you see in the following figure. You can see that lower temperatures produce less light and the light that is produced is shifted toward red (700 nm) in the visible range. As the object gets hotter, more light is emitted, and it shifts toward the blue (400 nm) end of the visible spectrum. The problem is that if light is a wave, this is not the graph we should get. Instead, the wave theory of light predicts way too much ultraviolet light. Thus, the problem was called the "ultraviolet catastrophe."

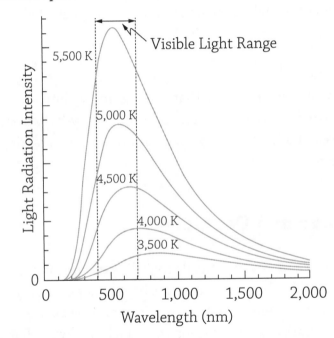

Max Planck proposed a solution to fix the ultraviolet catastrophe. Planck suggested that atoms could not emit light at any energy they wanted to. Instead, energy radiated by atoms comes in **quantized** chunks depending on the vibration frequency. At first, Planck viewed his solution as a "mathematical trick" not based on reality. But it turned out his wild assumption was right on the mark, and he won a Nobel Prize for his solution.

Today we call Planck's **quanta** of light *photons*. A nice way to think of a photon is a bundle or package of electromagnetic waves or energy. The amount of energy in that bundle is directly related to the electromagnetic wave's frequency:

$$E = hf = \frac{hc}{\lambda}$$

where:

- E is the energy of the photon of light, in units of joules (J) or electron-volts (eV)

- h is Planck's constant, which comes in several forms:

    $$h = 6.63 \times 10^{-34} \text{ J} \cdot \text{s} = 4.14 \times 10^{-15} \text{ eV} \cdot \text{s}$$

    $$hc = 1.99 \times 10^{-25} \text{ J} \cdot \text{m} = 1240 \text{ eV} \cdot \text{nm}$$

- f is the frequency of light in hertz (Hz)

- c is the speed of light, $c = 3.0 \times 10^8$ m/s

- λ is the wavelength in meters (m)

The next figure shows the entire electromagnetic spectrum of light from radio waves to gamma rays. Looking at the photon energy equation, we can see that gamma rays have the most energy because they have the highest frequency and shortest wavelength. Radio waves have the least energy and the longest wavelengths and the lowest frequencies.

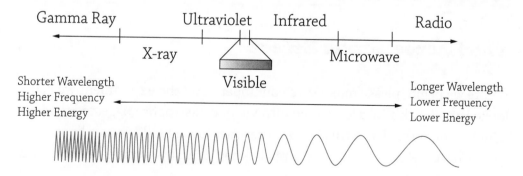

EXAMPLE

> Calculate the energy of the photons in a 650 nm red laser pointer. Give your answer in both joules and eV.

> Since we know the wavelength of the light, let's use the version of the photon energy equation that has wavelength instead of frequency. First, let's find the energy in joules. Remember to convert 650 nm into scientific notation and use the correct constant, hc:

$$E = \frac{hc}{\lambda} = \frac{1.99 \times 10^{-25}\,\text{J}\cdot\text{m}}{650\,\text{nm}} = \frac{1.99 \times 10^{-25}\,\text{J}\cdot\text{m}}{650 \times 10^{-9}\,\text{m}} = 3.1 \times 10^{-19}\,\text{J}$$

> Now we could convert our answer directly into electron-volts, but I'm going to show you how to use our second hc constant: $hc = 1{,}240\,\text{eV}\cdot\text{nm}$. Notice how this constant already has units of nanometers. Therefore, we do not need to convert our wavelength into scientific notation because the nanometers will already cancel out:

$$E = \frac{hc}{\lambda} = \frac{1{,}240\,\text{eV}\cdot\text{nm}}{650\,\text{nm}} = 1.9\,\text{eV}$$

IRL Gamma rays and X-rays have enough energy to pass through our bodies and mutate our DNA and possibly cause cancer. Ultraviolet light has less energy, but can still cause a chemical reaction in our skin, giving us a sunburn if we get too much. Infrared light causes our skin to vibrate, which we perceive as heat. Most microwaves and radio waves just pass right through us without doing much, which is why we use them to transmit cell phone calls, radio signals, and other forms of communication.

The Photoelectric Effect

Before we begin, please make sure you understand the experiment we just looked at. Planck's solution to the ultraviolet catastrophe suggested that light was quantized, but this is the experiment that really proves light has to be a particle, even though we already know that it is a wave, because light has wave properties like interference. This is the experiment that opened up the rabbit hole of **wave-particle duality**.

When light shines on a metal, you can get electrons to fly off the surface. Scientists call this the **photoelectric effect**. You might say to yourself, "What's so important about that?" There are so many important processes in nature that are governed by the photoelectric effect, you might be shocked. Photosynthesis, photographic sensors in cameras, solar panels, and cancer, for instance, are all related to the photoelectric effect.

To really understand the photoelectric effect, you have to understand the experiment that was used to study it. The figure below shows the experimental setup. Don't worry if it looks a little scary—it'll make sense in a few minutes. In the experiment, two metal plates are sealed in a vacuum container. One of the plates is either made of or coated with the metal being studied. A variable potential source is connected across the two metal plates. (Think of the variable potential source as a dimmer switch in your home that varies the voltage across your lights, making them brighter or dimmer.) Light is directed at the plate with the metal being tested, and under the right conditions, electrons are emitted from that plate and travel to the other plate, and a current can be seen on the ammeter.

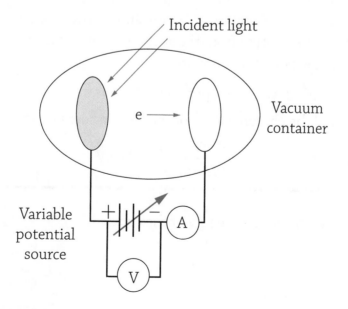

Notice that the variable potential source is oriented in such a way that the electrons are traveling toward the negative plate and away from the positive plate. This means that the voltage across the plates is acting to stop the electrons from traveling from one side to the other. That voltage is called the **stopping potential**; it lets us calculate the maximum kinetic energy of the electrons. For example, let's say we have light incident upon the left plate and electrons are flowing across to the plate on the right. This shows up as a current reading on our ammeter. We start turning up our voltage until all the electrons are stopped and our ammeter reads zero. For argument's sake, I'll say 2 volts stop all the current. The work done on the electron by the electric field is equal to the maximum kinetic energy of the electron:

$$K_{max} = \Delta U = q\Delta V = (1\,e)(2\,V) = 2\,eV$$

This tells us that the maximum kinetic energy of electrons is 2 eV. So, the ammeter tells us the rate the electrons are being emitted, and the voltmeter tells us the maximum kinetic energy of the electrons.

The easiest way to understand this is with an example.

BTW

Remember, when you are calculating your energy in electron-volts, your charge must be in multiples of the elementary charge, and an electron is exactly one elementary charge.

EXAMPLE

▶ Blue light with a wavelength of 443 nm shines on a potassium surface, and electrons are emitted from the surface. A current is measured until a 0.5 V stopping potential is applied.

What does this tell you about potassium?

▶ Let's first find the photon energy of the light incident on the surface:

$$E = hf = \frac{hc}{\lambda} = \frac{1{,}240\,eV \cdot nm}{443\,nm} = 2.8\,eV$$

Next, let's find the kinetic energy of the emitted electrons.

BTW

When using Planck's constant, be sure to pick the correct one. In my example, I am using the equation $E = \frac{hc}{\lambda}$ and the constant hc, with units of eV.

▶ It takes 0.5 volts to stop the electrons, so:

$$K_{max} = \Delta U = q\Delta V = (1\,e)(0.5\,\text{V}) = 0.5\,\text{eV}$$

▶ Wait…Something does not make sense. Why, when a 2.8 eV photon hits potassium, does an electron get emitted with only 0.5 eV of kinetic energy? You know enough about conservation of energy to ask: "Why is 2.3 eV of energy missing?"

▶ Let's think about this. The electron is negative and the nucleus is positive, so the two are attracted to each other. It takes work to "tear" that electron away from the nucleus. (For you chemistry students out there, it's similar to ionization energy—the amount of energy required to strip a molecule of an outer electron.) Physicists call this amount of energy the **work function**, ϕ. The work function is based upon the properties of the material the electron is being torn free from. Each material, like potassium or gold, will have a different work function. To find the work function:

$$K_{max} = hf - \phi$$

where:

- K_{max} is the maximum kinetic energy of the emitted electron.
- hf is the energy of the incident photon, from Planck's equation: $E = hf$.
- ϕ is the work function. The work function is the energy required to remove an electron from a specific material.

▶ Finally, completing our example, we can calculate the work function:

$$K_{max} = hf - \phi$$
$$0.5\,\text{eV} = 2.8\,\text{eV} - \phi$$
$$\phi = -2.3\,\text{eV}$$

▶ Okay, so now we have proven that it takes 2.3 eV to remove an electron from potassium. Thus, the work function of potassium is 2.3 eV.

What if we make the original 443 nm blue light that is striking the potassium brighter—how will this change things? One important thing to keep in mind is that the photons cannot gang up on an electron; two photons can't act together and add their energies. The interaction between photons and electrons is on a one-to-one basis. The photon energy of the blue light is still 2.8 eV, and the work function is still 2.3 eV, so electrons will still be emitted with an energy of the same 0.5 eV. Making the light brighter means there are more 2.8 eV photons of blue light incident on the metal, so there will be a greater number of 0.5 eV electrons ejected from the surface, which would be seen as higher current measured by the ammeter.

So, let's change things. Suppose red light with a photon energy of 1.9 eV is incident on potassium. What will happen? It takes 2.3 eV to remove an electron from potassium, so nothing happens—no electron emission.

Okay, so let's just send in more red light. Let's make that red light really bright—what will happen then? This is where the photoelectric experiment really proves that light has particle properties. If light was just a wave, we should be able to shine brighter red light on the potassium and force an electron to jump free. If light is just a wave, then brighter light means a larger amplitude with more energy. But that's not what happens. Making the red light brighter simply means more 1.9 eV photons are hitting the potassium. So, no matter how bright the red light is made, there is no way a 1.9 eV photon will make potassium, with a 2.3 eV work function, emit an electron. So, blinding, skin-burning, bright-red light will never eject an electron from potassium. Never. Think of it like a soda machine that only accepts dollar bills. It does not matter that you have a pocket full of quarters. The machine requires a minimum of $1 in one single transaction to spit out a soda. The quarters are not big enough to make the machine work.

This can be confusing, so let's see if we understand what we have just learned by working through an example.

EXAMPLE

▶ Let's shine some violet light with a wavelength of 400 nm on our potassium surface, which has a work function of 2.3 eV. Will any electrons be ejected from the potassium?

▶ That's simple enough. First, let's calculate the energy of the 400 nm violet light:

$$E = hf = \frac{hc}{\lambda} = \frac{1{,}240\,\text{eV} \cdot \text{nm}}{400\,\text{nm}} = 3.1\,\text{eV}$$

▶ Yes, electrons will be ejected from the surface. The photons have 3.1 eV of energy, and it only takes 2.3 eV to remove the electrons (the work function) from the potassium.

▶ What is the maximum kinetic energy of the ejected electrons?

▶ You can probably do this in your head, but I'm going to write down our photoelectric equation just to make it easier:

$$K_{max} = hf - \phi = 3.1\,\text{eV} - 2.3\,\text{eV} = 0.8\,\text{eV}$$

▶ What happens if the violet light is made brighter? Will the kinetic energy of the ejected electrons go up?

▶ Remember that photons act like particles and each photon particle interacts with an electron on a one-to-one basis. Making the violet light brighter only means that we have more photons with the same 3.1 eV of energy. Therefore, there will be more ejected electrons because more photons are striking the potassium, but the maximum energy of the electrons remains the same at 0.8 eV of energy.

When we shine many different wavelengths of light on the potassium surface and measure the kinetic energy of the ejected electrons, we get a graph as shown on the next page. Notice how the graph is linear because the kinetic energy equation represents a line with the slope equaling Planck's constant and the y-intercept representing the work function of the metal.

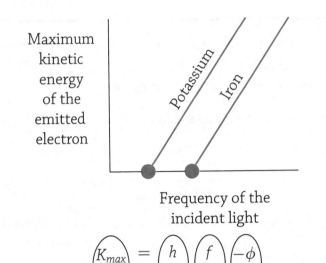

$$\begin{pmatrix} K_{max} \\ y \end{pmatrix} = \begin{pmatrix} h \\ m \end{pmatrix} \begin{pmatrix} f \\ x \end{pmatrix} \begin{pmatrix} -\phi \\ +b \end{pmatrix}$$

When we repeat this experiment for a different metal, the new graph will have the same slope, but its own unique work function, as seen in the previous figure. So, the slope of these lines always equals Planck's constant, and the intercept will equal the work function of the specific material. Notice the large points where the lines intersect the x-axis. This point is called the **threshold frequency**. It is the minimum frequency required for electron emission. Potassium has a lower threshold frequency than iron because it takes less energy to remove an electron from potassium. Another way to find the threshold frequency is to use the equation $E = hf$ and enter the work function for your energy. Since the work function is the minimum energy needed to cause electron emission, the resulting frequency will be the lowest frequency to cause photo-electron emission.

> **IRL** You are very familiar with the photoelectric effect. We use this technology in photovoltaic cells, or solar panels, to generate electricity from the Sun. Sunlight strikes the solar panel, causing electrons to be emitted. The electrons are collected by wires and funneled to the electronic devices we wish to power.

Remember the importance of this experiment: The photoelectric effect shows us the particle behavior of light. We already discussed the double-slit experiment in Chapter 16, which demonstrates interference and the wave nature of light. So, we say that light exhibits the properties of both waves and particles. It has **wave-particle duality**. When we are working with energies down near the photon $E = hf$ range, we see more of the particle nature of light. When light interacts with objects around the size of its wavelength and smaller, we will see more of its wave behavior.

Photon Momentum

We now know that the energy of a photon is $E = hf$ and that energy and mass are related by the equation $E = mc^2$. And we also know that photons of light have particle properties. Is it possible for photons to have other particle properties such as momentum? Yes! When a photon hits an atom and that atom emits an electron, as in the photoelectric effect, both energy and momentum are conserved in the process. This means that even though a photon does not have any mass, it still has momentum. The momentum of a photon can be calculated using the equation:

$$p_{photon} = \frac{h}{\lambda} = \frac{E_{photon}}{c}$$

where:

- E is the energy of the photon of light

- h is Planck's constant, which comes in several forms depending on the units being used:

  $$h = 6.63 \times 10^{-34} \, \text{J} \cdot \text{s} = 4.14 \times 10^{-15} \, \text{eV} \cdot \text{s}$$

- c is the speed of light, $c = 3.0 \times 10^8 \, \text{m/s}$

- λ is the wavelength in meters (m)

So, a photon of light can "collide" with an electron just like two cars collide, and the momentum will be conserved. Energy must also be conserved in the interaction. Let's take a look at a couple of examples.

▶ A photon collides head on with an electron, as seen in the following figure.

▶ This is a one-dimensional interaction. The photon hits a stationary electron, and afterward, the electron takes off with some of the photon's original energy and momentum.

▶ Thus, the photon lost both momentum and energy. Look back at our equation. To lose momentum and energy, the wavelength of the photon must have increased and the frequency decreased. This means the photon has changed "color" during the interaction.

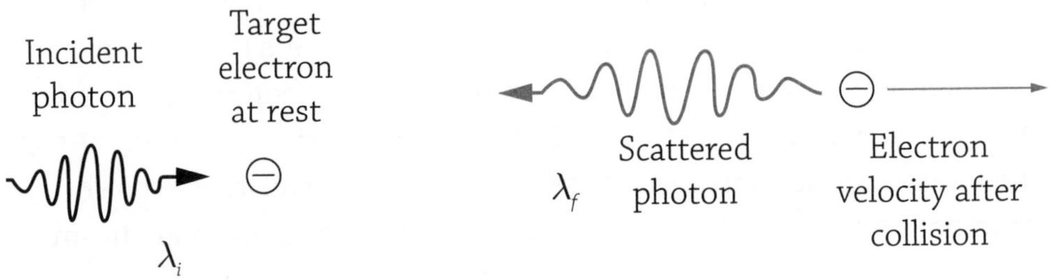

Incident photon   Target electron at rest

Scattered photon λ_f Electron velocity after collision

λ_i

Now let's look at a two-dimensional example.

▶ A photon strikes an electron and the electron moves off at an angle, as seen in the following figure.

▶ This is a two-dimensional momentum collision. Since the electron moves upward after the collision, the photon must move downward. The initial y-direction momentum of the system is zero; therefore, the final y-direction momentum must also equal zero.

▶ The initial and final x-momentums must also remain constant. The photon has again transferred some of its momentum and energy to the electron, which decreases its frequency and increases its wavelength.

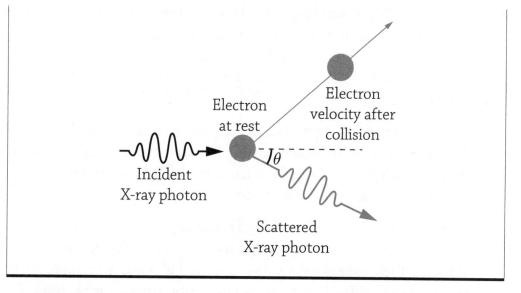

Incident X-ray photon → Electron at rest → Electron velocity after collision, θ, Scattered X-ray photon

De Broglie Wavelength and Wave Functions

When we get down into the nano world nitty-gritty, our old physics models don't adequately explain things anymore. This caused physicists to invent better explanations for how things behave. We have already seen this with light. We now know that light exhibits both wave *and* particle properties, and so the photon model of light was born. Turns out our vision of atoms, protons, and electrons behaving like little solid spheres was off as well, and a better model needed to be built.

You already know that light waves can act like a particle. Photon momentum and the photoelectric effect are examples of light acting like a particle or a photon. If that wasn't weird enough, physicist Louis de Broglie started to contemplate: "If waves can act like particles, then maybe particles can act like waves." This sounded ridiculous, but this crazy idea explains the odd behavior of nano-sized particles like electrons. As it turned out, de Broglie was correct in his groundbreaking assumption.

To understand this, you need a good grasp of how the electromagnetic spectrum behaves:

■ Long wavelengths of light, such as radio waves, demonstrate lots of wave behaviors like interference and diffraction. However, they have so little energy, they don't exhibit much in the way of particle properties like photon momentum or the photoelectric effect.

■ Short wavelengths of light, like X-rays and gamma rays, demonstrate a large amount of photon momentum and the photoelectric effect. However, their wavelengths are so short that it's hard to measure wave properties like diffraction and interference for gamma-ray photons.

Particles behave the same way. If the particle has a short wavelength, it should behave more like a particle, and if the particle has a longer wavelength, it should behave more like a wave. To find the wavelength for a particle, which is called the **de Broglie wavelength**, physicists took the equation for photon momentum and worked it in reverse:

$$\lambda = \frac{h}{p_{particle}} = \frac{h}{mv}$$

where:

■ λ is the de Broglie wavelength in m.

■ h is Planck's constant. (This time you have no choice; you must use $h = 6.63 \times 10^{-34}$ J·s because of the other units used in the equation.)

■ m is the mass the particle in kg.

■ v is the velocity of the particle in m/s.

Let's concentrate on calculating the wavelength of a particle first, and then we will discuss what the wavelength means. Here is an example.

> **EXAMPLE**
>
> Find the de Broglie wavelength of a 0.15 kg baseball moving at 40 m/s:
>
> $$\lambda = \frac{h}{p_{particle}} = \frac{h}{mv} = \frac{6.63 \times 10^{-34} \, \text{J} \cdot \text{s}}{(0.15 \, \text{kg})(40 \, \text{m/s})} = 1.1 \times 10^{-34} \, \text{m}$$

If this seems small, it certainly is. In fact, it is smaller than gamma rays, which means the baseball will behave like a particle. It won't diffract around objects or show interference as you might expect from a wave. When it collides with another object, it will act like a particle because its wave wavelength is just too small to exhibit any wave properties.

To get more wavelike behavior from a particle, I need to increase its wavelength. To do that, I have to decrease the momentum of the particle by decreasing the mass and/or the velocity of the particle.

> **EXAMPLE**
>
> Let's find the de Broglie wavelength of one of the smallest particles: an electron moving at 3.0×10^6 m/s.
>
> $$\lambda = \frac{h}{p_{particle}} = \frac{h}{mv} = \frac{6.63 \times 10^{-34} \, \text{J} \cdot \text{s}}{(9.11 \times 10^{-31} \, \text{kg})(3.0 \times 10^6 \, \text{m/s})} = 2.4 \times 10^{-10} \, \text{m}$$
>
> Now we are talking! This might seem very small, but it is right in the wavelength range of an X-ray.

Okay, so we can calculate the supposed wavelength of an electron, but what does it mean? How can electrons, which are particles with mass, show wave behavior? If we shoot electrons at a double slit, will they form an interference pattern? Well, in fact they do! If we use a gun to shoot electrons at a barrier with two openings, as shown in the following figure, the electrons don't just land anywhere. Most of them fall in alternating bands. After many electrons are fired, we clearly see what is happening. The electrons are forming an interference pattern just like photons of light do. The electrons are exhibiting classic wave properties! So de Broglie was right: particles can have wave properties and exhibit wave-particle duality just like photons of light.

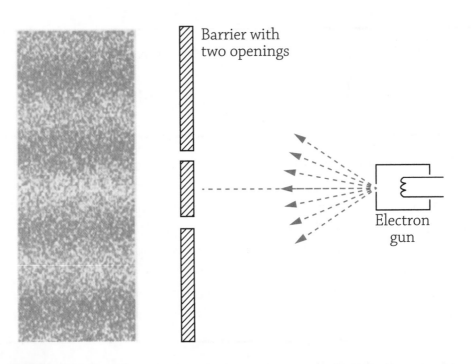

Barrier with
two openings

Electron
gun

IRL In Chapter 16, we talked about Rosalind Franklin using X-rays shot at DNA to form an interference pattern. We can do the same thing with electrons. When electrons with the same wavelength as X-rays are fired at an atomic structure, a very similar interference pattern occurs, which means the electrons are undergoing diffraction and interference. Electron microscopes take advantage of the wave properties of electrons to form amazingly clear images of super-small objects.

Photograph of ant taken with electron microscope.
By Greg Meeker, U.S. Geological Survey.

The Wave Function

If particles have wave behaviors, what does their wave look like and what does it mean? Particles will have a **wave function**, Ψ, as a function of location. Let's discuss what the wave function means with an example.

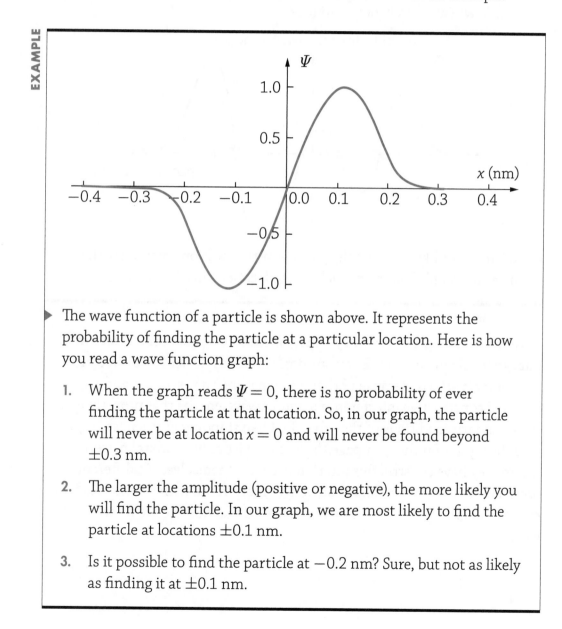

The wave function of a particle is shown above. It represents the probability of finding the particle at a particular location. Here is how you read a wave function graph:

1. When the graph reads $\Psi = 0$, there is no probability of ever finding the particle at that location. So, in our graph, the particle will never be at location $x = 0$ and will never be found beyond ± 0.3 nm.

2. The larger the amplitude (positive or negative), the more likely you will find the particle. In our graph, we are most likely to find the particle at locations ± 0.1 nm.

3. Is it possible to find the particle at -0.2 nm? Sure, but not as likely as finding it at ± 0.1 nm.

Here is another example.

▶ Look at the wave function in the following figure. Where will the particle most likely be found? Are there any locations where the particle will absolutely not be found?

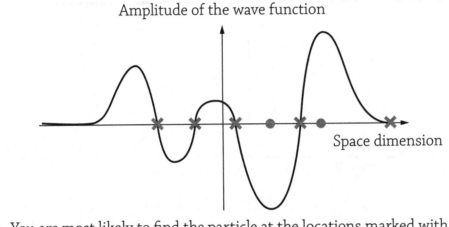

Amplitude of the wave function

Space dimension

▶ You are most likely to find the particle at the locations marked with a circle. You won't find the particle at the locations marked with an X.

Another example of the wavy behavior of electrons is in the **electron energy levels** in an atom. Electrons don't orbit the nucleus in a circular "planetary" pattern. Because of their wave nature, they set up standing waves of constructive interference in whole wavelengths that properly match the boundary conditions of the orbit. Just like standing waves in strings. (See the figures on the next page.) They can exist only in orbital patterns where they have constructive interference with themselves. At different orbital radii, they will exhibit destructive interference with themselves and cancel themselves out. They simply cannot exist at destructive interference locations.

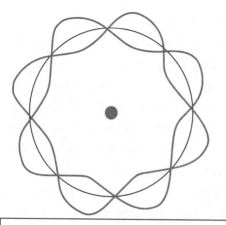

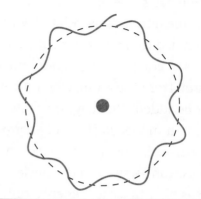

| Electron forming a constructive standing wave. The electron can exist in this orbit. | Electron in an orbit that causes destructive interference. The electron cannot exist in this orbit. |

Let's examine this orbital behavior closer.

Electron Energy Levels in an Atom

Niels Bohr discovered that electrons existed only in specific orbital energy locations. We now know that this is due to the constructive and destructive wave nature of electrons. For an electron to move from one of these **energy levels** to another, the electron has to either absorb or release a little bit of energy in the form of a photon. To make this easier to understand, physicists came up with a simple picture called an **energy level diagram** that helps us visualize what is going on.

In the following figure, you see an energy level diagram for a hypothetical atom, where n_1, n_2, and n_3 represent the first three energy levels of the electrons in this atom. Another way to look at it is that if the electron is in the n_1 level, the path the electron takes around the atom is like a standing wave of one wavelength. For n_2, the electron path would be a standing wave of two wavelengths, and so on. n_∞ means that the atom is **ionized** and the electron has had enough energy added to it to completely remove it from the atom. The numbers associated with each level tell you how much energy has to be added to the electron, at that particular energy level, to remove it from the atom. For example, in our hypothetical atom, to completely remove the electron from the atom, when the electron is in the n_1 level, 10 eV of energy must be added. The diagram shows $n_1 = -10$ eV; the negative sign is telling you the atom needs 10 eV of energy before it can be ionized. Remember that the nucleus is positive, the electron is negative, and it takes work to "pull" the electron away from the nucleus. If the electron is in a higher energy level, such as n_2, it takes less energy, only 7 eV, to remove the electron. Bohr called this little bundle of energy a "**quanta.**" These quanta of energy are delivered in the form of photons of light.

$$n_\infty = 0 \text{ eV} \quad \underline{\hspace{4cm}}$$

$$n_3 = -6.0 \text{ eV} \quad \underline{\hspace{4cm}}$$

$$n_2 = -7.0 \text{ eV} \quad \underline{\hspace{4cm}}$$

$$n_1 = -10.0 \text{ eV} \quad \underline{\hspace{4cm}}$$

Electrons can move from one energy level to another by either absorbing a photon or emitting a photon of light. Here is how this works:

- n_1 is called the ground state. It is the lowest possible energy level for the electron.

- Moving from a lower to a higher energy level requires the electron to absorb a photon with exactly the delta energy difference between the two energy levels. For instance, for an electron to jump up from energy level n_1 to n_2, it must absorb a photon of exactly 3 eV.

- Moving from a higher to a lower energy level requires the electron to emit a photon with exactly the delta energy between the two energy levels. For instance, for an electron to fall downward from energy level n_3 to n_2, it must emit a photon of exactly 1 eV.

- The electron can move only from one exact energy level to another. There are no intermediate levels, such as $n_{1.333}$. If you think of this like a set of stairs, when you move up and down the stairs, you have to put your foot on only one step. So, you can be standing on the first step, second step, third step, and so on, but you cannot land on the 1.333 step. In the same way, the electron can move up and down only directly from one energy level to another. Electrons don't ever exist at an intermediate level because of destructive wave interference.

- To find the energy of the absorbed or released photon, all you have to do is subtract the energies of the two energy levels.

- If an electron absorbs a photon with more than is needed to ionize the atom, the extra energy is carried away with the electron in the form of kinetic energy. For instance, let's say the electron starts in energy level n_3 and absorbs an 8 eV photon. This is 2 eV more than the electron needs to be ionized from the atom, so the electron departs the atom with 2 eV of kinetic energy.

Let's look at a couple of examples.

EXAMPLE

▶ What energy photon is absorbed or released in our hypothetical atom when an electron moves from $n = 3$ to $n = 1$?

▶ Easy! Simply calculate the difference in the two energy levels, which turns out to be 4 eV. Since the electron moved from a higher energy level to a lower energy level, a 4 eV photon must be released or emitted from the atom.

▶ Now let's assume our hypothetical atom is in the ground state when a 3.5 eV photon is incident upon it. Describe what happens.

▶ Well, we know that the electron can move only from one exact energy level to another. So, for the electron to jump from $n = 1$ to $n = 2$, the atom must absorb exactly 3 eV, and to jump from $n = 1$ to $n = 3$, the atom must absorb 4 eV.

▶ Since there is no exact jump equal to 3.5 eV, the photon passes through the atom and does nothing. It is impossible for the photon to be absorbed by the electron.

One more thing to remember about electron quantum jumps:

■ When the atom releases energy going from a higher to a lower energy level, every and any possible combination of downward energy level jumps may happen.

Let me explain what that means. If an atom is in the $n = 4$ level and is returning to the $n = 1$ level, how many different photons might be released on its downward path to the ground state?

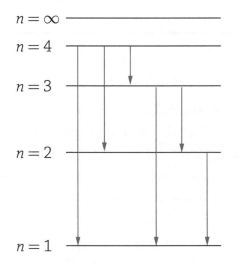

Electrons can jump directly from the $n = 4$ level directly to the $n = 1$ level or they can jump from $n = 4$ to 3 to 2 to 1, releasing three photons. Jumping down from $n = 4$ to 2 to 1 is another possibility.

So, every possible jump down from every level can happen, as shown in the previous figure. Six possible photons could be emitted in this process.

When electrons in a material fall downward into lower energy levels, as shown earlier, they produce what is called an **emission spectrum** that is unique to the material. (See the figure below.) Notice how each element emits a different set of wavelengths of light. The visible bands of hydrogen combine to produce a purplish light, and sodium produces a yellowish light. Scientists can use this unique "photon fingerprint" to determine what the material is.

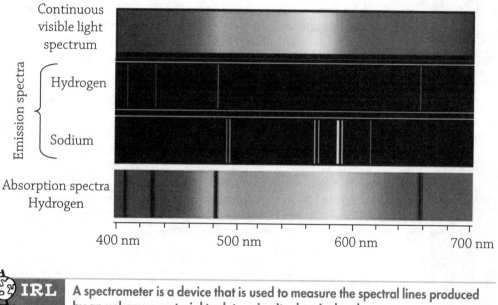

IRL A spectrometer is a device that is used to measure the spectral lines produced by an unknown material to determine its chemical makeup.

When electrons in a material jump upward by absorbing photons, unique bands of light are removed from the light striking the material. This is called an **absorption spectrum** and shows up as dark bands removed from the continuous spectrum. The previous figure shows an absorption spectrum for hydrogen. Notice how it matches the same color and frequency of the hydrogen in the emission spectrum.

▶ An astronomer collects light from a distant star. How could they use this light to determine what the star is made of?

▶ Astronomers can look for emission spectrum lines and match them up with the photoemission spectrum of the elements to discover what elements are present in the star.

IRL As a child, you might have had glow-in-the-dark stars on your ceiling. This is called *phosphorescence*, and is produced when electrons are excited by light, which moves the electrons up to a *meta-stable electron energy level*. Normally, electrons fall quickly back to the ground state after jumping to higher energy states. But in meta-stable states, the electrons "hang out" for a while before finally falling back down to the ground state. In a sense, the electrons have stored energy that they release over time. This is why your glow-in-the-dark stars are bright at first but slowly fade to black as all of the stored energy is released as photons over time.

IRL A *laser* (Light Amplification by Stimulated Emission of Radiation) is produced when the electrons in an atom become excited, or "pumped," from the ground state into a meta-stable state, where the electrons hang out waiting to be released. When one of the electrons in this state falls to the ground state, it stimulates others to fall with it. This creates a cascade of falling electrons, all emitting photons with the same single wavelength, and all in phase moving together, which creates a coherent beam of intense light.

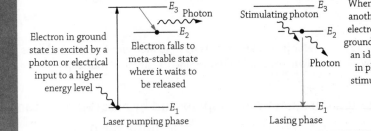

Electron in ground state is excited by a photon or electrical input to a higher energy level

Electron falls to meta-stable state where it waits to be released

Laser pumping phase

When stimulated by another photon, the electron drops to the ground state, releasing an identical photon in phase with the stimulating photon

Lasing phase

REVIEW QUESTIONS

Let's show what we have learned in the wonderfully weird world of the ultra-fast and the ultra-small by answering the following questions.

1. When you go for a jog, do you experience the effects of relativity?

You are in a fast-moving rocket traveling toward a distant space station, and I am on the ground at NASA monitoring your progress. Use this scenario to answer questions 2 to 5.

2. Will our watches run at the same speed?

3. From your perspective in the rocket, why don't you age much on the trip?

4. From my perspective on the ground, why don't you age much on the trip?

5. Your rocket is traveling at 80 percent the speed of light. You turn on the headlights on the spaceship. What speed do I see the light leaving your spaceship? What speed do you see the light leaving the spaceship?

6. Convert 200 eV into joules.

7. Which has more energy, orange light or green light?

8. Which has a longer wavelength, UV or IR light?

9. Which has a higher frequency, microwaves or gamma rays?

10. How much energy do the photons in a 532 nm green laser have? Give your answer in both joules and eV.

11. Calculate the energy of a 96.7 MHz radio wave. Give your answer in both joules and eV.

12. For sound waves, a louder sound means a larger amplitude of the wave. For light, what does a brighter light mean?

A particular metal is photosensitive to green light and higher energy photons. Use this information to answer questions 13 to 15.

13. Which produces more electricity from the metal: shining red light or violet light on the metal?

14. When we shine brighter green light on the metal, what happens to the electrons that are ejected from the surface?

15. When ultraviolet light shines on the metal, what happens to the energy of the ejected photons?

The work function of cesium is 2.1 eV. Use this information to answer questions 16 and 17.

16. Calculate the minimum frequency of light that will eject photoelectrons from the cesium.

17. A photon with a wavelength of 300 nm strikes the cesium. Calculate the kinetic energy of the ejected electrons.

18. For each of the following phenomena, indicate which are best understood by considering the particle properties of light and which are best understood by viewing light as a wave.
    a. Visible light shining on a multislit grating will create an alternating pattern of bright and dark bands.
    b. Ultraviolet light of a frequency below a minimum frequency will fail to cause electrons to be emitted from a metal.
    c. X-rays incident on a smooth crystalline surface create a diffraction pattern.
    d. An X-ray striking an electron causes a change in direction of the electron's motion.

When a stationary electron is struck by a photon of light, the electron takes off with kinetic energy and momentum. Use this information to answer questions 19 and 20.

19. How did the electron acquire its new momentum and energy?

20. What happens to the photon in the collision?

An electron has a de Broglie wavelength of 5.3×10^{-10} m. Use this information to answer questions 21 and 22.

21. How fast is the electron traveling?

22. How can we decrease the wavelength of the electron?

The following figure represents the wave function of an exotic new particle. Use this graph to answer questions 23 and 24.

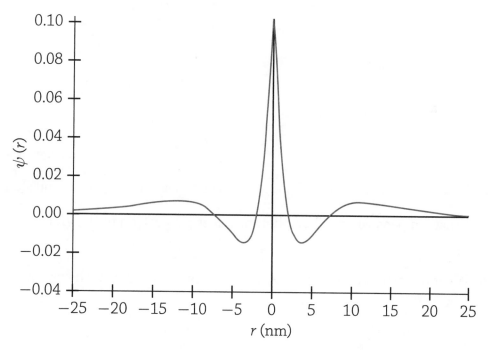

23. On the figure, circle the locations where the particle is most likely to be found. Rank the locations from most likely to least likely to find the particle.

24. Place an X on the horizontal axis where the particle will never be found.

The next figure shows an energy-level diagram of a theoretical atom. Use it to answer questions 25 to 28.

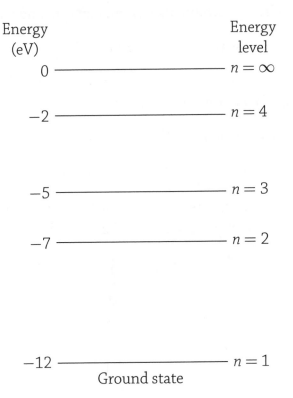

25. What happens when a 7 eV photon strikes an electron in the ground state?

26. What happens when a 9 eV photon strikes an electron in the ground state?

27. What happens when a 15 eV photon strikes an electron in the ground state?

28. What are all the possible photon emission energies from an electron starting in the $n = 4$ energy level? Sketch all the possible paths the electron can take to return to the ground state.

An electron transitions between the following energy levels:

    a. $n = 4$ to $n = 3$
    b. $n = 4$ to $n = 2$
    c. $n = 3$ to $n = 2$
    d. $n = 2$ to $n = 1$

Use this information to answer questions 29 and 30.

29. Rank the emitted photons from highest to lowest frequency.

30. Rank the emitted photons from longest to shortest wavelength.

Flashcard App

Nuclear Physics

MUST KNOW

⚡ Energy and mass are connected by the equation $E = mc^2$. All atomic nuclei have a smaller mass than the sum of their parts. This mass defect contributes to the binding energy of the nucleus and the nuclear strong force, which holds the nucleus together.

⚡ Four key quantities are always conserved during a nuclear reaction: conservation of mass/energy, conservation of nucleon number, conservation of charge, and conservation of momentum.

⚡ There are three main nuclear reactions: decay, fission, and fusion. The decay rate of a nucleus is measured in half-lives.

This chapter takes us to the very center of the atom itself: the nucleus. Our goal in this chapter is to investigate how the nucleus holds together and also how it can fall apart and **transmute** into something new.

Long ago, **alchemists** dreamed of turning lead into gold. Sir Isaac Newton himself tried to accomplish the feat of **transmutation**. Eventually, the women and men of science convinced themselves that alchemy was a myth akin to magic. But if you have learned anything at all from physics, it's that the universe is filled with many wonderous and unexpected things that continue to surprise.

In 1895 Wilhelm Roentgen discovered X-rays, which was a new part of the electromagnetic spectrum that had not yet been discovered. This is generally considered the beginning of the modern physics revolution, that over the next 30 or so years, transformed physics forever. Roentgen's discovery sparked a cascade of new research that led Henri Becquerel to accidently discover another new form of radiation from uranium, just a few months after X-rays had been discovered. This, in turn, led Marie Curie to investigate the phenomena and, in 1898, along with her husband Pierre, who had joined her research, began discovering the new radioactive elements of polonium, thorium, and radium. Marie Curie was the first to coin the term "**radioactivity**" to describe this strange new form of radiation.

At the same time all of these discoveries about the atom and radioactivity were going on, Einstein was fleshing out the details of relativity, the particle nature of light, and the wave nature of nano-particles (wave-particle duality). The first part of the twentieth century was an amazing time of discovery for physics!

Soon after, Ernest Rutherford, who is considered the father of nuclear physics, entered the game. In 1899, Rutherford separated radioactivity into three distinct varieties, which he called **alpha**, **beta**, and **gamma radiation**. In 1908, Rutherford bombarded a thin foil of gold with alpha particles to investigate what the atom's structure looked like. This is kind of like trying to figure out the shape of an invisible object by throwing baseballs at it! To his great surprise, he discovered that the atom was mostly empty space and that all of the positiveness of the atom was concentrated in a tiny nucleus.

Rutherford continued his bombarding ways by exposing other materials to alpha radiation. In 1919, he exposed nitrogen gas to alpha radiation and transmuted some of the nitrogen into oxygen. In the process, Rutherford became the world's first successful alchemist by turning one element into another.

IRL Today, it is possible to create gold using a bombardment technique similar to the one used by Rutherford. The only problem is it costs about a trillion times more to create gold in a particle accelerator than the actual gold produced is worth!

The last naturally occurring element is uranium. All elements heavier than uranium have been created by humans through bombardment.

Hang on, physics fans! We are diving into the nucleus to discover its mysteries.

Subatomic Particles

Atoms are made up of three smaller subatomic particles: **protons, electrons,** and **neutrons**. Each of these has unique properties, as shown in the following table.

Particle	Symbol(s)	Mass (kg)	Charge (C)
proton	p, p$^+$	$m_p = 1.673 \times 10^{-27}$ kg	$+e = 1.6 \times 10^{-19}$ C
neutron	n	$m_n = 1.675 \times 10^{-27}$ kg	zero
electron	e, e$^-$, β$^+$, β$^-$	$m_e = 9.11 \times 10^{-31}$ kg	$-e = -1.6 \times 10^{-19}$ C
photon*	γ	zero	zero

*This makes four particles in the table, when we just named three. More on photons coming up!

Protons and neutrons are almost the same mass and are much more massive than the tiny election. Protons and electrons have the exact same magnitude of charge, but the proton is positive and the electron has a negative charge. Neutrons are neutral and don't have any charge at all. The protons and neutrons reside in a dense nucleus, with the tiny electrons filling up a very large cloud of volume surrounding the nucleus, as seen in the following figure.

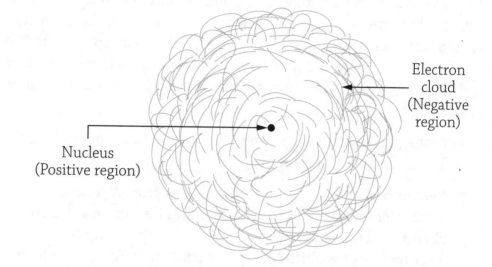

Almost all of the atom's mass is in the nucleus, but almost all of the volume of the atom is the electron cloud surrounding the nucleus, as seen in the figure. Individual atoms have different numbers of protons, neutrons, and electrons. The number of protons in an atom determines which **element** the atom is. Elements can have a different number of neutrons. Some carbon atoms have six neutrons, while others have eight. These are called **isotopes** of the same element. There are even more subatomic particles. In the 1,950s, **neutrinos** were discovered. In the 1,960s, **quarks**, which are the building blocks of protons and neutrons, were discovered. In our investigation of nuclear physics, we are going to stick with protons, neutrons, electrons, and **photons** (which are also listed in the previous table).

Nuclear Decay

When Becquerel, Curie, and other scientists first investigated elements that emitted radiation, they didn't really know what they were witnessing. The subatomic particles (protons, neutrons, and electrons) had not been definitively named—or even discovered yet. But physicists did notice that certain kinds of particles emerged repeatedly from their experiments. They categorized these by their different properties and called them alpha, beta, and gamma radiation. Years later, physicists found out what this radiation actually was:

- **Alpha particle (α)**   Two protons and two neutrons stuck together. This is the same thing as a helium nucleus.

- **Beta particle (β)**   An electron or a **positron**. A positron is the antimatter equivalent of an electron. It has the same mass as an electron and the same amount of charge, but the charge is positive. Since there are two different types of beta particles, we often write them as $\beta+$ for the positron and $\beta-$ for the electron.

- **Gamma ray (γ)**   A gamma-ray photon of electromagnetic radiation.

The alpha particle is by far the most massive and has a net charge of $+2e$. Alpha particles move the slowest of the three and hardly have any penetrating power. They can be stopped with a single sheet of paper. Beta particles are high-speed electrons and have more penetrating power. It takes a thin sheet of copper or aluminum to stop beta radiation. Gamma particles are massless, chargeless, and travel at the speed of light. It takes about a half of a meter of lead or two meters of concrete to stop most gamma radiation. You can see that gamma radiation is the most dangerous of the three.

By definition, any electromagnetic radiation that comes from the nucleus of the atom is called a gamma ray. Gamma rays have the highest energy and frequency in the electromagnetic spectrum.

Nuclear Notation

Before we can progress any further, we need to discuss nuclear notation. The two most important nuclear properties are the **atomic number** and its **mass number**. The atomic number, Z, tells us how many protons there are in the nucleus; this number determines what element we're dealing with, because each element has a unique atomic number. The atomic number also tells you the net positive elementary charge of the nucleus. The mass number, A, tells the total number of nuclear particles, or **nucleons**, a nucleus contains. It is equal to the number of protons plus the number of neutrons in the nucleus. Like the name implies, the mass number tells you the approximate mass of the nucleus in **atomic mass units,** "u" ($1\,u = 1.66 \times 10^{-27}\,kg$). Isotopes of an element have the same atomic number but different mass numbers.

A nucleus is represented using the following notation: $^A_Z X$, where X is the symbol for the element we're working with. For example, $^4_2 He$ represents helium (He) with two protons and four nucleons. (This means that there will be two neutrons.) A different isotope of helium that contains three neutrons would be represented by the symbol $^5_2 He$. You need to recognize the significance of the atomic number and the mass number. Both numbers are conserved in nuclear reactions. This is explained in detail in the following sections.

EXAMPLE

▶ Here is an isotope of gold: $^{197}_{79}\text{Au}$

▶ How many protons does gold have?

▶ The atomic number gives us the number of protons. Therefore, we have 79 protons.

▶ How many neutrons does this isotope of gold have?

▶ $197 - 79 = 118$ neutrons

▶ How many nucleons does this isotope of gold have?

▶ 197 nucleons = 79 protons + 118 neutrons = mass number

▶ What is the approximate mass of this isotope of gold?

▶ 197 u = mass number

Alpha Decay

Alpha decay happens when a nucleus emits an alpha particle. Since an alpha particle has two neutrons and two protons, the **daughter nucleus** (the nucleus left over after the decay) must have two fewer protons and two fewer neutrons than it had initially. Here is an example:

EXAMPLE

▶ Uranium-238 ($^{238}_{92}\text{U}$) undergoes alpha decay. Let's figure out the atomic number and mass number of the nucleus produced by the alpha decay.

▶ The answer is found very simply with basic arithmetic. The atomic number decreases by 2, to 90. The mass number decreases by 4, to 234. The element that is formed is thorium. This alpha decay can be represented by the following equation:

$$^{238}_{92}U \rightarrow \,^{234}_{90}Th + \,^{4}_{2}\alpha$$

▶ Think of this as an algebraic equation, but instead of an equal sign, there is an arrow to represent that the uranium is turning into something new.

▶ Notice how the mass number in the top row and the atomic number in the bottom row of the equation on the left equals the sum on the right side of the equation. This must always be true.

Beta Decay

In beta decay, a nucleus emits either a positron ($\beta+$ decay) or an electron ($\beta-$ decay). Because beta particles have very little mass and do not reside in the nucleus, the total mass number of the nucleus will stay the same after beta decay. The total charge must also remain the same, as charge is a conserved quantity.

EXAMPLE

▶ Let's consider an example of neon (Ne) undergoing $\beta+$ decay:

$$^{19}_{10}Ne \rightarrow \,^{19}_{9}F + \,^{0}_{1}e$$

▶ Here I have used $^{0}_{1}e$ to indicate the positron. Notice how the positron has a zero mass number. Therefore, the mass number of the daughter nucleus stayed the same. But look at the total charge present. Before the decay, the neon nucleus has a charge of +10. After the decay, the total net charge must still be +10, and it is: +9 for the protons in the fluorine (F) and +1 for the positron. Effectively, in $\beta+$ decay, a proton turns into a neutron and emits a positron.

▸ In β– decay, a neutron turns into a proton, as in this decay process shown for carbon-14 decay:

$$^{14}_{6}C \rightarrow \,^{14}_{7}N + \,^{0}_{-1}e$$

▸ Once again, the mass number of the daughter nucleus didn't change. But the total charge of the carbon nucleus was initially +6. Thus, a total charge of +6 has to exist after the decay as well. This is accomplished by the electron (charge −1) and the nitrogen (charge +7). In beta decay, a **neutrino** or an **antineutrino** must also be emitted to carry off some extra energy. But that doesn't affect the products of the decay—just the kinetic energy of the products. We are going to ignore neutrinos and leave that for you to discover in your next physics class!

Gamma Decay

A gamma ray is an electromagnetic photon, a particle of light. It has no mass and no charge. So, a nucleus undergoing gamma decay does not change its outward appearance:

$$^{238}_{92}U \rightarrow \,^{238}_{92}U + \gamma$$

The gamma ray simply carries away excess energy and momentum from the nucleus. Remember that photons have momentum just like particles with mass. So, when the photon leaves, the nucleus must recoil just like when a bullet leaves a gun.

Here is a decay we have not discussed: neutron decay.

EXAMPLE

See if you can fill in the blank in the following figure with what is missing.

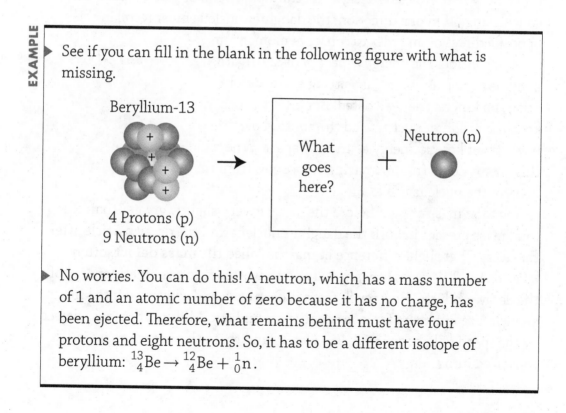

Beryllium-13

4 Protons (p)
9 Neutrons (n)

What goes here?

$+$

Neutron (n)

No worries. You can do this! A neutron, which has a mass number of 1 and an atomic number of zero because it has no charge, has been ejected. Therefore, what remains behind must have four protons and eight neutrons. So, it has to be a different isotope of beryllium: $^{13}_{4}\text{Be} \rightarrow {}^{12}_{4}\text{Be} + {}^{1}_{0}\text{n}$.

$E = mc^2$ and Conservation During Nuclear Reactions

In our discussion of nuclear decay, I have hinted at conservation rules that nature obeys during nuclear reactions. Let's take a look at them in more detail now.

Imagine that a nucleus is at rest and then an alpha particle shoots off from it during radioactive decay. This is just like what we learned way back in Chapter 3. When an object at rest explodes into pieces, the fragments must move off in opposite directions due to conservation of momentum. The exploding pieces are exerting equal opposite-direction forces on each other. So, when a nucleus decays and the alpha, beta, or gamma particles go whizzing off in one direction, the daughter nucleus must recoil in the opposite direction in order to conserve momentum.

Now consider the energy aspect of decay. Again, imagine a nucleus at rest just sitting there. Before decay, no kinetic energy existed. But after the decay, both the daughter nucleus and the emitted particle have lots of kinetic energy as they fly apart. Where did this energy come from? Amazingly, it comes from the mass of the nucleus itself.

The total mass present before the decay is very slightly greater than the total mass present in both the daughter nucleus and the decay particle after the decay. That slight difference in mass is called the **mass defect**, often labeled Δm. This mass is destroyed and converted into the kinetic energy of the decay products. How much mass is destroyed and converted into kinetic energy? Use Einstein's famous equation to find out. Multiply the mass defect by the speed of light squared: $E = (\Delta m)c^2$ and that is how much mass is converted into energy.

Okay, so now we have all the conservation laws that must be obeyed in all nuclear reactions. Let's review them before moving on. Things that are absolutely positively conserved during all nuclear reactions:

- **Conservation of mass/energy**   Before a nuclear reaction, the combination of energy in the form of mass, photon energy, kinetic energy, etc., must equal this same value after the nuclear reaction.

In every naturally occurring nuclear reaction, some of the mass of the original nucleus is lost to other forms of energy, such as photons or kinetic energy, and the final mass is less than the original mass. Less mass means more nuclear stability. We will discuss this in more detail later.

- **Conservation of momentum**   When a nucleus undergoes decay and shoots off a particle, there is a recoil in the daughter nuclei, just like when a bullet is fired from a gun.

- **Conservation of charge**   The total charge of the nuclear material before and after a nuclear reaction must be the same. The sum of the atomic numbers, which tells us the charge before, and the sum of the atomic numbers after the nuclear reaction are equal.

- **Conservation of nucleons**   Protons and neutrons reside in the nucleus and are referred to as nucleons. The sum of the protons and neutrons is equivalent to the mass number. The sum of the neutrons and protons before and after the nuclear reaction must be the same.

Half-Life

Every radioactive element will decay at a different rate. In fact, every radioactive isotope of the same element has its own unique decay rate. The decay rate is measured in **half-life**, which is the time it takes for half of the radioactive atoms to decay (transmute) into a new daughter nucleus, leaving only half of the original nucleus behind. Uranium-238 has a half-life of 4.5 billion years. Some elements have a half-life of less than a second! So don't blink, or it will already have transmuted into something new!

Which material is more dangerous: one with a long half-life like uranium or one with a short half-life? A short half-life = fast decay rate = lots of radiation in a short period of time and then it is all gone. So short-half-life materials are likely to kill you quickly and then disappear/transmute into something else. A long-half-life = slow decay rate = small radiation emissions = little danger to you. But this also means the material will be around for a long time.

EXAMPLE

▶ Carbon-14 has a half-life of 5,730 years. Look at the graph on the next page. Starting at time = 0, we have 100 percent of the original carbon-14. After 5,730 years, only 50 percent is left. After two half-lives, only 25 percent is left (that's ½ times ½).

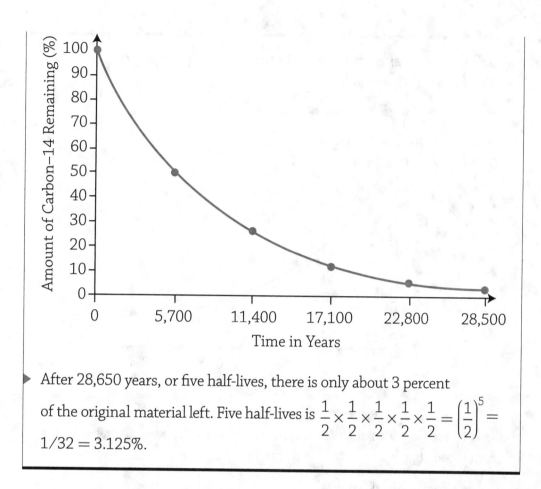

After 28,650 years, or five half-lives, there is only about 3 percent of the original material left. Five half-lives is $\frac{1}{2} \times \frac{1}{2} \times \frac{1}{2} \times \frac{1}{2} \times \frac{1}{2} = \left(\frac{1}{2}\right)^5 = 1/32 = 3.125\%$.

How do scientists measure half-life? Well, they can't actually count the atoms because they are tiny, and we can't directly see atoms anyway. So, scientists measure the radioactive activity with a **Geiger counter**. A Geiger counter literally counts the number of decay products like alpha and beta particles that are emitted by the radioactive material.

BTW

As far as we have been able to determine, decay rates appear to be independent of external conditions like temperature, electric or magnetic fields, or even chemical reactions. This means that we can use half-life as a consistent measure of time.

▶ Look at the Geiger activity count in the following graph. What is the half-life of this material?

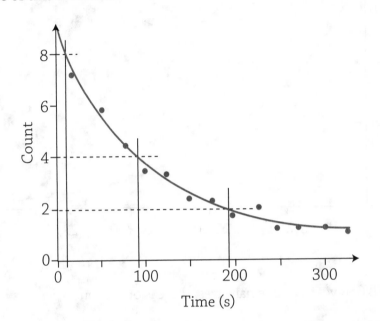

▶ Look to see how long it takes the count to be cut in half. The time to go from 8 counts per second to 4 counts per second is about 80 seconds.

▶ This is experimental data, which always has some errors or uncertainty, so let's check with another data point. The time it takes to go from 4 to 2 counts is about 100 s. So, we will take the average and say that the half-life is approximately 90 seconds for this sample.

▶ Scientists would use a more precise statistical model to get a better number than our quick estimate.

IRL Trying to experimentally determine the half-life of a radioactive material with my students is problematic. We need a material that has a short enough half-life that we can easily measure in the class time we have available. Unfortunately, materials with a short half-life are more highly radioactive and can be dangerous. But we can't use something safe like uranium, because its half-life of 4.5 billion years is way too long for us to notice any changes. So we measure the half-life of...root beer foam. We pour the root beer into a graduated cylinder and measure how much foam there is over time. Turns out root beer foam has a decay curve exactly like radioactive elements. Plus, we can drink it! There are videos online if you would like to see how it's done and drink some root beer—I mean "perform the experiment."

Carbon-14 Dating

Radiation from the Sun interacts with molecules in the upper atmosphere producing high-speed neutrons. Some of these neutrons collide with nitrogen-14, creating carbon-14, as shown in the following nuclear equation:

$$\ _0^1 n + \ _7^{14}N \rightarrow \ _6^{14}C + \ _1^1 p$$

Carbon-14 happens to be radioactive and it filters its way through the entirety of Earth's ecosystem. Remember that all life on Earth is carbon based. That means that when plants take in carbon dioxide as they grow, part of their structure is made from radioactive carbon. When animals eat these plants, part of their own structure becomes radioactive as well. But when a plant or animal dies, it is no longer ingesting this radioactive carbon and the carbon-14 eventually decays away into nitrogen in a beta decay. Knowing the background abundance of carbon-14 in the environment, scientists can get a good idea of when a living plant or animal died by checking to see how much carbon-14 remains in the dead plant or animal.

This is called carbon-14 dating and can be used to date organic material back to about 50,000 to 60,000 years, depending on the sample. Dead material older than this does not have enough carbon-14 remaining to get a reasonably accurate reading anymore.

IRL Carbon-14 is not the only radioisotope in the food chain. Bananas contain a lot of potassium, which your body needs to stay healthy. When you eat a banana, you are also eating a tiny fraction of potassium-40, which is a radioactive isotope with a half-life of 1.25 billion years. But no need to worry! You already have way more radioactive potassium in your body than you'll get from a single banana. You would need to eat over a million bananas—at once—to overdose on potassium radiation. You'll die much sooner from potassium poisoning while trying to eat those million bananas. So enjoy a tasty banana!

There are other forms of radioactive dating as well. The most famous is uranium-lead dating, which measures the radioactive decay of two different isotopes of uranium as it turns into lead. This method is used to measure the age of rocks that contain lead. It can measure the age of these rocks from about 10,000 years old to a couple of billion years old.

IRL Like all scientific tests, radioactive dating has experimental uncertainties. This is why when an object is radio-dated, the time range is given or the uncertainty is listed. For instance, a rock formation may be uranium-lead dated to be 2.40 billion years old ±150 million years. This means that, based on the rock formation or the precision of the equipment used, the age range of when the rock was formed falls between 2.25 billion and 2.55 billion years ago.

Other Types of Nuclear Reactions

Decay is not the only type of nuclear reaction. There are lots of different kinds, but they all abide by our four conservation rules. Here are some other nuclear events that you need to know about.

Fission

Fission occurs when a heavy nucleus is split into two relatively large chunks. Fission reactions are begun by shooting a neutron into the nucleus—this creates an unstable nucleus and initiates the reaction. The protons in the nucleus repel each other, which tries to tear the nucleus apart. The nuclear strong force links the nucleons together. The strong force is a short-range attractive force between nearby nucleons. When the high-speed neutron is absorbed, the nucleus deforms. Parts of the nucleus move beyond the reach of the strong force, and the long-range electromagnetic repulsion overpowers the strong force, and the nucleus breaks apart:

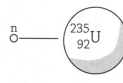   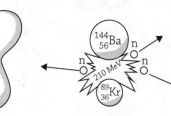

High-speed neutron approaches the nucleus.

Neutron is absorbed, creating an unstable nucleus.

Nucleus deforms beyond the ability of strong force to hold together.

Electrostatic repulsion breaks the nucleus into two big chunks and high-speed neutrons.

Large amounts of energy are created during fission as mass converts to energy in the reaction. Additionally, high-speed neutrons are part of the by-products of fission. These neutrons supply additional "neutron bullets" that can be used to split more atoms. By packing a large number of fissionable atoms together, a **chain reaction** can be achieved, as shown in the next figure. Chain reactions are used in nuclear power plants and nuclear weapons.

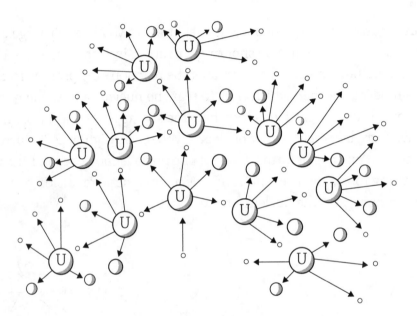

IRL Because fission is used in nuclear weapons, people have a natural and rightly placed fear of it. However, what is surprising to most people is that the electricity produced by a nuclear power plant is much safer than electricity produced by coal, oil, or natural gas due to their mining, pollution, and fire hazards. And this does not even consider environmental, CO_2, and climate change considerations. Nuclear power is even statistically safer than solar, wind, and hydroelectric power. So why are we so scared of nuclear power? Probably for the same reason people are scared of flying. Flying is the safest way to travel, but when there is an accident it is usually catastrophic. The same is true of nuclear power.

Next is an example of a plutonium fission reaction.

EXAMPLE

▶ Can you figure out what goes where the question mark is?

$$^{239}_{94}\text{Pu} + ^{1}_{0}\text{n} \rightarrow ^{100}_{40}\text{Zr} + ^{137}_{54}\text{Xe} + ? + \text{Energy}$$

▶ The mass number on the left adds to 240, and the atomic numbers add to 94. On the right, the zirconium and xenon have mass numbers that add to 237 and atomic numbers that add to 94.

▶ So we are short three nucleons without any charge. This would be accounted for by three neutrons ($3 \times ^{1}_{0}\text{n}$).

Fusion

Fusion occurs when two light nuclei combine to make a new heavier and more stable nucleus. In the figure shown next, you can see that two smaller isotopes of hydrogen collide to form a larger helium nucleus and a neutron. In order to accomplish fusion, the original nuclei must travel toward each other with vast amounts of kinetic energy to overcome the electrostatic repulsion of the two positive nuclei. Once the two nuclei are close enough, the strong force overpowers the electrostatic repulsion, and the two nuclei bind into something larger than before. Large amounts of energy are released in the reaction as some of the mass is converted into energy. This is the process that fuels the Sun.

Here is an example of fusion.

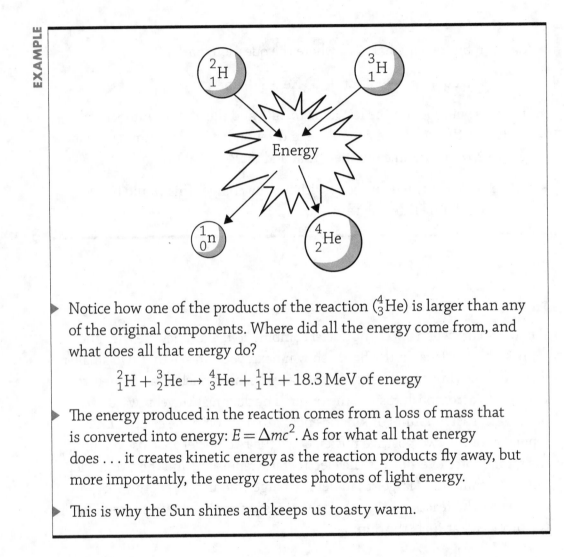

Notice how one of the products of the reaction ($^{4}_{3}$He) is larger than any of the original components. Where did all the energy come from, and what does all that energy do?

$$^{2}_{1}\text{H} + ^{3}_{2}\text{He} \rightarrow ^{4}_{3}\text{He} + ^{1}_{1}\text{H} + 18.3\,\text{MeV of energy}$$

The energy produced in the reaction comes from a loss of mass that is converted into energy: $E = \Delta mc^2$. As for what all that energy does . . . it creates kinetic energy as the reaction products fly away, but more importantly, the energy creates photons of light energy.

This is why the Sun shines and keeps us toasty warm.

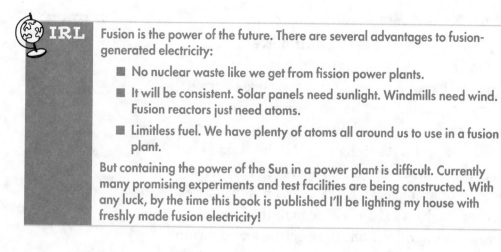

IRL Fusion is the power of the future. There are several advantages to fusion-generated electricity:

- No nuclear waste like we get from fission power plants.
- It will be consistent. Solar panels need sunlight. Windmills need wind. Fusion reactors just need atoms.
- Limitless fuel. We have plenty of atoms all around us to use in a fusion plant.

But containing the power of the Sun in a power plant is difficult. Currently many promising experiments and test facilities are being constructed. With any luck, by the time this book is published I'll be lighting my house with freshly made fusion electricity!

Induced Reactions

Sometimes scientists bombard a nucleus with high-speed particles to see what will happen. This is called an **induced reaction** because you are causing a reaction to happen. Rutherford was the first to do this in 1919 when he bombarded a stable isotope of nitrogen with a particle to induce a transmutation into oxygen and a proton. Here is the Rutherford reaction. What did he bombard the nitrogen with, and where do you think he got this particle from?

$$? + {}^{14}_{7}\text{N} \rightarrow {}^{17}_{8}\text{O} + {}^{1}_{1}\text{p}$$

The unknown in the nuclear equation must have a mass number of 4 and an atomic number of 2. Rutherford must have bombarded the nitrogen with an alpha particle or helium nucleus. Rutherford got his alpha particles from the alpha decay of other radioactive isotopes like uranium.

BTW

All the elements beyond uranium, element 92, are called transuranium elements. They are all radioactive, and they are not found in nature. Neptunium was the first transuranium element to be synthesized. In 1940, scientists from the University of California, Berkeley, bombarded uranium-238 with neutrons to produce uranium-239, which then underwent beta decay to form neptunium-239.

Antimatter and Annihilation

Every "normal" particle has an **antimatter** twin. For instance, an electron has an antimatter twin called a **positron**. Positrons are exactly like electrons in every way, except they have a positive charge. A curious behavior is that when matter and antimatter meet, they **annihilate** each other. All the mass of both particles is converted into energy in the form of photons. For example, an electron and positron greet each other and shake hands as seen in the following figure. Oh no, annihilation! The electron and positron disappear and turn into photon energy. How much energy will be converted into photon energy? Both particles turn completely into energy. Both have the same mass as an ordinary electron: $E_{electron} + E_{positron} = 2(m_{electron})c^2 = hf$.

Why does the electron-positron annihilation not just produce one photon? Think about all the conservation laws we have to abide by during a nuclear reaction.

Well, we have obeyed conservation of charge because the net charge of the two particles was zero and the photon has no charge either. We have obeyed conservation of mass/energy because all the mass has converted into energy. Hmmm...what about conservation of momentum? The electron-positron duo was at rest with a total momentum of zero when they shook hands. Due to conservation of momentum, we can't just have one photon as a product of the reaction because then there would be a net photon momentum after the reaction. Therefore, we must have at least two photons—of the same energy—traveling in opposite directions so that the final momentum is zero, just like the initial momentum.

Wait a minute.... If matter and antimatter annihilate to produce photons, can a photon be used to create matter? Sure! As long as we abide by the conservation laws that the universe demands. In fact, physicists have created matter-antimatter pairs using photons or light. But before you get excited about using light to make yourself a new car, just remember that it takes a tremendous amount of photon energy to produce mass. Remember Einstein's equation: $E = mc^2$. It would take about 10^{20} joules of light energy to make a car. That's equivalent to the electricity produced by a million power plants running for a year. It's probably cheaper to just buy a car instead!

Mass Defect, Binding Energy, and the Strong Nuclear Force

Carbon-12 is made up of six protons and six neutrons. We know the mass of a proton. We know the mass of a neutron. So, you would think that if we added the mass of six protons to the mass of six neutrons, we would get the mass of a carbon-12 nucleus. Remarkably, the mass of the carbon-12 is a little less than we expect! Scientists call this a **mass defect**, Δm. What has happened to the lost mass? The small amount of missing mass has been converted into energy ($E = \Delta mc^2$) that holds the nucleus together. The mass defect has become the nuclear binding energy and is equivalent to the strong nuclear force holding the nucleus together. If you wanted to break the carbon-12 apart into separate protons and neutrons, you would need to supply enough energy to the nucleus to re-create the mass that is missing: $\Delta m = \dfrac{E}{c^2}$.

Because physicists so often need to convert matter into energy and back again, they took Einstein's equation and a bunch of conversions to give us the following relationship that connects different units of mass:

$$1\,u = 1.66 \times 10^{-27}\,kg = 931\,MeV/c^2$$

where u is called the **unified atomic mass unit**. Remember that MeV is a **mega-electron-volt**. This conversion means that 1 u, or 1.66×10^{-27} kg of mass, is equivalent to 931 MeV when converted to energy.

Let's take a look at an example.

▶ Given the following masses shown in the table, what are the mass defect and binding energy for the helium atom?

Particle (Symbol)	Mass
proton (p)	1.0073 u
neutron (n)	1.0087 u
helium atom (He, α)	4.0016 u

▶ Let's first find the mass defect. It's really simple to do—just subtract the mass of the atom from the mass of its separate particles. The helium atom is made of two protons plus two neutrons, so:

$$2(1.0073\,u) + 2(1.0087\,u) = 4.0320\,u$$

▶ Next, subtract the mass of the helium nucleus from its separate parts:

$$4.0320\,u - 4.0016\,u = 0.0304\,u$$

▶ This is the mass defect of the helium nucleus. It is the amount of matter that's been converted into binding energy and is released from the atom when its component nucleons are put together during fusion.

▶ Now, we want to convert this amount of matter into energy. You could convert the mass into kilograms, use $E = \Delta mc^2$ to change the mass into energy, and then convert the energy from joules into mega-electron-volts, but it is so much easier to use the conversion listed earlier and in the appendix:

$$0.0304\,\text{u} \left(\frac{931\,\text{MeV}}{1\,\text{u}} \right) = 28.3\,\text{MeV}$$

▶ So, in order to break a helium nucleus apart into its pieces, we need to supply the nucleus with 28.3 MeV of energy to reconstitute the missing mass (mass defect).

The mass of a nucleus is less than the mass of its constituent parts. The bigger the mass defect, the harder it will be to break the nucleus apart and the more stable the nucleus will be. In essence, the mass defect is a measure of how large the nuclear force holding the nucleus together is. Remember that the nucleus is populated with protons that repel each other. The force holding it all together must be very strong. Thus, the name given to this binding force is the nuclear strong force.

Mass defect gives us a way to measure nuclear stability and to explain why fission and fusion occur. When we divide the mass of the nucleus by the number of nucleons and plot this against the atomic number, we get a curious graph, shown next. Notice that the mass per nucleon is highest at the ends, near hydrogen (H) and uranium (U), and lowest for iron (Fe). This means that iron has the lowest amount of mass per nucleon and would be the hardest element to break apart into its individual protons and neutrons. So, fission starts with large elements that break into smaller pieces closer to the iron minimum point. This means fission reactions will give up energy as the daughter products convert mass into energy and become more stable.

Likewise, fusion takes small nuclei and fuses them together, and in the process releases mass energy to create more stable nuclei closer to iron. In a similar way, all naturally occurring decay processes release energy to produce a lighter, more stable nucleus. So, in effect, all nuclear reactions are about moving the nucleus to a more stable configuration.

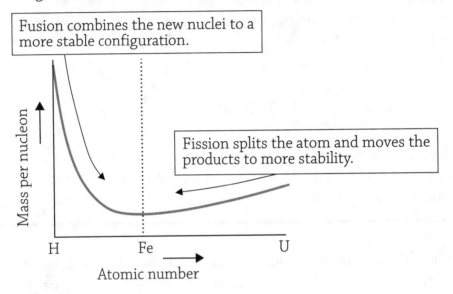

This graph has big implications for the evolution of stars. Stars are huge balls of mostly hydrogen and helium gas held together by gravity. When large enough, these balls of gas create enough pressure to ignite a fusion reaction at their core to begin fusing hydrogen and helium into larger atoms. This process creates a more stable nucleus and generates excess energy that is released as photons. This is why stars shine. As lighter-element fuel is used up, larger and larger nuclei are manufactured, all releasing energy.

Until the star gets to iron. Iron is the most stable nucleus in the universe. Fusing past iron requires a net input of energy to make the next heavier elements of cobalt, nickel, and copper, because now energy has to be converted into mass for the first time. This sucks energy out of the star's energy budget, and signals the death of the star.

Once fusion results in iron, the star begins to collapse on itself. This infalling of mass during the collapse of a dying star provides the energy required to produce all of the heavier elements on the periodic table. When the star explodes, this material is spread throughout the universe as the building blocks of planets and other things like you and me!

REVIEW QUESTIONS

Let's show how much we have learned about nuclear physics by answering the following questions.

1. What are the three primary methods of radioactive decay?

2. Which type of decay emits the most and least penetrating radiation?

3. How are carbon-12 and carbon-14 similar and different?

4. Do different isotopes of the same atom have the same chemical and nuclear properties? Explain.

Thorium-234 has a nuclear notation of $^{234}_{90}\text{Th}$. Use this information to answer questions 5 to 7.

5. How many protons does thorium have?

6. How many neutrons does thorium have?

7. What is the approximate mass of the thorium nucleus?

8. Discuss the forces that the nucleus experiences. Explain how the nucleus stays together.

9. In alpha decay, what happens to the number of protons in the daughter particle?

10. Why does the nucleus naturally transmute?

11. In the decay represented in the following equation, what is the nucleon number of the particle Y?

$$^{12}_{4}\text{Be} \rightarrow \text{Y} + \text{n}$$

12. Decide whether the following nuclear reactions are possible. If not, explain why not.

 a. $^6_3\text{Li} + ^4_2\text{He} \rightarrow ^{12}_7\text{N} + 2^0_{-1}\beta + \text{energy}$

 b. $^3_1\text{H} + ^2_1\text{H} \rightarrow ^4_2\text{He} + ^1_1\text{H} + \gamma$

13. Complete the following reactions and classify each as either decay (specify the type), fission, or fusion.

 a. $2^1_1\text{H} + 2^1_0\text{n} \rightarrow \text{X} + \text{energy}$

 b. $^{137}_{55}\text{Cs} \rightarrow \text{X} + ^0_{-1}\text{e}$

 c. $^{238}_{92}\text{U} \rightarrow ^{234}_{90}\text{Th} + \text{X} + \text{energy}$

 d. $^1_0\text{n} + ^{235}_{92}\text{U} \rightarrow ^{144}_{54}\text{Xe} + ^{90}_{38}\text{Sr} + \text{X}$

 e. $^{12}_6\text{C} \rightarrow ^{12}_6\text{C} + \text{X}$

14. Use the following graph of the number of radioactive emission counts by a Geiger counter versus time to estimate the half-life of the material.

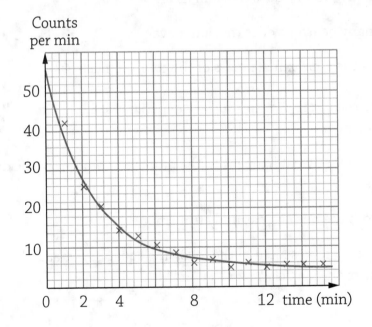

15. The half-life of carbon-14 is 5,730 years. Fresh wood has about 12.6 carbon-14 decays/min/g. A piece of wood from an archaeological dig has a radioactivity of 1.6 decays/min/g. How old is the wood?

16. Radon-222 has a half-life of 3.8 days. How long will it take for there to be only one-quarter of the radon left?

17. Describe the difference between fission and fusion.

18. What is mass defect, and how is it related to both the nuclear strong force and binding energy?

19. Why does nuclear decay occur?

20. An iron nucleus ($^{56}_{26}$Fe) has a mass of 92.86×10^{-27} kg (55.94 u). Find the mass defect and binding energy of this nucleus. Show your work.

Appendix

Periodic Table

Key:

10
Ne
Neon
20.18

10 — Atomic number
Ne
Neon
20.18 — Approximate average atomic mass.

1 / 1A	2 / 2A	3 / 3B	4 / 4B	5 / 5B	6 / 6B	7 / 7B	8 / 8B	9 / 8B	10 / 8B	11 / 1B	12 / 2B	13 / 3A	14 / 4A	15 / 5A	16 / 6A	17 / 7A	18 / 8A
1 **H** Hydrogen 1.008																	2 **He** Helium 4.003
3 **Li** Lithium 6.941	4 **Be** Beryllium 9.012											5 **B** Boron 10.81	6 **C** Carbon 12.01	7 **N** Nitrogen 14.01	8 **O** Oxygen 16.00	9 **F** Fluorine 19.00	10 **Ne** Neon 20.18
11 **Na** Sodium 22.99	12 **Mg** Magnesium 24.31											13 **Al** Aluminum 26.98	14 **Si** Silicon 28.09	15 **P** Phosphorus 30.97	16 **S** Sulfur 32.07	17 **Cl** Chlorine 35.45	18 **Ar** Argon 39.95
19 **K** Potassium 39.10	20 **Ca** Calcium 40.08	21 **Sc** Scandium 44.96	22 **Ti** Titanium 47.88	23 **V** Vanadium 50.94	24 **Cr** Chromium 52.00	25 **Mn** Manganese 54.94	26 **Fe** Iron 55.85	27 **Co** Cobalt 58.93	28 **Ni** Nickel 58.69	29 **Cu** Copper 63.55	30 **Zn** Zinc 65.39	31 **Ga** Gallium 69.72	32 **Ge** Germanium 72.59	33 **As** Arsenic 74.92	34 **Se** Selenium 78.96	35 **Br** Bromine 79.90	36 **Kr** Krypton 83.80
37 **Rb** Rubidium 85.47	38 **Sr** Strontium 87.62	39 **Y** Yttrium 88.91	40 **Zr** Zirconium 91.22	41 **Nb** Niobium 92.91	42 **Mo** Molybdenum 95.94	43 **Tc** Technetium (98)	44 **Ru** Ruthenium 101.1	45 **Rh** Rhodium 102.9	46 **Pd** Palladium 106.4	47 **Ag** Silver 107.9	48 **Cd** Cadmium 112.4	49 **In** Indium 114.8	50 **Sn** Tin 118.7	51 **Sb** Antimony 121.8	52 **Te** Tellurium 127.6	53 **I** Iodine 126.9	54 **Xe** Xenon 131.3
55 **Cs** Cesium 132.9	56 **Ba** Barium 137.3	57 **La** Lanthanum 138.9	72 **Hf** Hafnium 178.5	73 **Ta** Tantalum 180.9	74 **W** Tungsten 183.9	75 **Re** Rhenium 186.2	76 **Os** Osmium 190.2	77 **Ir** Iridium 192.2	78 **Pt** Platinum 195.1	79 **Au** Gold 197.0	80 **Hg** Mercury 200.6	81 **Tl** Thallium 204.4	82 **Pb** Lead 207.2	83 **Bi** Bismuth 209.0	84 **Po** Polonium (210)	85 **At** Astatine (210)	86 **Rn** Radon (222)
87 **Fr** Francium (223)	88 **Ra** Radium (226)	89 **Ac** Actinium (227)	104 **Rf** Rutherfordium (257)	105 **Db** Dubnium (260)	106 **Sg** Seaborgium (263)	107 **Bh** Bohrium (262)	108 **Hs** Hassium (265)	109 **Mt** Meitnerium (266)	110 **Ds** Darmstadtium (269)	111 **Rg** Roentgenium (272)	112 **Cn** Copernicium (272)	113 **Nh** Nihonium	114 **Fl** Flerovium	115 **Mc** Moscovium	116 **Lv** Livermorium	(117) **Ts** Tennessine	118 **Og** Oganesson

Lanthanide series:

58 **Ce** Cerium 140.1	59 **Pr** Praseodymium 140.9	60 **Nd** Neodymium 144.2	61 **Pm** Promethium (147)	62 **Sm** Samarium 150.4	63 **Eu** Europium 152.0	64 **Gd** Gadolinium 157.3	65 **Tb** Terbium 158.9	66 **Dy** Dysprosium 162.5	67 **Ho** Holmium 164.9	68 **Er** Erbium 167.3	69 **Tm** Thulium 168.9	70 **Yb** Ytterbium 173.0	71 **Lu** Lutetium 175.0

Actinide series:

90 **Th** Thorium 232.0	91 **Pa** Protactinium (231)	92 **U** Uranium 238.0	93 **Np** Neptunium (237)	94 **Pu** Plutonium (242)	95 **Am** Americium (243)	96 **Cm** Curium (247)	97 **Bk** Berkelium (247)	98 **Cf** Californium (249)	99 **Es** Einsteinium (254)	100 **Fm** Fermium (253)	101 **Md** Mendelevium (256)	102 **No** Nobelium (254)	103 **Lr** Lawrencium (257)

Legend:
- Metals
- Metalloids
- Nonmetals

Physics Constants

Physical Description	Symbol and Constant Value
Average atmospheric pressure near the Earth's surface	$1\ \text{atm} = 1.0 \times 10^5\ \text{Pa}$
Speed of light	$c = 3.0 \times 10^8\ \text{m/s}$
Magnitude of the electron/proton change	$e = 1.6 \times 10^{-19}\ \text{C}$
Electron volt	$1\ \text{eV} = 1.6 \times 10^{-19}\ \text{J}$
Acceleration caused by gravity near the Earth's surface	$g = 9.8\ \text{m/s}^2$
Universal gravitational constant	$G = 6.67 \times 10^{-11}\ \text{Nm}^2/\text{kg}^2$
Planck's constant	$h = 6.63 \times 10^{-34}\ \text{J} \cdot \text{s}$ $= 4.14 \times 10^{-15}\ \text{eV} \cdot \text{s}$
Planck's constant times the speed of light	$hc = 1.99 \times 10^{-25}\ \text{J} \cdot \text{m} = 1240\ \text{eV} \cdot \text{nm}$
Coulomb's law constant	$k = 9.0 \times 10^9\ \text{Nm}^2/\text{C}^2$
Boltzmann's constant	$k_B = 1.38 \times 10^{-23}\ \text{J/K}$
Light year	$1\ \text{ly} = 9.46 \times 10^{15}\ \text{m}$
Electron mass	$m_e = 9.11 \times 10^{-31}\ \text{kg}$
Proton mass	$m_p = 1.673 \times 10^{-27}\ \text{kg}$
Neutron mass	$m_n = 1.675 \times 10^{-27}\ \text{kg}$
Avogadro's number	$N_A = 6.02 \times 10^{23}\ \text{atoms/mol}$
Universal gas constant	$R = 8.31\ \text{J/mol} \cdot \text{K}$
Atomic mass unit	$1\ \text{u} = 1.66 \times 10^{-27}\ \text{kg} = 931\ \text{MeV}/c^2$
Vacuum permittivity (electric constant)	$\varepsilon_0 = 8.85 \times 10^{-12}\ \text{C}^2/\text{N} \cdot \text{m}^2$
Vacuum permeability (magnetic constant)	$\mu_0 = 4\pi \times 10^{-7}\ \text{T} \cdot \text{m/A}$

Physical Properties

Symbol	Description	Standard Physics Unit	Unit Name or Description
a	acceleration	m/s²	meters per second squared
A	amplitude	m	meter
A	area	m²	meters squared
b	length of a triangle's base	m	meter
B	magnetic field	T	tesla
C	capacitance	F	farad
d	distance	m	meter
d	separation between two things	m	meter
E	electric field	N/C, V/m	newtons/coulomb, volts/meter
E	energy	J, eV	joules or electron volts
f	frequency	Hz	hertz, cycles/second
f	focal length	m	meter
$F_{subscript}$	force (subscript indicates what kind)	N	newton
h	height or depth (y direction)	m	meter
I	current	A	amperes, amps
I	rotational inertia	kg · m²	kilograms meters squared
k	spring constant	N/m	newtons per meter
k	thermal conductivity	J/m · K · s	Joules/meter · Kelvin · second
K	kinetic energy	J	joules

Symbol	Description	Standard Physics Unit	Unit Name or Description
l	length	m	meter
L	angular momentum	kg · m²/s	kilograms meters squared per second
m	mass	kg	kilogram
m	multiplier or counter	no unit	number indicating how many multiples of
M	magnification	no unit	this is a ratio and has no unit
n	index of refraction	no unit	this is a ratio and has no unit
n	number of moles	mol	moles
N	number of atoms or molecules	no unit	just a number indicating how many
p	momentum	kg · m/s	kilograms meters per second
P	power	W	watt
P	pressure	Pa	pascal
q, Q	charge	C	coulomb
Q	heat, thermal energy transfer between objects	J	joule
r	radius or separation between objects	m	meter
R	resistance	Ω	ohm
t	time	s	second
T	time period or period	s	second
T	temperature	K or °C	kelvin or degrees Celsius
U	total internal energy of an object	J	joule

Symbol	Description	Standard Physics Unit	Unit Name or Description
$U_{subscript}$	potential or stored energy (subscript indicates what kind of energy)	J	joule
v	velocity or speed	m/s	meters per second
V	electric potential	V	volt
V	volume	m^3	meters cubed
w	width	m	meter
W	work (+ done on the system)	J	joule
x	position in x direction (horizontal)	m	meter
y	position in y direction (vertical)	m	meter
y	height	m	meter
z	position in z direction (in/out of the page)	m	meter
Greek letters			
α	angular acceleration	rad/s^2	radians per second squared
α	alpha particle	no unit	helium nucleus
β, β^-, e, e^-	beta particle	no unit	electron
β^+, e^+	beta particle	no unit	positron
γ	gamma ray	no unit	electromagnetic radiation from the nucleus
ε	emf	V	volt
θ	angle	° or rad	degrees or radians
κ	dielectric constant	no unit	this is a ratio and has no unit

Symbol	Description	Standard Physics Unit	Unit Name or Description
Greek letters			
λ	wavelength	m	meter
μ	coefficient of friction	no unit	this is a ratio and has no unit
ρ	density	kg/m^3	kilograms per meters cubed
ρ	resistivity	$\Omega \cdot m$	ohm meter
τ	torque	$N \cdot m$	newton meters
ϕ	work function	J	joule
Φ_M	magnetic flux	T/m^2, Wb	tesla per meter squared, weber
Ψ	wave function	no unit	this is a probability and has no units
ω	angular velocity or angular speed	rad/s	radians per second

Unit Prefixes

Symbol	Prefix Name	Multiplication Factor
T	tera-	$\times 10^{12}$
G	giga-	$\times 10^{9}$
M	mega-	$\times 10^{6}$
k	kilo-	$\times 10^{3}$
c	centi-	$\times 10^{-2}$
m	milli-	$\times 10^{-3}$
μ	micro-	$\times 10^{-6}$
n	nano-	$\times 10^{-9}$
p	pico-	$\times 10^{-12}$

Geometry Equations

Circle
Circumference $= 2\pi r$
Area $= \pi r^2$

Sphere
Surface Area $= 2\pi r^2$
Volume $= \dfrac{4}{3}\pi r^3$

Triangle
Area $= \dfrac{1}{2}bh$

Rectangle
Area $= lw$

Rectangular Solid
Volume $= lwh$

Right Triangle Trigonometry
$a^2 + b^2 = c^2$
$\sin\theta = \dfrac{a}{c} = \dfrac{\text{opposite}}{\text{hypotenuse}}$
$\cos\theta = \dfrac{b}{c} = \dfrac{\text{adjacent}}{\text{hypotenuse}}$
$\tan\theta = \dfrac{a}{b} = \dfrac{\text{opposite}}{\text{adjacent}}$

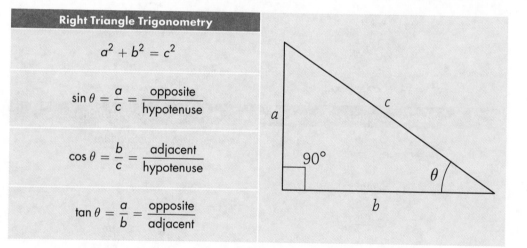

Physics Equations

Mechanics

Kinematics and Motion	Circular Motion and Orbits
$\Delta t = t_f - t_i$	$\sum F_{center} = ma_{center}$
$\Delta x = x_f - x_i$	$a_{center} = \dfrac{v^2}{r}$
$\Delta v = v_f - v_i$	$v_{circle} = \dfrac{2\pi r}{T}$
$v = \dfrac{\Delta x}{\Delta t}$	$F_G = G\dfrac{m_1 m_2}{r^2}$
$a = \dfrac{\Delta v}{\Delta t}$	$v_{orbit} = \sqrt{\dfrac{GM_{object\,being\,orbited}}{R_{orbit}}}$
$x = x_0 + v_0 t + \dfrac{1}{2}at^2$	
$v = v_0 + at$	$T^2_{orbit} = \left(\dfrac{4\pi^2}{GM_{object\,being\,orbited}}\right)R^3_{orbit}$
$v^2 = v_0^2 + 2a(x - x_0)$	

Newton's Laws and Forces	Oscillatory Motion
$\sum F = F_{net} = ma$	$f = \dfrac{oscillations}{\Delta t}$
$F_g = mg$	$T = \dfrac{1}{f}$
$F_G = G\dfrac{m_1 m_2}{r^2}$	$T_s = 2\pi\sqrt{\dfrac{m}{k}}$
$g = G\dfrac{M_{Earth}}{R^2_{Earth}}$	$T_p = 2\pi\sqrt{\dfrac{l}{g}}$
$F_f \le \mu F_N$	$x = A\cos(2\pi ft) = A\cos\left(\dfrac{2\pi t}{T}\right)$
$F_s = k\Delta x$	

Momentum and Impulse

$$p = mv$$

$$F\Delta t = \Delta p = m(v_f - v_i)$$

$$(mv_1 + mv_2 + \cdots)_i = (mv_1 + mv_2 + \cdots)_f$$

$$x_{center\ of\ mass} = \frac{\sum(xm)}{\sum m}$$

Energy and Work

$$K = \frac{1}{2}mv^2$$

$$K_{rotational} = \frac{1}{2}I\omega^2$$

$$\Delta U_g = mg\Delta y$$

$$\Delta U_s = \frac{1}{2}k(\Delta x)^2$$

$$W = \Delta E = Fd\cos\theta$$

$$P = \frac{W}{\Delta t} = \frac{\Delta E}{\Delta t}$$

$$E_1 + W = E_2$$

Rotational Motion

$$x_{center\ of\ mass} = \frac{\sum(xm)}{\sum m}$$

$$\omega = \frac{\Delta x}{\Delta t}$$

$$a = \frac{\Delta v}{\Delta t}$$

$$\theta = \theta_0 + \omega_0 t + \frac{1}{2}\alpha t^2$$

$$\omega = \omega_0 + \alpha t$$

$$\omega^2 = \omega_0^2 + 2\alpha(\theta - \theta_0)$$

$$I = \sum mr^2$$

$$\tau = rF\sin\theta = r_\perp F = rF_\perp$$

$$\sum \tau = I\alpha$$

$$K_{rotational} = \frac{1}{2}I\omega^2$$

$$L = I\omega = r_\perp mv$$

$$\tau\Delta t = \Delta L = I(\omega_f - \omega_i)$$

$$(I\omega_1 + I\omega_2 + \cdots)_i = (I\omega_1 + I\omega_2 + \cdots)_f$$

Nonsolid Behavior

Fluids (Behavior of Gases and Liquids)	Gases (Behavior of Gases Only)
$\rho = \dfrac{m}{V}$	$PV = nRT$
$P = \dfrac{F}{A}$	$K_{average} = \dfrac{3}{2}k_BT$
$P = P_0 + \rho g h$	$\Delta U = \dfrac{3}{2}nR\Delta T = \dfrac{3}{2}Nk_BT$
$F_b = \rho V g$	$v_{gas\,molecules} = \sqrt{\dfrac{3k_BT}{m_{gas\,molecules}}}$
$A_1 v_1 = A_2 v_2$	
$P + \rho g y + \dfrac{1}{2}\rho v^2 = \text{constant}$	

Thermal Physics

$$W_{gas} = -P\Delta V$$

$$\Delta U = \frac{3}{2}nR\Delta T = \frac{3}{2}Nk_BT$$

$$\Delta U = Q + W$$

$$\frac{Q}{\Delta t} = \frac{kA\Delta T}{l}$$

$$\frac{Q}{\Delta t} \propto T^4$$

$$\Delta S = \frac{Q}{T}$$

Electricity and Magnetism

Static Electricity (Forces and Energy)

$$F_E = \left| k \frac{q_1 q_2}{r^2} \right|$$

$$F_E = Eq$$

$$E = k \frac{Q_{charge\ producing\ the\ E-Field}}{r^2}$$

$$V = \frac{U_E}{q}$$

$$V = k \frac{Q}{r}$$

$$E = \left| \frac{\Delta V}{d} \right|$$

$$C_\parallel = \frac{\kappa \varepsilon_0 A}{d}$$

$$C = \frac{Q}{\Delta V}$$

$$U_C = \frac{1}{2} Q \Delta V = \frac{1}{2} C (\Delta V)^2$$

Magnetism and Electromagnetic Induction

$$B = \left(\frac{\mu_0}{2\pi} \right) \frac{I}{r}$$

$$F_M = IlB \sin \theta$$

$$F_M = qvB \sin \theta$$

$$r = \frac{mv}{qB}$$

$$\varepsilon = Blv$$

$$\Phi_M = BA \cos \theta$$

$$\varepsilon = \frac{\Delta \Phi_M}{\Delta t}$$

Circuits

$$I = \frac{\Delta q}{\Delta t}$$

$$R = \frac{\rho L}{A}$$

$$\Delta V = IR$$

$$P = I\Delta V = I^2 R = \frac{\Delta V^2}{R}$$

$$\sum I_{junction} = 0$$

$$\sum \Delta V_{closed\ loop} = 0$$

$$R_{series} = R_1 + R_2 + R_3 \ldots$$

$$\frac{1}{R_{parallel}} = \frac{1}{R_1} + \frac{1}{R_2} + \frac{1}{R_3} \ldots$$

$$\frac{1}{C_{series}} = \frac{1}{C_1} + \frac{1}{C_2} + \frac{1}{C_3} \ldots$$

Wave Behavior and Optics

Waves and Interference	Refraction, Lenses, and Mirror Optics
$$f = \frac{1}{T}$$	$$\theta_{incidence} = \theta_{reflection}$$
$$v_{wave} = \frac{\Delta x}{\Delta t} = \frac{\lambda}{T} = f\lambda$$	$$f = \frac{R}{2}$$
$$x = A\cos(2\pi ft) = A\cos\left(\frac{2\pi t}{T}\right)$$	$$n = \frac{c}{v}$$
$$f_{beat} = |f_1 - f_2|$$	$$n_1 \sin\theta_1 = n_2 \sin\theta_2$$
$$\Delta L = m\lambda$$	$$\sin\theta_{critical} = \frac{n_2}{n_1}$$
$$d\sin\theta = m\lambda$$	$$\frac{1}{f} = \frac{1}{d_i} + \frac{1}{d_o}$$
	$$M = \frac{h_i}{h_o} = -\frac{d_i}{d_o}$$

Quantum and Nuclear Physics

$$E = mc^2$$

$$E = hf = \frac{hc}{\lambda}$$

$$K_{max} = hf - \phi$$

$$\lambda = \frac{h}{p} = \frac{h}{mv}$$

Answer Key

1

Motion

1. Vectors (magnitude and direction): position, displacement, velocity, and acceleration. Scalars (magnitude only): distance and speed.

2. Distance = 70 m, displacement = 50 m at 37° south of east (remember to pythagorize), speed = 0.7 m/s, velocity = 0.5 m/s at 37° south of east.

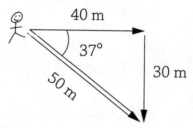

3. For the first 4 seconds, the distance between the dots is increasing. Therefore, the object is accelerating. After this, the spacing is uniform and the object is traveling at a constant velocity. See the following figure.

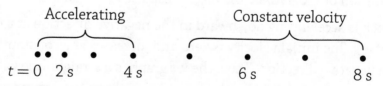

4. The object is moving to the right at a constant velocity, begins slowing down at 4 s as it accelerates to the left until 6 s, then travels at a constant slower speed until 14 s.

5. Remember that time is plotted on the x-axis and position is plotted on the y-axis.

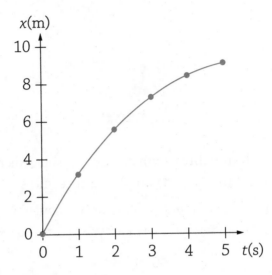

6. $v = \dfrac{\Delta x}{\Delta t} = \dfrac{(5.4\,\text{m/s}) - (0)}{2\,\text{s}} = 2.7\,\text{m/s}$

7. This is an x-t graph. The slope equals the velocity. The object starts at 2 m and travels to 4 m in 5 s at a constant velocity of 0.4 m/s.

8. This is a v-t graph. The slope equals the acceleration, and the area of the graph equals the displacement. At the start, the object is traveling with a positive velocity of 4 m/s but is slowing down with a constant acceleration of $-0.4\,\text{m/s}^2$. The object displaces 15 m in 5 s.

9. The rock is accelerating downward in the negative direction at a constant rate of $-g$. The initial velocity is zero and increases at a constant rate in the negative direction. Thus, the v-t graph is a straight line with a negative slope of g. Finally, the rock begins above the ground at x_0, accelerating downward, which shows up as a downward parabola on the x-t graph. See the graphs on the next page.

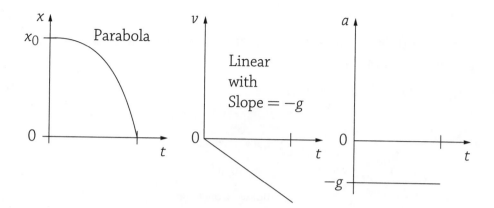

10. Unfortunately, no. Everything accelerates downward toward the Earth at 9.8 m/s². The puppy has a 1-second head start and will therefore always be traveling 9.8 m/s faster than our hero. The hero can never catch the puppy unless there is enough air resistance to reduce the puppy's acceleration.

11. Organize your information as shown:

y_0	0
y	What we want to find
v_0	19.6 m/s
v	0 (at the top)
a	−9.8 m/s²
t	What we want to find

$$v^2 = v_0^2 + 2a(x - x_0)$$
$$0 = (19.6\,\text{m/s})^2 + 2(-9.8\,\text{m/s}^2)(y - 0)$$
$$y = 19.6\,\text{m}$$

$$v = v_0 + at$$
$$0 = 19.6\,\text{m/s} + (-9.8\,\text{m/s}^2)t$$
$$t = 2\,\text{s}$$

But this is only the time to the top of the motion, when the velocity equals zero. Total time = 4 s.

12. Organize your data as shown:

x_0	0
x	What we want to find
v_0	10 m/s
v	What we want to find
a	4 m/s²
t	8 s

$$x = x_0 + v_0 t + \frac{1}{2} a t^2$$

$$x = 0 + (10\,\text{m/s})(8\,\text{s}) + \frac{1}{2}(4\,\text{m/s}^2)(8\,\text{s})^2$$

$$x = 208\,\text{m}$$

$$v = v_0 + at$$

$$v = 10\,\text{m/s} + (4\,\text{m/s}^2)(8\,\text{s})$$

$$v = 42\,\text{m/s}$$

13. This is circular motion: $v = \dfrac{2\pi r}{\Delta t}$

$$4.5\,\text{m/s} = \frac{2\pi r}{6.8\,\text{s}}$$

$$r = 4.9\,\text{m}$$

14. Both rocks will hit the ground at the same time. The y-motion of both rocks is identical. Both start with a y-velocity of zero and accelerate toward the ground at the same rate of g.

15. The object will accelerate downward toward the Earth at a rate of 9.8 m/s². Therefore, the y-velocity will be changing and the object slows down on its way upward and then accelerates on its way downward. The *x*-velocity remains constant and unchanged the entire time.

2

Forces

1. Forces always come in equal and opposite pairs (Newton's First Law). You move backward because your friend pushed you backward even if they didn't try to.

2. Newton's Third Law tells us that when the father pushes the girl forward, the daughter is also pushing the father backward with the exact same force. Newton's Second Law tells us that this backward force will accelerate the father backward, but since the father probably has more mass than the girl, he will have a smaller acceleration and final velocity than his daughter.

3. Sure! Gravity from the Earth, normal force, and friction from the road. Even air resistance, but since the car is traveling at a constant velocity, all of these forces must cancel out or sum to zero ($\Sigma F = 0$).

4. There are many examples: When a car rear-ends a truck, the truck is pushed forward while the car is slowed to a stop. In baseball, the bat hits the ball and causes it to change direction, and the ball slows the bat down in its swing. As I type this sentence, my fingers push the keys of the laptop downward while the keys push upward on my fingers, stopping their downward motion.

5. Your head has inertia and wants to stay at rest, but your body is being pushed forward by the car seat. We need the headrest to push your head forward with the rest of your body. Without a headrest, your body will move forward but your head would stay put until your neck stretches out and pulls your head forward. This causes whiplash.

6. When she fills the glass with wine, it adds more mass and therefore inertia. The glass will now be harder to move when she pulls the tablecloth out from underneath it. This makes it easier to perform the trick.

7. You are an object in motion with inertia. When the car turns to the left, you want to keep going straight forward. The only way you turn with the car is if something pushes or pulls you around the turn, like the door or seatbelt.

8. Gravity always pulls objects together. On Earth, that means everything is pulled toward the center of the Earth. Santa, at the North Pole, is pulled downward. At the South Pole, penguins are pulled upward toward the center of the Earth.

9. Normal force and friction only occur when there is contact between two objects. For example, when you jump off the floor, there is no normal or friction force between you and the floor. Normal force always points perpendicular to the surface, away from the surface. Friction is always parallel to the surface, trying to prevent sliding. Let's say you are leaning against a wall, as shown in the figure. There are two normal forces, one from the floor pushing upward and one from the wall pushing you away from the wall. Friction between your feet and the floor will be pushing to prevent your feet from sliding. (See the following figure.)

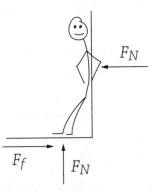

10. Gravity accelerates the football downward at 9.8 m/s². There are no horizontal forces on the ball, assuming air resistance to be small. Therefore, the ball continues to move horizontally at a constant x-velocity while it is accelerating downward.

11. The tension is greatest when accelerating upward because the tension must be larger than gravity to accelerate the yo-yo upward. It is least when accelerating downward and equal to gravity when the yo-yo is at rest. When accelerating upward at 3.0 m/s², $F_T - mg = ma$. This gives us $F_T = m(g + a) = (0.2\,\text{kg})(9.8\,\text{m/s}^2 + 3\,\text{m/s}^2) = 2.6\,\text{N}$.

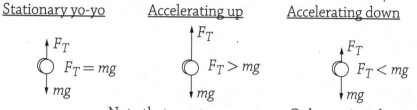

Stationary yo-yo   Accelerating up   Accelerating down

F_T F_T F_T

$F_T = mg$ $F_T > mg$ $F_T < mg$

mg mg mg

Note that mg stays constant. Only tension changes.

12. Let's choose an x-y coordinate system with left as positive so that the velocity is positive. Since we don't know the acceleration, we will need to find it. We will also need the normal force to find the coefficient of friction from the friction equation. See the following figure.

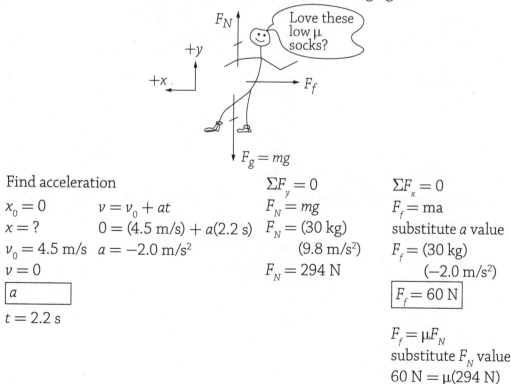

Find acceleration

$x_0 = 0$ $v = v_0 + at$

$x = ?$ $0 = (4.5\,\text{m/s}) + a(2.2\,\text{s})$

$v_0 = 4.5\,\text{m/s}$ $a = -2.0\,\text{m/s}^2$

$v = 0$

$\boxed{a}$

$t = 2.2\,\text{s}$

$\Sigma F_y = 0$

$F_N = mg$

$F_N = (30\,\text{kg})$

$(9.8\,\text{m/s}^2)$

$F_N = 294\,\text{N}$

$\Sigma F_x = 0$

$F_f = ma$

substitute a value

$F_f = (30\,\text{kg})$

$(-2.0\,\text{m/s}^2)$

$\boxed{F_f = 60\,\text{N}}$

$F_f = \mu F_N$

substitute F_N value

$60\,\text{N} = \mu(294\,\text{N})$

$\boxed{\mu = 0.20}$

13. Since the acceleration will be down the mountain, tilt the coordinate system as shown in the FBD and split up the gravity force into its components. See the following figure. Notice that only the parallel component of the gravity force causes the skier to accelerate and that the mass of the skier cancels out of the equation and does not matter. The formal force and the perpendicular part of gravity are equal and opposite.

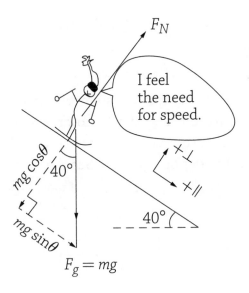

$$\Sigma F_{||} = ma$$
$$mg \sin \theta = ma$$

(Notice that we can cancel the mass out from both sides of the equation.)

$$a = g \sin \theta$$
$$a = (9.8 \text{ m/s}^2) \sin 40°$$
$$\boxed{a = 6.3 \text{ m/s}^2}$$

14. The child is not accelerating, so the x- and y-forces cancel out. See the following figure.

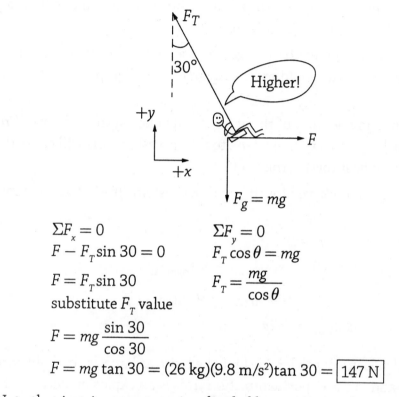

$$\Sigma F_x = 0 \qquad\qquad \Sigma F_y = 0$$
$$F - F_T \sin 30 = 0 \qquad F_T \cos \theta = mg$$

$$F = F_T \sin 30 \qquad\qquad F_T = \frac{mg}{\cos \theta}$$
substitute F_T value

$$F = mg\,\frac{\sin 30}{\cos 30}$$
$$F = mg \tan 30 = (26\,\text{kg})(9.8\,\text{m/s}^2)\tan 30 = \boxed{147\,\text{N}}$$

Note that in trigonometry, sine divided by cosine equals tangent:
$$\tan \theta = \frac{\sin \theta}{\cos \theta}.$$

Momentum and Impulse

1. $p_i = 32{,}000\,\text{kg} \cdot \text{m/s}^2$, $p_f = 72{,}000\,\text{kg} \cdot \text{m/s}^2$, $\Delta p = 40{,}000\,\text{kg} \cdot \text{m/s}^2$

2. I am assuming up to be positive in this problem: $\Delta p = m(v_f - v_i) = 0.056\,\text{kg}((12\,\text{m/s}) - (-15\,\text{m/s})) = 1.5\,\text{kg} \cdot \text{m/s}$

$$F\Delta t = \Delta p,\ F(0.03\,\text{s}) = 1.5\,\text{kg} \cdot \text{m/s},\ F = 50\,\text{N}$$

3. The water balloon will pop if you catch it with too much force. To reduce the force, we need to extend the time it takes to bring it to rest. Therefore, move your hands in the same direction the balloon is going to increase the time to slow it to a stop.

4. Bouncing off would require a larger change in momentum due to the change in direction, would produce a larger overall impulse, and would hurt worse.

5. The original momentum of the astronaut–wrench system is zero. Throw the wrench in a direction opposite to the spaceship. This will cause the astronaut to float toward the ship.

6. Momentum is conserved for the two-cart system. The initial momentum is zero.

$$p_i = p_f$$
$$0 = (mv_{\text{left cart}} + mv_{\text{right cart}})_f$$
$$0 = (mv_{\text{left cart}} + 2mv_{\text{right cart}})_f$$
$$v_{\text{left cart}} = -2v_{\text{right cart}}$$

The final velocity of the left cart will be twice as large as the right cart's velocity and in the opposite direction. This is an explosion because there was no motion before the string was cut. After it was cut, the energy stored in the spring turns into kinetic energy for the carts.

7. Momentum is conserved for the truck–car system. I assumed north was the positive direction:

$$p_i = p_f$$
$$(mv_{\text{truck}} + mv_{\text{car}})_i = (m_{\text{truck}} + m_{\text{car}})v_f$$
$$((1{,}900\,\text{kg})(15\,\text{m/s}) + (1{,}000\,\text{kg})(-30\,\text{m/s}))_i = ((2{,}900\,\text{kg}))v_f$$
$$v_f = -0.52\,\text{m/s}$$

The final velocity of the wreckage is south because the final velocity is negative.

8. Momentum is conserved for the granddaughter–grandfather system. Assume the original direction of the child to be positive:

$$p_i = p_f$$

$$(mv_{child} + 0)_i = (mv_{child} + mv_{grandparent})_f$$

$$(20\,kg)(4.5\,m/s)_i = ((20\,kg)v_f + (80\,kg)(1.0\,m/s))_f$$

$$v_f = 0.5\,m/s$$

The granddaughter's final velocity is in the same direction she was originally running.

9. Unfortunately, it looks like she will scratch and put the cue ball in the bottom right pocket. Due to conservation of momentum, when the eight-ball moves up and to the right with a positive y-momentum, the cue ball must move down and to the right with a negative y-momentum.

10. The two will move off together along path C. The easterly momentum of the truck and the northerly momentum of the car add to produce a northeastward momentum along path C.

4

Energy and Work

1. While running, the pole vaulter–Earth–pole system has kinetic energy. This turns into spring/elastic potential energy when the pole is bent, and finally into gravitational potential energy when the vaulter is at the top of their motion.

2. The work is the same because they both acquire the same gravitational potential energy. The runner generates more power because they accomplish the same work in less time.

3. $P = \dfrac{W}{\Delta t} = \dfrac{\Delta K}{\Delta t} = \dfrac{\frac{1}{2}(1{,}600\,kg)30\,m/s^2}{3.4\,s} = 210{,}000\,W$

4. $W = \Delta E = \Delta U_g = mg\Delta y$. Since they both run to the same height, Jack does more work. Jill generates more power because her work is done quicker: $P_{Jill} = 1{,}680\,\text{W}$, $P_{Jack} = 1{,}568\,\text{W}$.

5. Tarzan must start with some kinetic energy to convert into the extra potential energy he needs to get up to where Jane and the cat are.

$$E_1 + W = E_2$$

$$\left(\frac{1}{2}mv^2 + mgy\right)_1 + 0 = (mgy)_2$$

$$\left(\frac{1}{2}mv^2\right)_1 = (mgy)_2 - (mgy)_1$$

$$\frac{1}{2}mv^2 = mg\Delta y$$

$$\frac{1}{2}v^2 = g\Delta y$$

$$v = 8.9\,\text{m/s}$$

6. Assuming no loses of energy due to friction: $v_B = 20\,\text{m/s}$, $v_C = 37\,\text{m/s}$.

7. Both balls start with the same kinetic and potential energies. Both balls fall the same distance and convert the same amount of potential energy into kinetic energy. Therefore, they both hit the ground at the same speed.

8. The roller coaster will reach a maximum height that is lower. The roller coaster is traveling on a parabolic trajectory. It will have a horizontal velocity when it reaches its maximum height. Therefore, all of the kinetic energy does not convert completely back into gravitational potential energy, and it never regains its original height.

9. Energy is conserved for the falling person and trampoline system.

$$E_1 + W = E_2$$

$$mg\Delta y + 0 = \frac{1}{2}kx^2$$

$$x = 2.1\,\text{m}$$

10. With no air resistance:

$$mg\Delta y = \frac{1}{2}mv^2$$

$$v = 44 \text{ m/s}$$

With air resistance:

$$mg\Delta y - W_{air\ resistance} = \frac{1}{2}mv^2$$

$$mg\Delta y - (F_{air\ resistance})d = \frac{1}{2}mv^2$$

$$(80 \text{ kg})(9.8 \text{ m/s}^2)(100 \text{ m}) - (200 \text{ N})(100 \text{ m}) = \frac{1}{2}(80 \text{ kg})v^2$$

$$v = 38 \text{ m/s}$$

11. Spring potential energy is being converted to gravitational potential:

$$\frac{1}{2}kx^2 = mgh$$

$$h = 12 \text{ m}$$

12. Since the box is coming to rest, the final energy is zero:

$$E_1 + W = E_2$$

$$K - W_{friction} = 0$$

$$\frac{1}{2}mv^2 - F_f d = \frac{1}{2}mv^2 - \mu F_N d$$

$$= \frac{1}{2}mv^2 - \mu mgd = \frac{1}{2}v^2 - \mu gd = 0$$

$$d = \frac{1}{2}\frac{v^2}{\mu g}$$

$$d = 4.6 \text{ m}$$

13. Although they won't all get to the bottom at the same time, they will all be going the same speed. All the crates start with the same amount of gravitational potential energy and will therefore end with the same kinetic energy at the bottom of the ramp.

14. Assuming the ball is caught at the same height it was released, it will be going 22 m/s again. The height of the ball at the top of the arc is 8.2 m. See the following equation.

$$E_1 + W = E_2$$

$$K + 0 = K + U_g$$

$$\left(\frac{1}{2}mv^2\right)_{start} = \left(\frac{1}{2}mv^2 + mg\Delta y\right)_{top}$$

$$\left(\frac{1}{2}v^2\right)_{start} = \left(\frac{1}{2}v^2 + g\Delta y\right)_{top}$$

$$\left(\frac{1}{2}v^2\right)_{start} - \left(\frac{1}{2}v^2\right)_{top} = g\Delta y$$

$$\Delta y = 8.2\,\text{m}$$

Three Special Situations: Oscillations, Circular Motion, and Rotational Motion

1. The time period of pendulums is independent of mass: $T_p = 2\pi\sqrt{\dfrac{l}{g}}$.

   Quadrupling the mass will increase the time period of the spring mass by a factor of 2: $T_s = 2\pi\sqrt{\dfrac{m}{k}}$. So no, they won't be the same.

2. Amplitude $= 4$ m, time period $= 4$ s, frequency $= 0.25$ Hz.

3. $x = A\cos\left(\dfrac{2\pi t}{T}\right) = (4\,\text{m})\cos\left(\dfrac{2\pi t}{4\,\text{s}}\right)$

4. Maximum gravitational potential energy will be at maximum amplitude times of 0, 2 s, 4 s, and 6 s.

5. Maximum kinetic energy is at equilibrium, which occurs at 1 s, 3 s, and 5 s.

6. Equilibrium occurs at 1 s, 3 s, and 5 s.

7. The velocity of the pendulum is zero at maximum amplitude times of 0, 2 s, 4 s, and 6 s.

8. The planet closest to the sun: Earth. $v_{orbit} = \sqrt{\dfrac{Gm_{sun}}{r}}$

9. The planet with the largest radius: Mars. $T_{orbit}^2 = \left(\dfrac{4\pi^2}{Gm_{sun}}\right)r_{orbit}^3$

10. The only two forces will be the normal force and gravity. See the following figure.

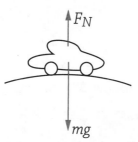

11. The car is accelerating downward toward the center of the circle. Therefore, gravity must be bigger than the normal force.

12. I assumed down to be positive toward the center of the circle.

$$\Sigma F_y = ma_c$$
$$mg - F_N = m\dfrac{v^2}{r}$$

13. The plane is traveling in a circle and must have a force toward the center of the circle to accelerate it around the turn. See the following figure.

14. The plane is traveling in a circle and must be accelerating toward the center of the circle, as shown in the figure.

15. Velocity is always tangent to the path of the object, as shown in the figure.

16. Remember Newton's Second Law. When the string breaks, the central force is gone and the plane will travel in a straight line, as shown in the figure.

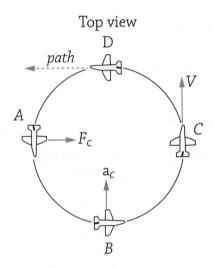

17. Organize our data as follows:

θ_0	0
θ	What we want to find
ω_0	0
ω	What we want to find
α	1.1 rad/s²
t	3.2 s

Use the equation $\omega = \omega_0 + \alpha t = 3.5 \, \text{rad/s}$.

18. Use the equation $\theta = \theta_0 + \omega_0 t + \dfrac{1}{2}\alpha t^2 = 5.6$ radians. Remember: 2π radians $= 1$ revolution. Therefore, $\theta = 0.89$ revolutions.

19. Convert rotational displacement to translational displacement using our connector equation. Be sure to use radians:

$$\theta = \frac{x}{r}$$

$$x = \theta r = (0.89 \, \text{rad})(0.12 \, \text{m}) = 0.11 \, \text{m}$$

20. The ball is starting from rest and experiencing a constant positive acceleration of 1.1 rad/s²:

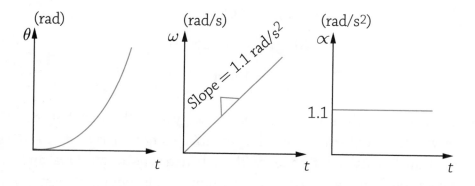

21. It is easier to rotate about the center because the rotational inertia is smaller about the center of mass than at the end of the bat.

22. Extend the length of the lug wrench by sliding the pipe over the end of the lug wrench and apply a force to the pipe. The extra length will increase the torque.

23. I'm going to use the perpendicular force times the radius to find the torque: $\tau = rF_\perp = (0.2 \, \text{m})(120 \, \text{N})\sin 45° = 17 \, \text{N} \cdot \text{m}$.

24. The clockwise and counterclockwise torques about the edge of the building must balance so that the worker doesn't fall. The center of mass of the board is 1 m from the edge of the building, as seen in the following figure.

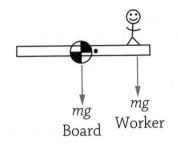

$$(rF)_{clockwise} = (rF)_{counterclockwise}$$

$$r(mg)_{worker} = r(mg)_{board}$$

$$r(g)_{worker} = r(g)_{board}$$

$$r(40\,\text{kg}) = (1\,\text{m})(55\,\text{kg})$$

$$r = 1.4\,\text{m}$$

25. They both have the same total kinetic energy at the bottom, but the ice wins the race because all of its potential energy turns completely into translational kinetic energy. The sphere has both rotational and translational kinetic energy at the bottom, so it is moving slower.

26. The hoop climbs higher. The hoop has a larger rotational inertia. Therefore, while rolling, the hoop has more rotational kinetic energy. Both have the same translational kinetic energy. This means the hoop has a greater total kinetic energy while rolling toward the ramp to convert into gravitational potential energy and a greater height.

27. Angular momentum is conserved. Therefore, the velocity at perigee is 4 times that of apogee:

$$L_i = L_f$$

$$(rmv)_{apogee} = (rmv)_{perigee}$$

$$((4r)v)_{apogee} = (r(4v))_{perigee}$$

28. Angular momentum is conserved. Therefore, when the rotational inertia doubles, the rotational velocity will be cut in half:

$$L_i = L_f$$

$$(I\omega)_i = (I\omega)_f$$

$$(I\omega)_i = \left[(2I)\left(\frac{\omega}{2}\right)\right]_f$$

6

Fluids

1. Both flow, and neither has a definite shape. Both exert pressure on whatever they touch.

2. Vastly different densities. Only gases can change volume and are compressible. Gas molecules are far apart and fly free of each other. Liquid molecules are close and hold together.

3. The pressure is the same because the size of the fluid does not matter. Only the depth below the surface affects pressure.

4. Remember that atmospheric pressure is 100,000 Pa and the density of water is $1{,}000\ \text{kg/m}^2$.

$$P = P_0 + \rho g h$$

$$200{,}000\ \text{Pa} = 100{,}000\ \text{Pa} + (1{,}000\ \text{kg/m}^2)(9.8\ \text{m/s}^2)h$$

$$h = 10\ \text{m}$$

5. The atmospheric pressure is acting on both you and the IV bag and will cancel out, as you will see in the following equation.

$$P = P_0 + \rho g h$$

$$100{,}000\ \text{Pa} + 10{,}600\ \text{Pa} = 100{,}000\ \text{Pa} + (1{,}050\ \text{kg/m}^2)(9.8\ \text{m/s}^2)h$$

$$h = 1.03\ \text{m}$$

6. The atmosphere outside the barometer pushes the fluid up into the tube.

7. Use Pascal's principle:

$$P_1 = P_2$$

$$\frac{F_1}{A_1} = \frac{F_2}{A_2}$$

$$\frac{F_1}{0.01\,\text{m}^2} = \frac{1{,}200\,\text{N}}{0.20\,\text{m}^2}$$

$$F_1 = 60\,\text{N}$$

8. The object is floating. Therefore, the buoyancy force must equal the force of gravity on the block:

$$F_b = F_g = mg = (7.0\,\text{kg})(9.8\,\text{m/s}^2) = 69\,\text{N}$$

9. Use the buoyancy force equation:

$$F_b = \rho V g$$

$$V = \frac{F_b}{\rho g} = \frac{69\,\text{N}}{(1{,}000\,\text{kg/m}^2)(9.8\,\text{m/s}^2)} = 0.0070\,\text{m}^3$$

10. You float higher in the water when your lungs are full because your chest expands when you inhale air. This increases your volume so that you displace more water, which increases the buoyancy force.

11. When a ship springs a leak, the hull no longer displaces as much volume of water and the buoyant force is no longer large enough to support the ship.

12. Use continuity. Cover part of the hose opening with your thumb. The water must shoot out faster to maintain conservation of mass.

7

Gases

1. Since the gas is heating up, the speed of the molecules will increase and the graph will shift to the right. The area under the new hotter distribution must be the same at the original distribution because the number of molecules has remained the same. This means that the new curve must be lower than the original curve:

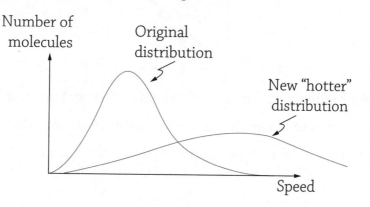

2. Oxygen has a lower molar mass; therefore, it is moving faster. $v \propto \sqrt{\dfrac{1}{m}}$

3. Both have the same average kinetic energy because they are at the same temperature. $K_{average} = \dfrac{3}{2} k_B T$

4. Use the ideal gas law:

$$PV = nRT$$

$$P = \frac{nRT}{V}$$

$$P = \frac{(1.5 \, \text{mol})(8.31 \, \text{J/mol} \cdot \text{K})(400 \, \text{K})}{(0.006 \, \text{m}^3)} = 831{,}000 \, \text{Pa}$$

$$= 8.3 \, \text{atmospheres of pressure}$$

5. The rooms have the same volume because they are identical. They also have the same pressure because the door is open between them. This leaves us with nT = constant. Therefore, the colder room must have more moles of gas inside.

6. Volume is directly proportional to the temperature: $V \propto T$. See the following figure.

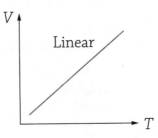

7. The number of moles and pressure are being held constant. This leaves us with $\dfrac{V}{T}$ = constant. Therefore, as the temperature goes up, the balloon will expand and as the temperature goes down, the balloon will shrink in size.

The First Law of Thermodynamics

1. The internal energy of ice increases. The ice is changing state from solid to liquid. This requires an input of energy equal to the heat of fusion.

2. Gas molecules are not connected to each other and therefore don't have any internal potential energy. They only have internal kinetic energy.

3. Steam, because it is in a higher energy state. The steam has been vaporized into a gas, so it has more energy than the water by an amount equal to the heat of vaporization.

4. The hot water added heat to the lid. This made the molecules move faster and caused the lid to expand. The expanded lid separated from the jar and made it easier to unscrew.

5. $T_K = T_C + 273 = 37 + 273 = 310\,\text{K}$

6. Heat flows from hot objects to cold objects. If they were running a fever, I would feel it as heat flowing into my forehead. This only works as long as I'm not running a fever too!

7. Use the ideal gas law:

$$PV = nRT$$

$$T = \frac{PV}{nR} = \frac{(500{,}000\,\text{Pa})(0.1\,\text{m}^3)}{(5.0\,\text{mol})(8.31\,\text{J/mol}\cdot\text{K})} = 1{,}200\,\text{K}$$

8. Any path that will increase the pressure, volume, or both. The key is that the PV value must increase. The following figure shows three possible paths.

9. To accomplish this, we need to decrease the temperature. Any path that will decrease the pressure times volume value will work. The figure shows three possible paths.

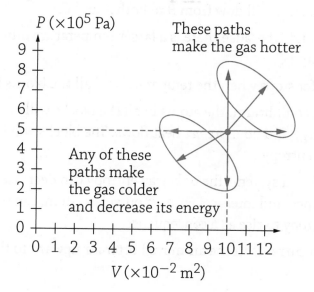

10. $T_D > T_C > T_A > T_B$. Temperature is directly proportional to the product of the pressure and volume (PV). The highest PV value is at point D, so it has the highest temperature. The lowest PV and temperature are at point B.

11.

Name of process	ΔT	ΔU	W	Q	Notes
Adiabatic	–	–	–	0	ΔU and W are the same value.
Isothermal	0	0	+	–	Q and W are the same magnitude.
Isochoric (Isovolumetric)	+	+	0	+	ΔU and Q are the same value.
Isobaric	–	–	+	–	Q has a larger magnitude than W.

The Second Law of Thermodynamics

1. The blocks that are in contact will transfer heat between them by conduction. Heat will flow from B to both A and C.

2. Between B and C because there is a larger temperature difference between B and C.

3. Heat transfer stops when the temperature of all the blocks is the same.

4. At thermal equilibrium, the atoms in all the blocks will have the same kinetic energy and no more *net* heat flows. The blocks will have reached maximum entropy.

5. Not for a closed system. The only way entropy can decrease is if the system is open and interacts with an outside system, causing the outside system entropy to rise as a consequence.

6. Convection currents in the air move hotter air upward to the second floor.

7. Keep the lid on to prevent heat from escaping via convection.

8. The reflective wrapping will reflect away radiation from outside on hot days and reflect the heat back into the house on cold days. This makes it easier to maintain the normal house temperature.

9. The crust is a poor conductor of heat but the tomato sauce is a great conductor of heat. Plus, the crust cools faster due to radiation and convection. The tomato sauce is insulated by the cheese and toppings and stays hotter longer.

10. You are speeding up the convection.

10

Electric Force and Electric Fields

1. From the triboelectric series chart, we can see that human skin will give up electrons to the polyester shirt. So the shirt is getting its static charge from you.

2. $(5,000,000 \text{ electrons})(1.6 \times 10^{-19} \text{ C/electron}) = 8.0 \times 10^{-13} \text{ C}$. You and the shirt will have the same magnitude of charge. You will be positive because you lost electrons. The shirt will be negative because it gained electrons.

3. All I would need to do is touch you, and by conduction, we would share the negative charge.

4. I would be grounded, and all the excess charge will move to the Earth, and I would end up neutral.

5. It polarizes the two spheres. The two spheres are touching; therefore, charge can flow between them. Thus, sphere Y becomes negative and sphere Z becomes positively charged.

6. If the spheres stay in contact and the rod is removed, the two spheres will no longer be polarized and the charges will be spread out evenly again. The sphere will return to being neutral.

7. If the spheres are moved apart while the negative rod is still polarizing them, the charges will be stranded where they are. Sphere Y will have electrons trapped on it and will remain negatively charged. Sphere Z will have a net positive charge trapped on it.

8. Charges 1 and 3 are positive because the electric field points away from them. Charge 2 is negative because the electric field points toward it.

9. A proton placed at A will receive a force to the right in the same direction as the electric field.

10. An electron placed at B will receive a force in the opposite direction as the field downward.

11. $\vec{F}_E = \vec{E}q = (50{,}000 \, \text{N/C})(1.6 \times 10^{-19} \, \text{C}) = 8.0 \times 10^{-15} \, \text{N}$

12. First convert everything to the standard units:

$$r = 4 \, \text{cm} = 0.04 \, \text{m}$$

$$q_1 = 2 \, \mu\text{C} = 2.0 \times 10^{-6} \, \text{C}$$

$$q_2 = -8 \, \mu\text{C} = -8.0 \times 10^{-6} \, \text{C}$$

$$F_E = \left| k \frac{q_1 q_2}{r^2} \right|$$

$$= \left| (9.0 \times 10^9 \, \text{Nm}^2/\text{C}^2) \left(\frac{(2.0 \times 10^{-6} \, \text{C})(-8.0 \times 10^{-6} \, \text{C})}{(0.04 \, \text{m})^2} \right) \right| = 90 \, \text{N}$$

13. The two charges attract, so the force is inward toward the other charge.

14. $F_E \propto q_1$, so as the charge doubles, the force will also double.

15. $F_E \propto \dfrac{1}{r^2}$, when the distance is cut in half, the force will be quadrupled.

11

Electric Potential Energy and Electric Potential

1. Since the equipotential lines around the top charge are increasingly positive the closer you get to the charge, the top charge must be positive. The two bottom charges are negative because the equipotential lines around them are increasingly negative.

2. $E_B > E_A > E_C$. How quickly the electric potential is changing indicates where the electric field is strongest $E = \left| \dfrac{\Delta V}{d} \right|$.

3. See the following figure. Electric field lines are always perpendicular to the equipotential lines and point from more positive to more negative.

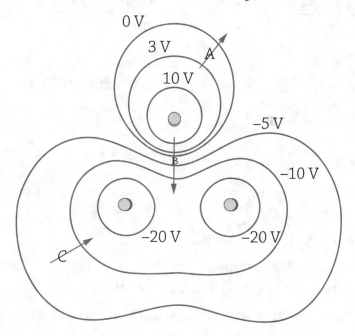

4. $V_A > V_B = V_D > V_C$. The potential at $A = 20$ V, at B and D it is zero, and at C it is -10 V.

5. $W = \Delta U_E = \Delta Vq$. Moving from B to A would require the most work. Moving the proton to D would require no work at all, because there isn't any change in potential. Moving the charge to C also wouldn't require any work, because the proton already wants to move to a lower potential.

6. $K = -\Delta U_E = -\Delta Vq = -(0\,\text{V} - (10\,\text{V}))(1.6 \times 10^{-19}\,\text{C}) = 1.6 \times 10^{-18}\,\text{J}$

7. $K = -\Delta U_E = -\Delta Vq = -(10\,\text{V} - (-10\,\text{V}))(1.6 \times 10^{-19}\,\text{C}) = 3.2 \times 10^{-18}\,\text{J}$

8. No. There is no change in electric potential when we move from point D to point B.

9. $\vec{E} = \left| \dfrac{\Delta V}{d} \right| = \dfrac{100\,\text{V}}{0.006\,\text{m}} = 4{,}000\,\text{V/m} = 17{,}000\,\text{N/C}$ To the left. The electric field will point away from the positive plate toward the negative plate.

10. $C_{\parallel} = \dfrac{\kappa \varepsilon_0 A}{d} = \dfrac{(1.0)(\varepsilon_0 = 8.85 \times 10^{-12}\,\text{C}^2/\text{N} \cdot \text{m}^2)(0.03\,\text{m}^2)}{(0.006\,\text{m})}$

    $= 4.4 \times 10^{-11}\,\text{F} = 44\,\text{pF}$

11. $C = \dfrac{Q}{\Delta V}$ or $Q = C\Delta V = (4.4 \times 10^{-11}\,\text{F} = 18\,\text{pF})(100\,\text{V}) =$
    $4.4 \times 10^{-9}\,\text{C} = 4.4\,\text{nC}$

12. $U_C = \dfrac{1}{2}Q\Delta V = \dfrac{1}{2}(4.4 \times 10^{-9}\,\text{C})(100\,\text{V}) = 4.4 \times 10^{-7}\,\text{J} = 440\,\text{nJ}$

13. To the right. The electron will receive a force in the opposite direction of the electric field.

14. Negative charge flows from A to B.

15. Charge will stop flowing when equilibrium is established. This is when both objects have the same electric potential: $V_B = V_A$.

16. Sphere B is three times smaller, $r_A = 3r_B$, so it will end up with one-third the charge of sphere A, $Q_B = \dfrac{1}{3}Q_A$.

17. Charge must be conserved. The original charge of sphere A is now spread out between both spheres A and B.

12

Circuits

1. The smaller-diameter resistor has a higher resistance because it has a smaller cross-section and the same length: $R = \dfrac{\rho L}{A}$.

2. $I = \dfrac{\Delta q}{\Delta t} = \dfrac{17\,\text{C}}{3.4\,\text{s}} = 5.0\,\text{A}$

3. Switch 1 controls the current to resistor 3 because it is in series with resistor 3.

4. Switch 2 controls the current to all three resistors because it is in the main current line.

5. $P = \dfrac{\Delta V^2}{R}$, $R = \dfrac{\Delta V^2}{P} = \dfrac{(12\,\text{V})^2}{(27\,\text{W})} = 5.3\,\Omega$

6. $I = \dfrac{\Delta V}{R} = \dfrac{(12\,\text{V})}{(5.3\,\Omega)} = 2.3\,\text{A}$

7. $P = I\Delta V = (3.34\,\text{A})(19.5\,\text{V}) = 65.1\,\text{W}$

8. Remember that parallel components must have the same voltage but be in separate pathways. See the boxed areas in the following figure.

9. Remember that series circuit elements have the same current and therefore must have only one current pathway without any branching wires. See the circled areas in the figure:

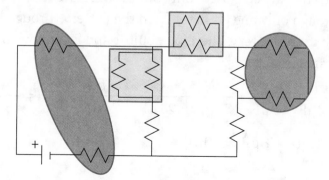

10. Batteries get hot due to internal resistance.

11. $R_{series} = R_1 + R_2 + R_3 \ldots = 40\,\Omega + 60\,\Omega + 120\,\Omega = 220\,\Omega$

12. $\dfrac{1}{R_{parallel}} = \dfrac{1}{40\,\Omega} + \dfrac{1}{60\,\Omega} + \dfrac{1}{120\,\Omega} = \dfrac{3}{120\,\Omega} + \dfrac{2}{120\,\Omega} + \dfrac{1}{120\,\Omega} = $

$\dfrac{6}{120\,\Omega} = \dfrac{1}{20\,\Omega}, R_{parallel} = 20\,\Omega$

13. See the following table.

	ΔV	I	R
R_1	60 V	5 A	12 Ω
R_2	20 V	1.7 A	12 Ω
R_3	20 V	1.7 A	12 Ω
R_4	40 V	3.3 A	12 Ω
Total	100 V	5 A	20 Ω

14. Ammeter 1 reads 5 A. Ammeter 2 reads 1.7 A.

15. The voltmeter indicates 20 V.

16. 1 = 4 > 2 = 3. The total current passes through both bulb 1 and 2. The current splits evenly to go through bulbs 2 and 3.

17. Bulbs 3 and 4 go out because the short circuit wire allows a zero-resistance path back to the battery that bypasses them.

18. 1 = 2. With the switch closed, only bulb 1 and 2 are left, and the current travels through only one pathway from the battery through bulb 1 and 2 and back to the battery. Therefore, bulb 1 and 2 are now in series and have the same brightness.

19. $\dfrac{1}{C_{series}} = \dfrac{1}{12\,\mu F} + \dfrac{1}{8\,\mu F} = \dfrac{2}{24\,\mu F} + \dfrac{3}{24\,\mu F} = \dfrac{5}{24\,\mu F}, C_{series} = \dfrac{24}{5}\,\mu F = 4.8\,\mu F$

20. $C_{parallel} = 12\,\mu F + 8\,\mu F = 20\,\mu F$

13
Magnetism

1. Use "the curly finger" right-hand rule:

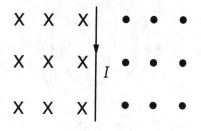

2. Use "the curly finger" right-hand rule:

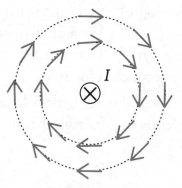

3. Use "the curly finger" right-hand rule:

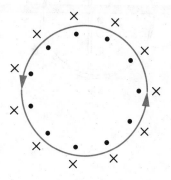

4. The magnetic field always exits out of the north pole and enters the south pole:

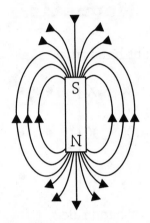

5. Iron, because it is ferromagnetic and will greatly enhance the magnetic field produced by the solenoid. Aluminum will only enhance the B-field a little because it is paramagnetic. Graphite is diamagnetic and will reduce the B-field produced by your coil of wire.

6. The compass will align with the magnetic field produced by the current in the wire. Use your "curly fingers" right-hand rule to determine the direction of the B-field:

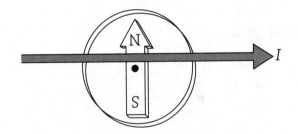

7. The compasses will turn to align with the *B*-field produced by the magnet:

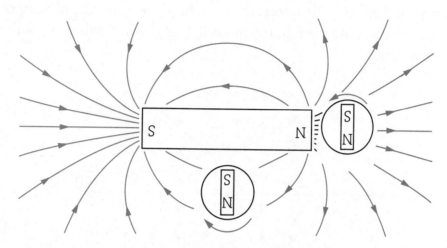

8. Remember to use the "flat fingers" right-hand rule to determine the direction of the force on each charge. Be sure to take your time to line up each part of your hand correctly!

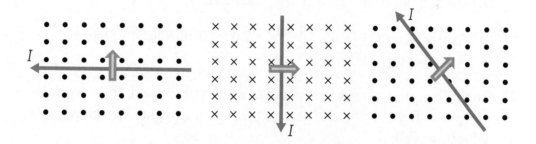

9. Note that the neutron travels along a straight path because it has no charge and is not affected by the *B*-field.

10. Using the "flat fingers" right-hand rule, the positive proton will curve toward the top of the page.

11. Using the "flat fingers" left-hand rule, the negative electron curves toward the bottom of the page. The proton and electron have the same charge and velocity. Therefore, they will have the same magnetic force acting on them. As a result, the much lighter electron will curve with a tighter radius.

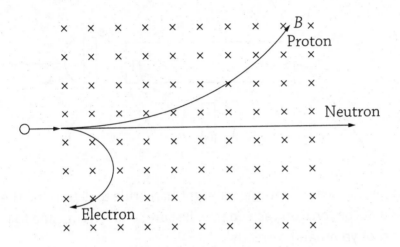

12. Use the magnetic force on a charge equation:

$$F_M = qvB\sin\theta = (1.6 \times 10^{-19}\,\text{C})(2.0 \times 10^6\,\text{m/s})(0.45\,\text{T})\sin 90°$$
$$= 1.4 \times 10^{-13}\,\text{N}$$

13. Sure! When the charge is traveling parallel to the B-field, there won't be a magnetic force on the charge and it will pass straight through without deflecting at all.

14. Use the magnetic force on a current carrying wire equation:

$$F_M = IlB\sin\theta = (0.50\,\text{A})(3.0\,\text{m})(0.88\,\text{T})\sin 90° = 1.3\,\text{N}$$

Induction

1. This will not produce current because the area, orientation, and magnetic field strength passing through the loop all remain the same.

2. This will produce a current because the magnetic field strength passing through the loop changes to zero when the loop is moved away from the loop of wire.

3. As long as the loop stays inside the magnetic field, the flux stays the same and no current will be produced.

4. Only locations 2 and 4 have a change in flux because the loop is entering the B-field at 2 and leaving the field at 4.

5. The magnetic flux is increasing through the loop into the page at location 2. Lenz's law says we need to oppose this increase in flux into the page by creating a magnetic field out of the page. Using the right-hand rule, we need a counterclockwise current in the wire to produce a magnetic field out of the page at location 2. See the figure below. At location 4 the loop is leaving the B-field. Thus, the magnetic flux into the page is decreasing. Lenz says we need to oppose this decrease in flux into the page by producing a magnetic field into the page to shore up the decrease. Using the right-hand rule, we can see that we need a clockwise current in the loop to produce a magnetic field into the page, as shown in the figure.

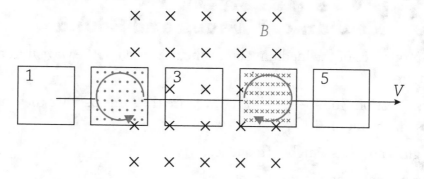

6. When the metal bar is moved to the right, the magnetic flux increases, which produces an emf and current.

7. The area between the resistor and the metal bar is increasing. Therefore, the flux out of the page is increasing. Lenz's law says we need to oppose this increase in flux out of the page by creating a magnetic field into the page. Using the right-hand rule, we need a clockwise current in the wire to produce a magnetic field into the page. A clockwise current around the closed conducting loop means the induced current passes through the resistor upward from the bottom of the page to the top of the page.

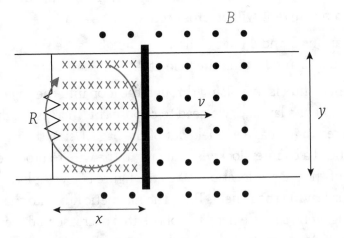

15

Mechanical Waves and Sound

1. The properties of the medium determine the speed of the wave in that medium.

2. Trick question! The speed of both waves are the same because they are in the same medium.

3. The girl's note has a higher frequency.

4. The boy's note has a longer wavelength.

5. Mechanical waves require a physical medium to travel through. Electromagnetic waves can travel through a vacuum because they supply their own medium.

6. Longitudinal wave

7. The wavelength is 9 m long.

8. Can't really tell, but it must be to the right or left.

9. The medium is vibrating parallel to the velocity of the wave, right and left.

10. Transverse

11. The amplitude is 0.9 m.

12. The wavelength is 1.4 m long.

13. $v = f\lambda = \left(\dfrac{12 \text{ waves}}{4\,\text{s}}\right)(1.4\,\text{m}) = 4.2\,\text{m/s}$

14. Can't really tell, but it must be to the right or left.

15. The medium is vibrating perpendicular to the velocity of the wave, up and down.

16. $T = \dfrac{1}{f} = \dfrac{1}{512\,\text{Hz}} = 0.0020\,\text{s}$

17. $\lambda = \dfrac{v}{f} = \dfrac{330\,\text{m/s}}{512\,\text{Hz}} = 0.64\,\text{m}$

18. 3.0 s

19. $f = \dfrac{1}{T} = \dfrac{1}{3.0\,\text{s}} = 0.33\,\text{Hz}$

20. 0.20 m

21. $v = f\lambda = (0.33\,\text{Hz})(3.7\,\text{m}) = 1.2\,\text{m/s}$

22. $x = A\sin\left(\dfrac{2\pi t}{T}\right) = (0.2\,\text{m})\sin\dfrac{2\pi t}{3.0} = (0.2\,\text{m})\sin 2.1t$

23. No. Humans can only hear up to 20,000 Hz.

24. Hearing loss begins at 85 dB over long periods of exposure. At 100 dB, hearing damage occurs within minutes.

25. Student A is three times closer and will receive nine times the sound intensity.

26. The transmitted portion of the wave will be upright with the same frequency but a new velocity that will be determined by the rope. The wavelength will also change.

27. The reflected portion of the wave will be inverted but will maintain the same speed, frequency, and wavelength.

28. The frequency will be higher than normal.

29. Shorter because the waves in front of the fire truck will be compressed.

30. The wave will diffract as it passes through the opening:

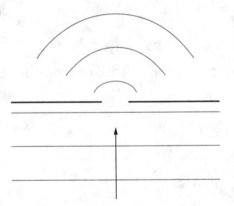

31. Diffraction is the property that says waves tend to bend around obstacles.

32. When the waves overlap, they have destructive interference because each is trying to move the medium in the opposite direction:

33. This is destructive interference because the two waves are attempting to move the medium in opposite directions, creating a wave that has a decreased amplitude when they meet.

34. 3 Hz

35. The fundamental wavelength is equal to twice the length of the guitar string: 1296 mm or 1.296 m.

36. $v = f\lambda = (196\,\text{Hz})(1.296\,\text{m}) = 254\,\text{m/s}$

37. 392 Hz and 588 Hz

Electromagnetic Waves

1. Light is a transverse electromagnetic wave that travels through electric and magnetic fields. It can travel through a vacuum.

2. Light travels at 3.0×10^8 m/s in a vacuum. It travels at different speeds in other mediums, depending on the index of refraction of the medium.

3. Light can be polarized, and only transverse waves can be polarized.

4. Electromagnetic waves are produced when a charged object oscillates.

5. Since everything above absolute zero has vibrating atoms inside and atoms are made up of charged particles, everything is emitting electromagnetic radiation.

6. 323 years

7. $(323\,\text{ly})\left(\dfrac{9.46 \times 10^{15}\,\text{m}}{1\,\text{ly}}\right) = 3.06 \times 10^{18}\,\text{m}$

8. gamma ray, visible light, radio wave

9. microwave, ultraviolet, X-ray

10. ultraviolet, visible light, infrared

11. For a perfect polarizer, 50% of the light passes through, and the other 50% will be absorbed because it is oscillating in the perpendicular direction.

12. Waves diffract because each point on a wave is the beginning of the next wave. This is called Huygens's principle. The waves near the boundary produce a circular wave that bends around the corners of objects that the wave passes.

13. Light has a much smaller wavelength than sound. The smaller the wave is compared to the opening, the less we see the diffraction effect.

14. Light demonstrates diffraction and interference, as seen in the double-slit experiment.

15. Use the double-slit interference equation with $m = 1$:

$$d\sin\theta = m\lambda$$

$$d\sin(7.5°) = (1)(650 \times 10^{-9})$$

$$d = 5.0 \times 10^{-6}\,\text{m} = 5.0\,\mu\text{m}$$

16. The total must add up to 100%. Therefore, 5% of the light is reflected by the sunglasses.

17. This is a regular reflection. This means that the surface is smooth on a microscopic level. When you wax a car, you fill in the little bumps and ridges with wax that makes the car smooth and shiny.

18. Water has a smaller index of refraction than glass, so light travels faster in water than in glass.

19. Use the index of refraction equation:

$$n = \frac{c}{v}$$

$$v = \frac{c}{n} = \frac{3.0 \times 10^8 \text{ m/s}}{1.60} = 1.9 \times 10^8 \text{ m/s}$$

20. The frequency always stays the same. The index of refraction tells us that the light slows down entering the glass. Therefore, the wavelength must also be getting shorter.

21. Light reflects off the front and back surface of the bubble. The two reflected waves interfere with each other, creating constructive and destructive interference in the light and causing some colors to be amplified and others to be reduced. This is called thin film interference.

Reflection and Mirrors

1. Remember that the angles are measured with respect to the normal line. See the following figure.

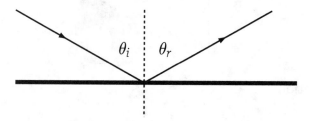

2. Remember that images formed by a plane mirror will appear to be behind the mirror. The image will be virtual and be the same size and distance from the mirror as the object. See the following figure.

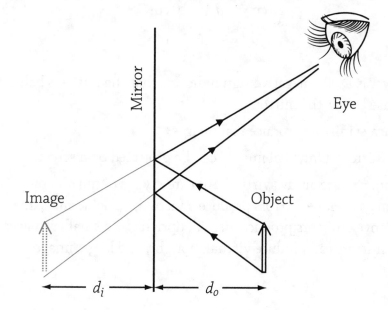

3. I have drawn one of the rays with a solid line and the other with a dashed line so that you can see it a little better.

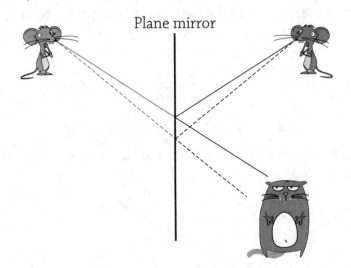

Plane mirror

4. Use the mirror equation:

$$\frac{1}{f} = \frac{1}{d_i} + \frac{1}{d_o}$$

$$\frac{1}{25\,\text{cm}} = \frac{1}{d_i} + \frac{1}{10\,\text{cm}}$$

$$d_i = -16.7\,\text{cm}$$

The answer is D. The negative sign indicates that the image is behind, or in this case below, the mirror.

5. Only concave mirrors produce real images.

6. Concave mirrors. Only real images can be projected on a screen.

7. Upright images are always virtual. All three types of mirrors produce virtual images. Plane mirrors produce a virtual, upright, and same-size image. Convex mirrors produce virtual, upright, and smaller images. Concave mirrors can produce virtual, upright, and larger images.

8. Be careful when you draw these ray diagrams. There are both concave and convex mirrors. I also put the objects in different places than the ones earlier in the chapter so that you would get a more diverse practice set and build a more robust understanding of mirrors. See the following figures.

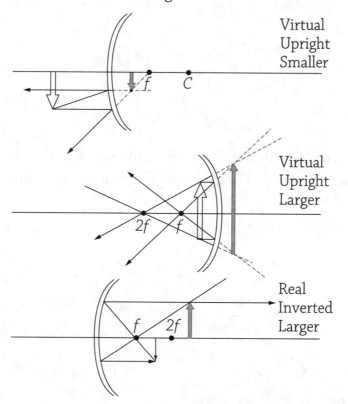

Virtual
Upright
Smaller

Virtual
Upright
Larger

Real
Inverted
Larger

9. $f = \dfrac{R}{2} = \dfrac{40\,\text{cm}}{2} = 20\,\text{cm}$. Since this is a convex mirror with a virtual focus, $f = -20\,\text{cm}$.

10. Use the mirror equation:

$$\frac{1}{f} = \frac{1}{d_i} + \frac{1}{d_o}$$

$$\frac{1}{-20\,\text{cm}} = \frac{1}{d_i} + \frac{1}{30\,\text{cm}}$$

$$d_i = -12\,\text{cm}$$

The negative image distance tells us that the image is a virtual image located behind the mirror.

11. Use the magnification equation:

$$M = \frac{h_i}{h_o} = -\frac{d_i}{d_o}$$

$$M = \frac{h_i}{20\,\text{cm}} = -\frac{(-12\,\text{cm})}{30\,\text{cm}}$$

$$M = 0.40 \text{ and } h_i = 8.0\,\text{cm}$$

The image will be 8.0 cm tall, which is only 40% as tall as the object. Both magnification and image height are positive, indicating that the image will be upright.

12. Use the mirror equation:

$$\frac{1}{f} = \frac{1}{d_i} + \frac{1}{d_o}$$

$$\frac{1}{1.0\,\text{m}} = \frac{1}{d_i} + \frac{1}{1.5\,\text{m}}$$

$$d_i = 3.0\,\text{m}$$

13. Use the magnification equation:

$$M = \frac{h_i}{h_o} = -\frac{d_i}{d_o}$$

$$M = -\frac{3.0\,\text{m}}{1.5\,\text{m}}$$

$$M = -2.0$$

The image is twice the size of the bunny.

14. The image of the bunny is inverted, real, and larger than the object.

18

Refraction and Lenses

1. Medium 1. The angle with the normal line is larger in medium 1.

2. Medium 2. Light is traveling slower in medium 2.

3. Medium 2. The wavelength is shortest in the slower substance.

4. Trick question! The frequency is the same in both mediums.

5. Use the index of refraction equation:

$$n = \frac{c}{v}$$

$$v = \frac{c}{n} = \frac{3.0 \times 10^8 \, \text{m/s}}{1.61} = 1.86 \times 10^8 \, \text{m/s}$$

6. Use the wavelength equation:

$$n = \frac{\lambda}{\lambda_n}$$

$$\lambda_n = \frac{\lambda}{n} = \frac{405 \, \text{nm}}{1.61} = 252 \, \text{nm}$$

7. Use Snell's law to determine that the unknown medium is polycarbonate:

$$n_1 \sin\theta_1 = n_2 \sin\theta_2$$

$$1.00 \sin 33.0° = n_2 \sin 19.9°$$

$$n_2 = 1.60$$

8. Use Snell's law:

$$n_1 \sin\theta_1 = n_2 \sin\theta_2$$

$$1.33\sin 25° = 1.00\sin\theta_2$$

$$\theta_2 = 34°$$

9. Use the critical angle equation:

$$\sin\theta_{critical} = \frac{n_2}{n_1} = \frac{1.00}{1.33} = 0.752$$

$$\theta_{critical} = \sin^{-1} 0.752 = 49°$$

10. B, D, and F are convex. Convex lenses are wider in the middle. Convex mirrors bow outward.

11. A, C, and E are concave. Concave lenses are skinnier in the middle. Concave mirrors bend inward.

12. B, D, and E are converging. Convex lenses are converging. Concave mirrors are converging.

13. A, C, and F are diverging. Concave lenses are diverging. Convex mirrors are converging.

14. Convex lenses are converging and have a real focal point. Therefore, the focal length is positive.

15. Between the focal point and the lens.

16. The apple is out beyond $2f$. Therefore, the image will be real, inverted, and smaller.

17. Notice how the ray passes through the focal point and travels out parallel, missing the lens in my diagram. That is okay. Just draw the ray from the bat to the lens plane and bend the ray horizontal to the principal axis like we always do. (See the figures of the bat and light bulb images on the next page.) Also notice that I have drawn an additional third ray that passes straight through the center of the lens without bending because the sides of the lens are parallel. This is just an extra ray that is not really necessary because we only need two rays to locate the image.

18. Look at the figures below. Remember the rules for drawing our rays. Notice that I have drawn an additional third ray that passes straight through the center of the lens without bending because the sides of the lens are parallel. This is just an extra ray that is not really necessary because we only need two rays to locate the image.

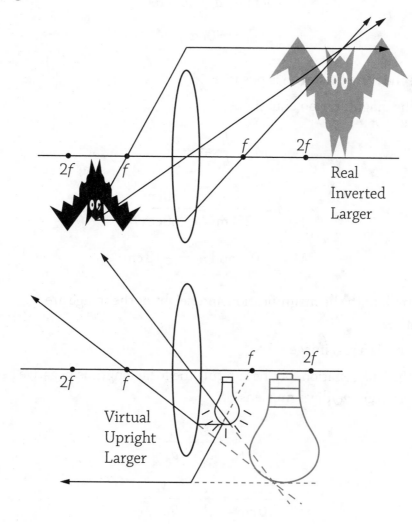

19. Use the lens equation:

$$\frac{1}{f} = \frac{1}{d_i} + \frac{1}{d_o}$$

$$\frac{1}{20\,\text{cm}} = \frac{1}{d_i} + \frac{1}{60\,\text{cm}}$$

$$d_i = 30\,\text{cm}$$

20. Real. The image distance is positive.

21. Use the magnification equation:

$$M = \frac{h_i}{h_o} = -\frac{d_i}{d_o}$$

$$M = \frac{h_i}{26\,\text{cm}} = -\frac{30\,\text{cm}}{60\,\text{cm}}$$

$$M = -0.5 \text{ and } h_i = -13\,\text{cm}$$

22. Inverted. Both the magnification and height of the image are negative.

23. Yes! This is a real image.

24. Use the lens equation, but be careful! The focal length is negative because this is a concave/diverging lens.

$$\frac{1}{f} = \frac{1}{d_i} + \frac{1}{d_o}$$

$$\frac{1}{-10\,\text{cm}} = \frac{1}{d_i} + \frac{1}{23\,\text{cm}}$$

$$d_i = -7.0\,\text{cm}$$

25. Virtual. The image distance is negative.

26. Upright because the image is virtual.

27. Smaller. This type of lens only produces upright, virtual, smaller images. You can also use the magnification equation to determine that the image is smaller.

28. No! This is a virtual image.

29. The focal length is positive, so I will assume that this is a convex/ converging lens. Using the lens equation:

$$\frac{1}{f} = \frac{1}{d_i} + \frac{1}{d_o}$$

$$\frac{1}{20\,\text{cm}} = \frac{1}{d_i} + \frac{1}{30\,\text{cm}}$$

$$d_i = 60\,\text{cm}$$

30. Using the magnification equation:

$$M = -\frac{d_i}{d_o}$$

$$M = -\frac{60\,\text{cm}}{30\,\text{cm}}$$

$$M = -2.0$$

The image is two times bigger.

31. Since the magnification is negative, the image is inverted.

19

Relativity and Quantum Behavior

1. Technically yes, but the effects are not noticeable until you travel much faster.

2. No! Time is not constant.

3. From your perspective, space shrinks in the direction you are going and the distance to the space station is less. Therefore, you get there in no time at all and you hardly age.

4. From my perspective, you seem to slow down and age much slower. It still takes you a long time to get to the space station, but you hardly age because time has slowed down for you.

5. The speed of light must always be c. Therefore, I see the light moving at c and leaving your spaceship at a relative speed of $0.2c$. You see the light leaving your rocket at c.

6. $1\text{ eV} = 1.6 \times 10^{-19}\text{ J}$. Therefore, $200\text{ eV} = (200)(1.6 \times 10^{-19}\text{ J}) = 3.2 \times 10^{-17}\text{ J}$.

7. Green light because it has a higher frequency.

8. IR has a longer wavelength.

9. Gamma rays have a higher frequency.

10. Use the photon energy equation:

$$E = \frac{hc}{\lambda} = \frac{1{,}240\,\text{eV} \cdot \text{nm}}{532\,\text{nm}} = 2.3\,\text{eV} = 3.7 \times 10^{-19}\,\text{J}$$

11. Use the photon energy equation:

$$E = hf = (6.63 \times 10^{-34}\,\text{J} \cdot \text{s})(96.7 \times 10^{6}\,\text{Hz})$$
$$= 6.4 \times 10^{-26}\,\text{J} = 4.0 \times 10^{-7}\,\text{eV}$$

12. A brighter light means more photons.

13. Because the metal is sensitive to only green light and higher energy light. Red has less energy than green. Therefore, red will not cause any electrons to be emitted. Violet light will eject electrons because it has more photon energy than the green light.

14. Brighter green light simply means more photons of green light. The photons still have the same energy. Therefore, brighter green light will eject more electrons, but the electrons will still have the same kinetic energy as with the dimmer green light.

15. UV light has more energy than green light. Therefore, the ejected electrons will have more kinetic energy.

16. Use the photoelectric equation with the kinetic energy equal to zero because this is the minimum frequency to eject photoelectrons:

$$K_{max} = hf - \phi$$

$$0 = (4.14 \times 10^{-15}\,\text{eV} \cdot \text{s})f - 2.1\,\text{eV}$$

$$f = 5.1 \times 10^{14}\,\text{eV} \cdot \text{s}$$

17. Use the photoelectric equation:

$$K_{max} = hf - \phi$$

$$K_{max} = \frac{hc}{\lambda} - \phi$$

$$K_{max} = \frac{1,240\,\text{eV} \cdot \text{nm}}{300\,\text{nm}} - 2.1\,\text{eV} = 2.0\,\text{eV}$$

18. a. This is an interference pattern, which is a wave property.

   b. This is the photoelectric effect, which proves that light has particle properties.

   c. Diffraction patterns are a wave property.

   d. Momentum and collisions are a particle property.

19. The electron acquired its new momentum and kinetic energy from the photon.

20. The photon loses energy and momentum to the electron. This means that the photon must decrease in frequency and its wavelength must increase. It still travels at the speed of light.

21. Use the de Broglie wavelength formula:

$$\lambda = \frac{h}{mv} = \frac{6.63 \times 10^{-34}\,\text{J} \cdot \text{s}}{(9.11 \times 10^{-31}\,\text{kg})v} = 5.3 \times 10^{-10}\,\text{m}$$

$$5.3 \times 10^{-10}\,\text{m} = \frac{6.63 \times 10^{-34}\,\text{J} \cdot \text{s}}{(9.11 \times 10^{-31}\,\text{kg})v}$$

$$v = 1.4 \times 10^{6}\,\text{m}$$

22. To get a smaller wavelength, we would need to increase the velocity of the electron. Wavelength is inversely proportional to the velocity: $\lambda \propto \dfrac{1}{v}$.

23. Where the wave function is at a relative maximum, either positive or negative, is the most likely locations to find the particle. See the following figure.

24. The particle can never exist where the value of the save function is zero. See the figure on the next page.

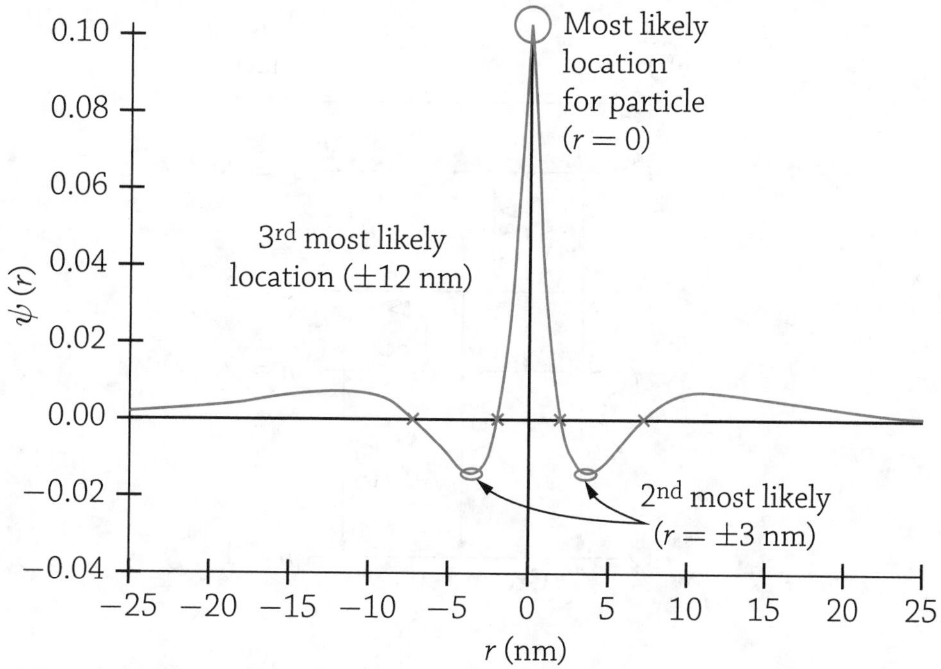

25. The electron will absorb the photon and jump up to an energy of -5 eV, which is the $n = 3$ energy level.

26. The electron cannot absorb the photon because absorbing it would take it to an energy of -3 eV. The electron cannot exist at this location.

27. The electron will absorb the photon and leave the atom with 3 eV of excess kinetic energy.

28. Starting in the $n = 4$ energy level, the electron can follow any path that eventually takes the electron back to the ground state. All the possible transitions are shown in the following figure. Each of these transitions will be accompanied by a photon emission, with energy equal to the difference in energies of the two states:

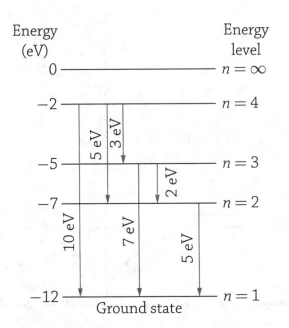

29. Since, $E = hf$, the frequency is directly related to the energy change in the transition. Therefore, the greater the change in energy, the greater the f of the emission. The frequency ranking from highest to lowest is $B = D > A > C$.

30. The wavelength is inversely related to the energy change in the transition: $E = \dfrac{hc}{\lambda}$. Therefore, the smaller the change in energy, the greater the λ of the emission. The wavelength ranking is opposite that of frequency: $C > A > B = D$.

Nuclear Physics

1. Alpha particle, beta particle, and gamma rays.

2. Gamma rays are the most penetrating and therefore the most dangerous. Alpha particles can be stopped by a single sheet of paper, so alpha emitters are generally not dangerous to humans unless you ingest them.

3. Carbon-12 has six protons and six neutrons. Carbon-14 has six protons and eight neutrons. Both are the element carbon because they both have six protons. They are different isotopes of carbon and will have different nuclear properties.

4. Isotopes have the same number of protons and therefore the same chemical properties. Since the nucleus is different, they will have different nuclear properties. Some are stable, while others are radioactive.

5. 90 protons

6. $234 - 90 = 144$ neutrons

7. 234 u

8. The protons in the nucleus try to force the nucleus apart. The nuclear strong force is stronger than the electrostatic force and holds the nucleus together. The nuclear weak force causes beta decay by converting a proton into a neutron and a positron or a neutron into a proton and an electron. Gravity between the nuclear particles is so small by comparison that we ignore it.

9. Alpha particles are helium nuclei so they take two protons away from the nucleus.

10. A nucleus will naturally decay in order to become more stable by converting some of its mass into energy.

11. Careful. I didn't give you the nuclear notation numbers of the neutron. There will be 11 nucleons in the daughter Y.

12. a. Not possible! The number of nucleons is not conserved. b. Not possible! Conservation of charge is violated.

13. a. Fusion: $2^1_1\text{H} + 2\,^1_0\text{n} \rightarrow\ ^4_2\text{He} + \text{energy}$

   b. Beta decay: $^{137}_{55}\text{Cs} \rightarrow\ ^{137}_{56}\text{Ba} + ^{\ 0}_{-1}\text{e}$

   c. Alpha decay: $^{238}_{92}\text{U} \rightarrow\ ^{234}_{90}\text{Th} + ^4_2\alpha + \text{energy}$

   d. Fission: $^1_0\text{n} + ^{235}_{92}\text{U} \rightarrow\ ^{144}_{54}\text{Xe} + ^{90}_{38}\text{Sr} + 2^1_0\text{n}$

   e. Gamma decay: $^{12}_6\text{C} \rightarrow\ ^{12}_6\text{C} + \gamma$

14. The half-life is approximately 2.5 minutes.

15. 1.6 decays/min/g is approximately one-eighth of the original activity: $(12.6/2/2/2 = 1.57)$. This is three half-lives. Therefore, the wood is approximately 17,000 years old: 5,730 years × 3 half-lives = 17,190 years.

16. This will take two half-lives: $\frac{1}{2} \times \frac{1}{2} = \left(\frac{1}{2}\right)^2 = \frac{1}{4}$. 3.8 days times 2 equals 7.6 days.

17. Fission is when a large nucleus is struck by a high-speed neutron that causes the nucleus to split into two large chunks. Fusion is when two smaller nuclei are joined together to form a larger nucleus than what you started with.

18. The mass defect is the missing mass that a nucleus should have based on what it is constructed of. The missing mass has been converted into binding energy that holds the nucleus together. This is equivalent to measuring the strong force's effect on the nucleus.

19. Nuclear reactions naturally occur when the new nucleus becomes more stable than the original by converting mass into binding energy to hold the nucleus together.

20. Find the total mass of the 26 protons and 30 neutrons that make up this iron nucleus:

$$26(1.0073\,\text{u}) + 30(1.0087\,\text{u}) = 56.4508\,\text{u}$$

Subtract this from the actual mass of the iron nucleus:

$$55.94\,\text{u} - 56.4508\,\text{u} = 0.5108\,\text{u} = 0.5108\,\text{u}\left(\frac{931\,\text{MeV}}{1\,\text{u}}\right) = 476\,\text{MeV}$$

Mass defect = $0.5108\,\text{u}$, and binding energy = $476\,\text{MeV}$